A Synopsis
of Human
Biochemistry
WITH MEDICAL APPLICATIONS

W.C.McMurray, *Ph.D.*

Professor, Department of Biochemistry,
Faculties of Medicine and Dentistry,
University of Western Ontario,
London, Ontario

A Synopsis
of Human
Biochemistry
WITH MEDICAL APPLICATIONS

HARPER & ROW, PUBLISHERS

PHILADELPHIA

Cambridge
New York
Hagerstown
San Francisco

London
Mexico City
São Paulo
Sydney

1817

A SYNOPSIS OF HUMAN BIOCHEMISTRY: WITH MEDICAL APPLICATIONS. Copyright © 1982 by Harper & Row, Publishers, Inc. All rights reserved. No part of this book may be used or reproduced in any manner whatsoever without written permission except in the case of brief quotations embodied in critical articles and reviews. Printed in the United States of America. For information address Harper & Row Publishers, Inc., East Washington Square, Philadelphia, PA 19105.

10 9 8 7 6 5 4 3 2 1

Library of Congress Cataloging in Publication Data

McMurray, W. C.,   DATE
  A synopsis of human biochemistry.

  Bibliography
  Includes index.
   1. Biological chemistry. I. Title. [DNLM:
1. Biochemistry.   QU4 M478f]
QP514.2.M44   612'.015   81-6767
ISBN 0-06-141642-8   AACR2

Dedicated to my friend and colleague,
JOHN C. RATHBUN, 1915 – 1972
A Frontiersman of Pediatric Biochemistry

Contents

Preface

Several years ago I undertook the task of elucidating metabolism for pre-clinical students in the health sciences. The outcome of that effort was my book *Essentials of Human Metabolism*, in which I presented in concise terms the major interconversions of carbohydrates, lipids, and nitrogenous substances from the standpoints of the enzymes involved, the energetics, and the control mechanisms of the organs and tissues in the body.

The process of writing is as much a matter of omission and elimination as it is of inclusion. In defining "essentials of human metabolism" I inevitably had to exclude many no less important topics. This book seeks to redress those omissions which I feel to be a "synopsis" of the present frontiers of knowledge in the molecular approach to medicine.

The first book was designed to provide the basic information suitable for an introductory course; the present work contains more applied aspects relating to biochemistry in humans and could serve as a subject outline for a more advanced course. Although I have assumed, therefore, a familiarity with basic biochemical concepts and a nodding acquaintance with biochemical terminology as set out in the preceding book, I hope to have met my original intention—namely to render a current account of the subject with broad brushstrokes.

Chapter 1 examines the main constituents of the blood and the chemical properties of blood proteins in relationship to their functions in the transport systems and defense mechanisms of the circulation. Chapter 2 provides a survey and critical analysis of human nutritional requirements in health and disease states. Chapter 3 discusses the actions of hormones in target tissues from a molecular point of view, emphasizing the controlling influences upon gene expression and metabolism. Chapter 4 is an overview of cell membrane properties and our present concepts of the nature of extra- and intracellular signals. Chapter 5 elaborates some key biochemical features of the human genetic apparatus and those factors in the environment which may give rise to genetic diseases or neoplasia.

The frontiers of biochemical investigation have a habit of continually advancing and shifting. It is indeed a bewildering enterprise to keep abreast of the findings and changes of emphasis. Although there is no true substitute for an ongoing evaluation of the current periodical literature it is my hope that *A Synopsis of Human Biochemistry,* by presenting one person's overview, may serve to highlight the critical features of human biochemical studies.

W. C. McMurray, Ph.D.

Acknowledgments

I would like to acknowledge the kind assistance and critical comments from my colleagues who read the following chapters: Dr. Val Donisch (Chapters 1 and 3); Dr. Bill Magee (Chapter 2); Dr. Chris Grant (Chapter 4); Drs. George Mackie and Ian Walker (Chapter 5).

A Synopsis
of Human
Biochemistry
WITH MEDICAL APPLICATIONS

Chapter One *Blood*

The bloodstream is truly a river of life, and it reflects much about our present and past existence. In its salt content blood is an echo of the sea water from which unicellular organisms adapted the extracellular fluids in their evolution to metazoan creatures. The metabolites in blood mirror those of body cells and thus provide an index of nutritional and physiologic states within tissues. The soluble protein constituents convey information about the tissues where they originate, as well as information about the foods and wastes which they carry, or their protective–maintenance functions in the circulation. Studies of the various blood cells tell us much about the status of our respiratory mechanisms, and our defense responses to injury, infection, and inflammation. Chemical analysis of the blood (and other fluids from the body—cerebrospinal fluid, synovial fluid, intestinal fluid, urine) is the simplest, most direct means for the assessment of metabolic, pathologic, or forensic aspects of the human body, and has provided the starting point for clinical biochemistry and laboratory medicine.

Functions of the blood may be grouped into three broad categories: (1) the *maintenance* roles which include balancing fluid compartments between cells and extracellular spaces; maintenance of body temperature; control of electrolyte and acid–base balance; (2) the *carrier* roles which involve respiratory gas exchanges by the erythrocytes; transportation of dissolved salts, nutrients, or waste materials in the plasma; binding and transport of many insoluble substances such as bilirubin, lipids, fat-soluble hormones, and certain minerals as complexes with plasma proteins; (3) the *protective* roles whereby antibodies and phagocytic cells attack invading organisms; clotting proteins and platelets prevent the loss of blood from the circulation by hemorrhage.

In analytic work blood is usually treated to remove the cells either by allowing a clot to form producing the cell-free *serum*, or by inhibiting coagulation and centrifuging to yield the blood *plasma*. Most components are present in the same concentrations in the two fluids, although those plasma proteins such as fibrinogen which are involved in production of the clot are obviously absent from serum.

HEMOGLOBIN

Our survival as breathing, aerobic organisms depends upon rapid exchanges of respiratory gases between tissues and the atmosphere by the blood. Hemoglobin plays the main role in these exchanges in the body, both directly for oxygen delivery, and indirectly for carbon dioxide removal. For example, the aqueous liquid portion of blood could carry no more than 0.3% of its volume as dissolved oxygen; by contrast each gram of hemoglobin is capable of binding 1.35 ml of O_2, and thus each 100 ml portion of blood containing approximately 15 g of hemoglobin may carry over 20% (15 × 1.35) of its volume as bound oxygen. Although the mechanism is different, hemoglobin may also bind substantial amounts of carbon dioxide. However, the latter gas is carried in the blood mainly in its hydrated form, carbonic acid, or the dissociated form, bicarbonate anion, in solution. As noted below, hemoglobin also facilitates these soluble carbon dioxide transfers in blood by coordinated balancing of the acid–base changes associated with its reaction with oxygen. Work of recent decades has revealed intimate details about the organization of the hemoglobin molecule and about its interactions with gases and modifiers of binding. As is so often the case with scientific inquiry the dividends have been profitable, and at times unexpected: on the clinical level the molecular studies have led to understanding of the control of respiratory gas exchange, and have provided clues for the development of therapeutic strategies in disease states; on the basic level the hemoglobin molecule with its dynamic conformational changes during gas exchanges, sometimes referred to as "molecular breathing," has served as the model par excellence for investigations into the mechanisms of action of regulatory proteins and allosteric enzymes.

Human hemoglobin has a molecular weight of 64,400, and contains 574 amino acid residues in the apoprotein or globin component, and 4 heme prosthetic groups. Each heme is attached to a separate polypeptide chain, or subunit (Fig. 1–1). Several variants have been detected in human blood. In the normal adult over 95% of the total is hemoglobin A, a tetramer containing 2α-polypeptide chains (141 amino acid residues each) plus 2β chains (146 residues); thus hemoglobin A is sometimes designated $\alpha_2\beta_2$. Minor species of hemoglobin that usually constitute only a few percent are hemoglobin $A_2(\alpha_2\delta_2)$ in which the β chains are replaced by closely similar δ chains, hemoglobin $F(\alpha_2\gamma_2)$, the type found in fetal blood which is replaced by adult type in the first few months after birth, and hemoglobin

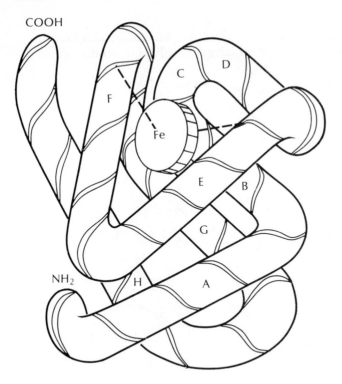

COOH

F

Fe

C

D

E

B

G

NH₂   H   A

FIG. 1–1.   Structure of a subunit (β chain) of hemoglobin.

$A_{1c}(\alpha_2(\beta-N-glucose)_2)$ in which the β chains have glucose molecules attached to their NH_2-nitrogens, and which may be elevated twofold in diabetes when blood sugar levels are high. In addition, abnormal types of hemoglobin have been detected in individuals suffering from hereditary hemoglobinopathies as noted further below.

Complete primary structures (amino acid sequences), and details of the three-dimensional configurations of human hemoglobins are now well established (Fig. 1–2). Overall shape of the tetrameric protein is almost spherical (50 × 55 × 65 Å) despite the fact that about 80% of its residues are present as straight α-helices. The globular structure results because the eight α-helical regions (designated A–H) in each of the subunits are twisted and folded at the nonhelical bends of the polypeptide chains; the subunits nestle together in the tetramer with many Van der Waals interactions and several salt-linkages coupling the α- and β-polypeptides together into a fairly tight ball. These interactions, particularly the electrostatic linkages, are strikingly altered during reactions of hemoglobin with oxygen or with carbon dioxide. The heme groups are attached through their Fe atoms to imidazole N atoms of histidine residues which are at the eighth position of α-helical region F (F8) of each of the α and β subunits. The four

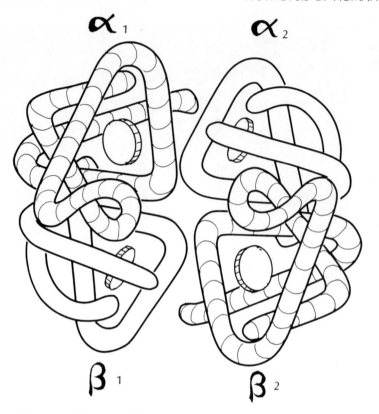

FIG. 1–2. The hemoglobin tetramer.

hemes lie in pockets at the surface of the molecule nearly equidistant from one another. Heme groups of the two β subunits appear to be shielded within narrow crevices of the protein, and are less accessible for reaction with oxygen than the hemes of the two α chains.

OXYGENATION

The oxygenation of hemoglobin involves stepwise binding of O_2 to the Fe atoms of each of the heme groups. Although the iron combines with oxygen it is not oxidized but remains in the Fe^{2+} (ferrous) state; hemoglobin in which the iron becomes oxidized to the Fe^{3+} (ferric) state is referred to as methemoglobin (hemoglobin M), and is incapable of binding oxygen. The terms deoxyhemoglobin and oxyhemoglobin are used to differentiate hemoglobin from its fully oxygenated form, designated Hb and $Hb(O_2)_4$ respectively. The partially oxygenated forms $Hb(O_2)$, $Hb(O_2)_2$, and $Hb(O_2)_3$ will also exist depending upon how much oxygen is available to combine with each of the four heme groups. Oxygenation of the hemes is a nonran-

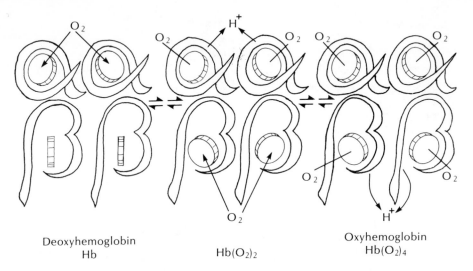

FIG. 1–3. Oxygenation and conformation changes in hemoglobin. Adding oxygen to the two α subunits "opens" the β subunits which facilitates further oxygen addition.

dom, interdependent sequence of events. Because the hemes of the α sub-units are more exposed they will react most readily when amounts of oxygen are low. As the oxygen concentration is increased and the α-hemes become oxygenated, they trigger a conformational change in neighboring subunits which opens the tight pockets surrounding the β-hemes enabling them also to bind oxygen (Fig. 1–3). This remarkable relaxation of the hemoglobin molecule is brought about by a slight change in shape of the heme group on oxygenation. In deoxyhemoglobin the iron atom is too large to lie flat in the heme disc, but projects slightly out of the plane of the porphyrin ring. When it combines with oxygen the ionic radius of the iron decreases sufficiently to permit it to pop down flush with the porphyrin ring. This small movement of the iron becomes amplified into larger movements of the attached polypeptide chain affecting several charged groups that link to adjacent subunits. As a result, the electrostatic bonds between subunits break, hydrogen ions are released, and the subunits slide apart into a more open or relaxed configuration. By this molecular breathing mechanism the addition of oxygen on one heme enhances addition on another, a phenomenon known as positive cooperativity, and results in the sigmoid-shaped binding curve that is typical of allosteric phenomena. This interaction between the heme groups has profound implications for the function of the hemoglobin molecule in delivery of oxygen to the tissues.

The preceding discussion relates to isolated hemoglobin molecules in solution. In reality hemoglobin is tightly packaged into the pliable biconcave corpuscles of blood, and is subjected to extreme variations in its odyssey through the circulatory system. Binding of oxygen in the blood is a

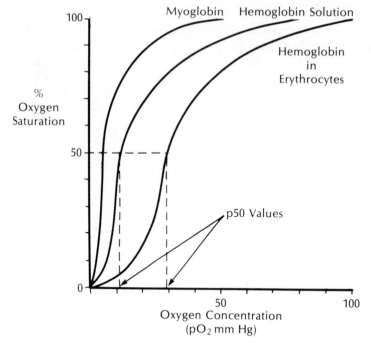

Fig. 1–4.   Comparison of oxygen dissociation curves for hemoglobin and myoglobin.

reversible process, the extent of which is dependent upon a number of factors in the erythrocyte, surrounding plasma, and extracellular fluids, including the temperature and the concentrations of hydrogen ions, organic anions, CO_2, and O_2. The gas concentrations of biologic fluids are generally expressed in terms of their contribution to the total pressure of the gas phase. Thus oxygen, which is 20% of the volume of air at atmospheric pressure (760* *Torr*), exerts a partial pressure (PO_2 of roughly 150 *Torr*). Inside the lung alveolus the PO_2 is somewhat lower (about 100 *Torr*), but it is still high enough to ensure that virtually all heme groups of hemoglobin in the alveolar capillaries will have an oxygen molecule bound to them (Fig. 1–4). This condition is referred to as 100% saturation, and represents the maximal capacity of blood for the transport of oxygen as it passes through the pulmonary circulation. When the arterial blood reaches the capillary beds of respiring tissues where mitochondrial oxidations will deplete the cellular oxygen, the PO_2 of surrounding fluids will drop (below 50 *Torr*). Under these circumstances oxyhemoglobin in the erythrocytes will partially dissociate to unload the O_2 required for cellular oxidations. In the converse of the oxygenation process heme–heme interactions ensure that the release of oxygen from one heme group facilitates the release by other hemes.

   * *Torr* = 1 mm Hg

The importance of the subunit interactions for oxygen unloading is apparent from a comparison of the oxygen dissociation curve for hemoglobin with that for myoglobin. The latter is a hemoprotein in muscle cells which stores oxygen for short peak periods of rapid contraction. Myoglobin is a monomer, roughly equivalent to one of the chains of hemoglobin. Since each heme reacts independently there is a hyperbolic relation between the binding and concentration of O_2 (Fig. 1–4). It requires a lower PO_2 corresponding to oxygen levels in tissue capillaries to saturate myoglobin, a feature that suits it well for its role in tissue storage, but would be inappropriate in blood oxygen transport. Myoglobin will give up its oxygen only when muscle is working and respiring so rapidly that the PO_2 in the cells drops drastically below 1 to 3 *Torr*. In hemoglobin on the other hand, because of the heme–heme interactions between subunits it requires only a small drop in PO_2 for the unloading of a substantial portion of bound oxygen. This effect is greatest in the steep region of the curve where the hemoglobin is 50% oxygenated; the oxygen concentration corresponding to this 50% saturation point is referred to as the P_{50} value, and for hemoglobin in neutral solutions (*p*H 7.0) is a PO_2 value of about 10 *Torr*. However, in the red cell at the conditions found in venous blood the P_{50} is increased to 25–30 *Torr*, indicating an enhanced tendency to give up oxygen. Several physiological factors contribute to this shift to the right of the oxyhemoglobin dissociation curve including increased concentrations of CO_2, H^+, and organic phosphates, principally 2,3-diphosphoglycerate (DPG), in the erythrocytes.

REGULATION OF O_2 DELIVERY

Carbon dioxide is the most important physiological regulator of oxygen delivery to the tissues. At the level of the tissues the concentration of carbon dioxide will rise owing to catabolic processes. The resulting increase of PCO_2 will shift the O_2^- dissociation curve for hemoglobin to the right thus elevating the P_{50} value (Fig. 1–5). This increase in oxygen release by carbon dioxide is referred to as the Bohr effect (after its discoverer), and results from two distinct reactions. First, some of the CO_2 may bind directly to hemoglobin, not to the heme group, but rather to the amino-terminal residues of the globin chains to produce carbamino compounds. Second, the reaction of CO_2 with water to produce carbonic acid will be accelerated by the red cell enzyme, carbonic anhydrase, causing a lowering of the *p*H inside the erythrocyte. These two effects of CO_2 reduce the affinity of hemoglobin for O_2. The *p*H effect is quantitatively most important and best understood. The addition of increased amounts of H^+ ions to oxyhemoglobin leads to protonation of residues within the polypeptides (histidine rings, NH_2 groups of lysine or arginine) which thus attain positive charges and form salt linkages to the negative charges of residues such as aspartate of other subunits. This favors a shift in configuration from the relaxed form of oxyhemoglobin into the tight association of subunits

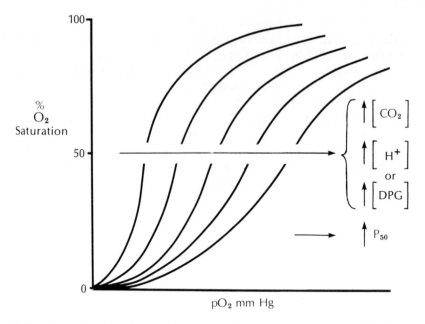

FIG. 1–5. The Bohr effect and DPG shift. Increasing concentrations of carbon dioxide, hydrogen ions, or 2,3-diphosphoglycerate (DPG) shift the oxygen dissociation curve of hemoglobin to the right to increase the P_{50} value.

characteristic of deoxyhemoglobin, with the consequence that the heme groups can no longer retain oxygen so tightly. Thus the production of carbonic acid or other acidic metabolites at the level of tissue capillaries which lower the pH will facilitate the release of O_2. The effect from binding of CO_2 is less substantial since only 10% of total carbon dioxide transported is in the form of carbamino groups. Deoxyhemoglobin binds CO_2 more readily than oxyhemoglobin, and formation of the N-terminal carbamino groups seems to accentuate the Bohr proton effect to promote the relaxed to tight configurational shift representing the "exhalation" stage of hemoglobin breathing.

At the level of the lungs O_2– CO_2 interactions are also important. Most of the CO_2 carried by venous blood is in the plasma as dissolved bicarbonate anions produced when carbonic acid gives up its hydrogen ions; the latter are transported in the erythrocyte complexed to deoxyhemoglobin. On oxygenation the hemoglobin, in undergoing the conformational changes of molecular "inhalation," becomes more acidic, discharging

both bound CO_2 from carbamino groups, and hydrogen ions. $Hb \underset{\diagdown}{\overset{\diagup H}{}} + CO_2$

$O_2 \rightarrow HbO_2 + CO_2 + H^+$. The protons then combine with bicarbonate to form carbonic acid which in the presence of carbonic anhydrase rapidly generates carbon dioxide gas for expiration: $HCO_3^- + H^+ \rightarrow H_2CO_3 \rightarrow H_2O + CO_2$. In this way it can be seen that hemoglobin carries out a cycle of physiologic and molecular breathing to carry CO_2 and O_2 in opposite directions through the circulation while buffering the H^+ concentrations in order to maintain the pH of the tissues close to neutrality.

Another important modulator of hemoglobin function is 2,3-diphosphoglycerate, usually abbreviated simply as DPG. This intermediate of glycolysis is found in trace amounts in most cells of the body, but in the erythrocyte it is the major phosphate compound, and is present approximately in amounts equivalent to the concentration of hemoglobin in the cell (5 mM). As in the case of carbon dioxide, DPG binds to hemoglobin reciprocally with oxygen: $Hb - DPG + 4O_2 \rightarrow Hb(O_2)_4 + DPG$. There is only one DPG binding site for each hemoglobin tetramer, and it lies in a cavity facing the two β chains. The negative charges of the phosphate anions form salt bridges to positive charges at the N-terminal valines and between histidine and lysine residues, thus producing cross-linkages between the β subunits. This enhances the tight configuration of deoxyhemoglobin, and thus lowers its affinity for oxygen. The reciprocal effect is due to the fact that in the relaxed conformation of oxyhemoglobin the N termini are too far apart to allow DPG to form the electrostatic bridge between β chains. At the level of the lungs, therefore, most DPG will be free in the erythrocytes, and will exert little effect upon oxygenation; in the tissue capillaries, however, significant amounts of DPG may be bound as hemoglobin dissociates its oxygen, and changes in the DPG level of the cell may play an important regulatory role in oxygen unloading. Increases in DPG, by stabilizing the deoxy form of hemoglobin will shift the P_{50} to higher PO_2 values. The acidity of the DPG molecule itself will accentuate this shift through the Bohr effect by lowering the intracellular pH.

The levels of DPG in erythrocytes will rise under a number of physiologic conditions *in vivo* where the supply of oxygen to tissues is decreased. The compensatory increase in DPG will shift the O_2 dissociation curve to the right, and by facilitating the discharge of oxygen will ensure the oxygenation of tissues in cases of anemia or cardiac failure. Two principal mechanisms in the erythrocyte seem to be involved in this adaptive response: (1) decreased oxygen, leading to increased deoxyhemoglobin, leading to more binding of DPG; (2) increased pH, leading to increased glycolysis. The two mechanisms are complementary and often occur simultaneously to promote a reinforcing effect. For example, upon an ascent to high altitudes the lower PO_2 of inspired air will result in accumulation of deoxyhemoglobin, which ties up proportionately more DPG, thus lowering the free DPG concentration in the cell. Since the key enzyme responsible for conversion of the glycolytic intermediate 1,3-diphosphoglycerate into 2,3-diphosphoglycerate (DPG mutase) is strongly feedback-inhibited by its product, a lowering of free DPG will enhance its own synthesis. The second factor will arise in response to hyperventilation at high altitude, an attempt by the body to adjust to lowered oxygen by accelerated breathing. In addition to providing more inspired air per unit time, of course there will also be an increase in the rate of gas exhalation, and thus more rapid removal of CO_2 from the body. In effect this corresponds to removal of H^+ by neutralization of the HCO_3^- reservoir for carbon dioxide gas, and a rise in body pH will ensue (hyperventilation alkalosis). In the red cell it is clear that glycolysis is pH-controlled especially at the level of phosphofructokinase, which is inhibited by H^+ ions and stimulated by phosphate ions. Thus when the pH rises glycolysis is accelerated providing more 1,3-DPG substrate for the mutase enzyme, and hence more 2,3-DPG. These shifts in DPG concentrations are important in a number of clinical situations where the acid–base balance is impaired. It is notable that acidosis, which shifts P_{50} values to the right, will lower DPG and thus move the P_{50} back to normal; as described above, alkalosis is also compensated in the red cell by the increased synthesis of DPG, thus offsetting the direct effect of decreased H^+ ion upon the P_{50}, and providing a normal O_2 dissociation.

In blood-banking practices it is important to provide storage conditions which maintain the DPG concentrations in the erythrocyte. Modern improvements include the additions of inorganic phosphate to the medium in order to stimulate the mutase enzyme, plus the addition of dextrose, pyruvate, dihydroxyacetone, and other substrates to activate the glycolytic pathway. Without such additives DPG levels of stored erythrocytes will drop to zero, and the transfusion into hypoxic patients would be of marginal benefit because of the impaired ability of the DPG-depleted cells to discharge their O_2 to the tissues. Although the newly transfused cells will eventually resynthesize the DPG *in vivo*, it is obviously better to circumvent the acute effects by appropriate maintenance measures *in vitro* in the banked blood.

INHERITED HEMOGLOBINOPATHIES

It has become apparent over the past thirty years that there are numerous variants of the normal hemoglobin composition in humans. In some instances the hemoglobin variants differ markedly from their normal counterparts with respect to their solubility, stability, or oxygen-binding properties, and their presence in affected individuals is grossly obvious from pathological effects upon the structure, survival, and function of the erythrocytes. In others, the alteration is minimal in terms of the physiological reactions of hemoglobin, and the consequences for the individuals are correspondingly minor. A comparison of the structural alterations with functional defects in human hemoglobin variants has provided a wealth of information concerning the crucial regions of the molecule and the requirements for normal mechanisms of action in the erythrocyte. In addition the patterns of inheritance, the determinants and controlling factors in biosynthesis of the polypeptide chains in these "experiments of nature" on our genetic machinery, have told us a great deal about the complex workings of human genes. The identification of the molecular lesions and much of the follow-up investigation have been made possible by the dedication of physicians and their patients; there is no better example in the fields of medicine where studies at a clinical level have furnished profound insights into molecular biology.

Almost three hundred different variants of human hemoglobin have been detected, but the list may be enlarged many-fold since there are undoubtedly numerous neutral alterations of the hemoglobin molecule whose detection is limited by the sensitivity of present methods of analysis. Some of the variants are common within particular racial groups and concentrated in geographical regions. This may explain the practice which has evolved of denoting some variants by the place names of locales of discovery (Hb Constant Spring, Hb Köln, Hb Chesapeake, Hb Little Rock, Hb Kansas, Hb M Saskatoon). On the whole, the incidence of any given variant in the general population may be quite low, but because of the large numbers of different variants it is estimated that one person in three hundred may bear a variant hemoglobin of some sort.

The genetically determined variations of hemoglobin structure may be divided into several categories. In all cases that have been elaborated the variation may be traced to the globin portion; the heme prosthetic group apparently has been conserved throughout evolution. Generally only one type of globin chain (α or β of adult hemoglobin) is altered, and this, plus other genetic evidence, has indicated that there is a discrete pair of allelic genes to specify each globin polypeptide. A majority of variants fall in the category of replacement of a single amino acid in one polypeptide chain by another. Thus Hb Bethesda, in which only the tyrosine residue at position 145 of the β chain has been altered by replacement with a histidine residue, is designated as $\alpha_2\beta_2^{145\ \text{Tyr}\rightarrow\text{His}}$. Such a single amino acid replacement can be traced to point mutation of a single nucleotide base in the DNA sequence of the gene coding for that polypeptide. During transcription of that gene in the developing erythrocyte the complementary mRNA will contain a modified triplet codon corresponding to a different tRNA anticodon, and a different amino acid will be substituted during translation of the particular globin message (see also Chapter 5). Since the genes are allelic pairs, inheritance of the abnormal hemoglobin will follow a codominant pattern whereby the heterozygotes containing one mutant and one normal gene will be expected to produce both the mutant and normal hemoglobins, while homozygotes containing both mutant genes will produce only the mutant hemoglobin species. Clinical consequences to the heterozygous and homozygous individuals will depend upon the position in the polypeptide chain where substitution occurs (*e.g.*, close to heme attachment or DPG-binding sites), and the degree of difference in properties between the normal and substituted residues (*e.g.*, relative sizes, electrostatic charges, abilities to form hydrogen bonds or hydrophobic interactions). Another category of variant arises when the mutation affects the DNA bases which specify the codon (UAA or UAG) that normally tells the ribosome to stop reading the mRNA and to discontinue synthesis of the globin polypeptide. The modified codon may now be used in translation to insert an additional amino acid. Since there is a substantial sequence of bases in globin mRNAs extending beyond the termination codon (see Chap. 5), the ribosome will continue to add amino acids at the C terminus, until another UAA or UAG triplet is encountered. This mechanism is consistent with the structures of "chain–termination" variants such as Hb Constant Spring in which the α chains are extended by an additional C-terminal sequence of 31 amino acids. Another type of elongated variant may arise as the result of deletion of a DNA base to produce a shift in the register of triplet condons within the messenger sequence. Once again the termination codon would be altered to that of another amino acid and the chain would continue, but a number of other codons would also be altered by the shift and the internal amino acid sequence would be changed from the point of the deletion. This type of "frame-shift" mechanism has been invoked to explain several elongated variants of α or β chains which have a few new amino acids just prior to the normal C terminus plus an extended region beyond (*e.g.*, Hb Wayne). In addition to the above lesions arising from point mutation of a single DNA base, there are other more complicated variants involving alterations of larger segments of the genome to delete sequences of several amino acid residues by nonhomologous crossing-over at meiotic division, or to fuse different subunit sequences producing hybrids of two different chains by mispairing of gene loci during chromosome crossover.

The most common and most thoroughly studied of the hemoglobinopathies is sickle cell anemia, a disease characterized by chronic hemolytic anemia, episodes of acute pain in various organs notably in the chest, abdomen, and musculoskeletal system, and the presence of bizarre, deformed erythrocytes in the shape of a crescent moon or sickle. The disease is particularly common in central Africa and in people of African descent. Afflicted individuals are homozygous for the mutant gene; the heterozygotes may show the existence of the mis-shaped erythrocytes in their peripheral blood (sickle-cell trait), but they do not suffer from the hemolytic anemia or painful crises (sickle-cell disease) of the homozygous state. The molecular basis of this inherited syndrome may be attributed entirely to the production of a mutant hemoglobin variant, HbS, $\alpha_2\beta_2^{6\ \text{Glu}\rightarrow\text{Val}}$.

Two consequences arise in the properties of HbS as the result of the amino acid substitution: first, the molecule has a more positive charge resulting from the replacement of negatively charged glutamic acids by neutral valine residues, and hence it may be readily separated from HbA

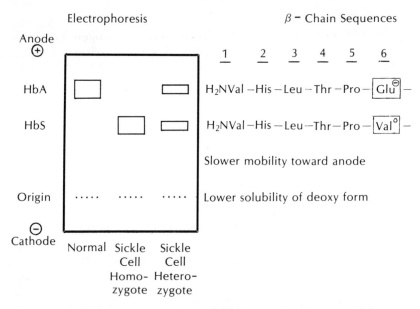

FIG. 1–6. Comparison of normal (HbA) and sickle-cell (HbS) hemoglobins.

by electrophoresis (Fig. 1–6); second, the valine residues are capable of forming hydrophobic linkages to residues on other HbS molecules to produce long, insoluble polymers. The first consequence allows the identification of HbS in blood samples and the detection of heterozygotes as well as homozygotes, while the second accounts for the phenomenon of sickling and the clinical symptoms of the disease.

Erythrocytes containing HbS exhibit normal binding of O_2 in the lungs, and as long as the HbS is in the oxygenated form they have the normal biconcave disc appearance. It is only when a high proportion of HbS undergoes deoxygenation that the molecules join together. The long fibers of deoxy HbS deform the erythrocyte membrane and induce the characteristic sickle shape. This process is normally reversed when the cell is re-exposed to oxygen. Accordingly, the sickling may be observed in smears of blood drawn from either heterozygotes or homozygotes when the O_2 levels fall on standing. Sickling would not be expected to occur in the peripheral circulation of heterozygotes until very severe hypoxia is experienced since there is HbA as well as HbS in the erythrocytes. Individuals with the heterozygous trait generally remain asymptomatic unless exposed to very low PO_2 values (*e.g.*, at high altitudes or in nonpressurized aircraft). However, in the homozygotes extensive sickling may occur with milder hypoxemia, particularly during the severe crises that may follow infections. A vicious cycle becomes established when the deformed erythrocytes occlude small vessels and thus potentiate the hypoxic effects upon sickling (Fig. 1–7). Vascular occlusion of pulmonary vessels is common,

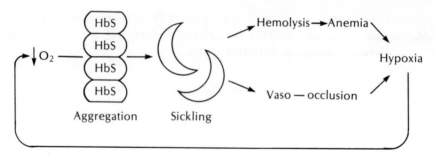

FIG. 1–7. The vicious cycle of sickle cell anemia. Lowered solubility of HbS in the deoxygenated state leads to further lowering of peripheral O_2.

and leads to the characteristic acute pleural pain and susceptibility to chest infections. The chronic effects of vaso-occlusion are seen in damage to a number of different organ systems: cardiac, skeletal, renal, hepatic, and neural involvement may be seen in addition to pulmonary damage.

The anemia is often severe, and results from the distortion of the erythrocyte membrane by the sickling process. In fact, repeated cycles of sickling may cause irreversible damage to the cell membrane with the loss of intracellular potassium and increased osmotic fragility. These damaged cells are accordingly more susceptible to hemolysis, and are removed from the circulation rapidly by the spleen. Hematopoiesis is accelerated in the bone marrow to compensate for the increased destruction of erythrocytes which have mean survival times of only 10 to 20 days compared with the average of 120 days in normal persons. Many of the erythropoietic stem cells are also induced to synthesize the fetal variant of hemoglobin, a response noted in other defects of HbA synthesis. The proportion of HbF may reach as high as 20% of the total. It seems that cells containing HbF show a lower tendency to sickle, and individuals with elevated HbF values have a better prognosis.

To date there is no cure for the disease. The anemia is often well-tolerated since the erythrocytes have compensatory increases of 2,3-DPG, and an increase in the P_{50} value, so that tissues may extract oxygen quite well. Good nutrition to maintain high intake of folic acid in the face of accelerated hematopoiesis is important. In the painful vaso-occlusive crises high pressure oxygen has been used effectively, along with antibiotic therapy and increased fluid intake to offset the deficient concentrating ability of the kidneys. Erythrocytes from patients with sickle cell anemia may be prevented from sickling by treatment with cyanate. The reaction with HbS results in modification of the $-NH_2$ terminal groups by addition of a carbamyl residue. The oxygen dissociation curve is shifted to the left since the amino groups of the N-terminal valines cannot form salt bridges to stabilize the deoxy form; thus carbamylation increases the oxygen affinity of HbS and reduces the Bohr effect. This is great in theory and in practice under *in vitro* conditions; unfortunately, cyanate has produced neuropathic effects upon patients in clinical trials. Other antisickling agents and extracorporeal techniques for modification of HbS are currently under active investigation.

Another type of hereditary hemolytic anemia is seen with certain hemoglobin variants that are readily denatured, and undergo precipitation either spontaneously or in the presence of oxidant drugs. These characteristic deposits of denatured hemoglobin are termed Heinz bodies, and may be detected in erythrocytes from blood smears when appropriately stained. As a result of deformation of their membranes the abnormal erythrocytes are trapped and removed from the circulation by the spleen. Severity of the anemia is quite variable dependent upon the degree of hemolysis, exposure to drugs, infections, and other stresses. The mode of inheritance in these disorders is invariably by autosomal dominant transmission, and it is the heterozygous individuals who are affected; presumably the homozygous conditions would be incompatible with survival. More than half of the circulating hemoglobin is normal HbA since the unstable variant hemoglobin is selectively removed. This class of hemoglobinopathy is typified by Hb Köln ($\alpha_2\beta_2^{98\,\text{Val}\rightarrow\text{Met}}$), the most common of the unstable hemoglobins. The amino acid substitution occurs in the hydrophobic pocket where the heme is normally inserted, with the consequences that heme binding is impaired in the β chains, and the molecule dissociates into unstable, heme-depleted dimers. In other instances the mutant molecule contains an amino acid substituent that is inappropriate for normal contacts at the faces between subunits, or that does not allow the normal α-helix structure within subunits, and hence the stable $\alpha_2\beta_2$ tetramer cannot hold together. Severe hemolytic cases have been treated by removal of the spleen, but most of these disorders need no therapy except for supplementation with folic acid to support the excess erythropoiesis, and avoidance of oxidant drugs.

In several of the hemoglobinopathies the primary effect of amino acid substitution in the mutant hemoglobin is an altered affinity for binding of oxygen. The conditions with increased oxygen affinity are more common, though only 100-odd cases have been reported. The pattern of inheritance is as a Mendelian dominant character, and affected heterozygous individuals have equal amounts of normal HbA and the variant Hb. The latter molecules have a very much lower P_{50} (10–15 *Torr*) so that in the normal course through the circulation they fail to give up their bound O_2 to the tissues. Consequently the only functional hemoglobin in the erythrocytes is the 50% which is HbA, and the body responds as though the individual were anemic. The kidney is stimulated to produce more erythropoietin thus inducing the bone marrow to synthesize more red cells in order to offset the symptoms of anemia. The result is that patients with these high-affinity variants develop an erythrocytosis (dramatic increase in red cell count) so that the total Hb levels in the blood may rise to twice normal values in excess of 20 g/100 ml. Typically the variant Hb has an amino acid substitution which either interferes with interfaces between the α and β subunits with the result that heme–heme interactions are prevented (*e.g.*, Hb Chesapeake, $\alpha_2^{92\,Arg\rightarrow Leu}\beta_2$; Hb Kempsey, $\alpha_2\beta_2^{99\,Asp\rightarrow Asn}$), or interferes with certain critical regions of the β chain with the result that DPG binding is prevented (*e.g.*, Hb Rahere, $\alpha_2\beta_2^{82\,Lys\rightarrow Thr}$) or the salt bonds required for the Bohr effect are prevented (*e.g.*, Hb Hiroshima, $\alpha_2\beta_2^{146\,His\rightarrow Asp}$). Although only half of the circulating hemoglobin is functional in oxygen transport in these cases the erythrocytosis is adequate to compensate for the defect, and apart from a ruddy complexion the conditions are asymptomatic and require no treatment. Diagnosis by determination of oxygen affinity is important only to prevent inappropriate therapy, such as the administration of radiomimetic drugs or radiation of bone marrow which are interventions used to treat the cancerous erythrocytosis of polycythemia vera.

The opposite situation, increased deoxy Hb in arterial blood with resulting blue–purple complexion (cyanosis) is seen in some individuals with low-affinity variants (*e.g.*, Hb Kansas, $\alpha_2\beta_2^{102\,Asn\rightarrow Thr}$). Here the blood is incompletely oxygenated in the lungs, but it does unload a normal quantity of O_2 to tissues, and the conditions are generally benign. Familial cyanosis, with blood coloration appearing brown to almost black, is also the obvious feature in individuals containing congenital methemoglobin variants (*e.g.*, HbM Saskatoon, $\alpha_2\beta_2^{63\,His\rightarrow Tyr}$). Here the histidine which normally binds to heme and keeps the iron in the ferrous state is replaced by tyrosine which favours binding to the oxidized ferric state in which the iron is incapable of binding oxygen. Again the conditions seem to be quite benign, and the major reason for correct diagnosis is to reassure the patient and to prevent incorrect therapy on the mistaken premise that the cyanosis is due to a cardiac or pulmonary disorder.

Finally, there is a group of hereditary disorders with defective total synthesis of one of the normal globin chains. Because of the large numbers of afflicted individuals with Greek or Italian ancestry the term thalassemia (Mediterranean anemia) has been applied to these syndromes. In α-thalassemia, production of α chains is affected so that the HbF ($\alpha_2\gamma_2$) and HbA ($\alpha_2\beta_2$) are decreased; the free γ and β chains may form the abnormal Hb Barts (γ_4) or HbH (β_4). In β-thalassemia HbA production will be impaired because of deficient β chain synthesis but HbF or HbA$_2$ ($\alpha_2\delta_2$) may be formed; the free α chains cannot combine to produce a stable tetrameric molecule, and instead they will aggregate into large inclusion bodies which stain like Heinz bodies, and damage the erythrocytes to cause a hemolytic anemia.

The genetics and expression of the defect in the thalassemia syndromes is quite complex. The mutation appears to affect regulatory genes that are separate from, but closely linked to, the structural genes for the respective globin chains. The major symptoms of hemolytic anemia with erythrocytes of low hemoglobin content (hypochromic) and small cellular size (microcytic) are most notable in homozygotes for β-thalassemia (Cooley's anemia); heterozygotes show lesser blood changes such as increases of HbF and HbA_2, and for this reason are designated as cases of thalessemia minor. In the case of α-thalassemia major symptoms seem to be related to deletion of two α-chain genes, and many of the heterozygous carriers show no clinical symptoms. The most severe homozygous condition is Hb Bart's hydrops fetalis syndrome in which the infant dies *in utero* or within minutes of birth with virtually all the hemoglobin as Hb Bart's (γ_4). With a less severe defect in the α-chain loci there is survival to adulthood with variable severity of the anemia, and a shift from γ_4 to β_4 production after birth (Hemoglobin H disease). The latter condition may result as a combination of two heterozygous defects at separate α-chain loci.

The molecular defect in the erythroid cells of thalassemia patients has been clarified in recent work by isolating the components involved in globin chain synthesis. These studies have demonstrated that the defective gene in thalassemia has the capacity to transcribe messenger RNA, but when the latter is added to purified ribosomes plus the components necessary for protein synthesis, it does not function optimally in translation of the specific globin chain. For example, mRNA isolated from nonthalassemic bone marrow cells directs the synthesis of equal amounts of α and β chains, while the mRNA from β-thalassemia cells promotes very little or no β chain synthesis relative to that of the α chains; similar but opposite imbalances were found with mRNA's from α-thalassemia cells. There is evidence indicating that the mRNAs are present, but in greatly reduced amounts in homozygous β-thalassemia, though in some instances the β-chain mRNA is not functional. In α-thalassemia of the severe homozygous variety (Hb Bart's hydrops) there is no evidence for α-chain mRNA and the α-chain gene appears to be completely deleted; the less severe HbH disease is characterized by reduced or less efficient α-chain mRNA in the cells.

PLASMA PROTEINS

Historically, the proteins in blood plasma or serum were divided into two categories: those which precipitated in the presence of half-saturated ammonium sulfate—the globulin fraction, and those which remained in solution—the albumin fraction. The latter constituted over half of the total (3.5–5.5 g/100 ml plasma), and was subsequently shown to consist of a single, simple protein (*i.e.*, composed entirely of amino acid residues). However the globulin fraction (2.0–3.0 g/ml) has since been found to contain many components with diverse conjugated groups attached. Early attempts to subfractionate this heterogeneous group of proteins based upon their differences in charge by the process of electrophoresis led to their classification into three groups: the α-glycoproteins which were rich in carbohydrates; β-lipoproteins which contained half or more of their weight as lipids; γ-immunoglobulins which represented circulating antibodies; present in plasma, but absent from blood serum, was an additional protein, Φ-fibrinogen which acted as the precursor of the clot protein, fibrin. Further refinements of separation techniques based upon molecular size as well as charge differences have revealed extensive heterogeneity and overlaps among the α, β, γ categories, and although vestiges of these designations persist in the medical literature it is probably better to group the blood proteins according to the systems or functions in which they participate. Examples of the systems include the immunoglobulins (IgG, IgA, IgM, IgD, IgE), the complement system of antigen–antibody reactions (C1, C2, C3, ... C9), the coagulation proteins (prothrombin, plasminogen, fibrinogen, and various activating or stabilizing factors), the protease inhibitors (antitrypsin, antichymotrypsin, antithrombin), and lipoproteins (HDL, LDL, VLDL). In addition there are the various transport proteins; for vitamins (retinol-binding globulin—vitamin A, transcobalamin—vitamin B_{12}); for minerals (transferrin—iron, ceruloplasmin—copper) for hormones (thyroxine-binding globulin—thyroid hormones, transcortin—steroid hormones); for heme (hemopexin); for free hemoglobin (haptoglobin); as noted below albumin also binds and transports many organic and inorganic molecules. Several proteins found in the blood plasma may be considered as passengers and include the polypeptide hormones (insulin, glucagon, growth hormone, corticotrophin, gonadotrophins); the tissue-derived enzymes (acid and alkaline phosphatase, lactate dehydrogenase, amylase, aminotransferases); the pregnancy-associated glycoproteins (SP_1, SP_2, SP_3); the tumour-associated proteins (myeloma proteins, α_1-fetoglobulin of hepatoma patients, lysozyme in monocytic leukemia). The ability to identify and measure fluctuations of particular plasma proteins is of obvious importance to our understanding of human physiology and to clinical medicine. Modern technical innovations utilizing gel polymers as supports for electrophoresis and immunoprecipitation methods allow even small regional hospitals to carry out rapid, quantitative studies of the diverse proteins in blood serum and other body fluids on a routine basis.

ALBUMIN

The primary metabolic function of albumin is the transport of nonesterified fatty acids in the blood plasma. This is exemplified by the rare genetic disease, analbuminemia, which is characterized by an inborn defect of albumin synthesis in the liver and a gross disturbance of fat utilization in the body. Normally, the free fatty acids that are liberated from adipose tissue by intracellular lipase action upon stored triacylglycerol, are carried with the albumin molecules in the bloodstream to the tissues of utilization such as the liver or muscles. Albumin has the capacity to bind six molecules of fatty acid per protein molecule; the first two molecules are more tightly bound ($K_a \sim 10^8 M$) than the other four ($K_a \sim 10^6$). In addition to providing a water-soluble complex for efficient transport through the circulation, the albumin facilitates the rapid uptake of fatty acids into peripheral tissues by acting as a reservoir to permit a much higher concentration of fatty acids near the surface of cells than could be attained if they were in an aqueous solution. Albumin also binds a number of hydrophobic metabolies (*e.g.*, bilirubin, tryptophan) and drugs (*e.g.*, salicylates, sulfonamides). The binding of bilirubin is critical in the neonatal period, when an excess of the bile pigments arising from accelerated hemolysis may lead to uptake in the nervous system with consequent brain damage (kernicterus). Protection from bilirubin toxicity may be afforded by infusion of extra albumin, which carries the bile pigments to the liver for conjugation and excretion in the bile. Competitive interactions may be seen for the binding of different hydrophobic agents: thus one drug may displace another depending upon their relative affinities for the same binding sites, or by altering the conformation of the albumin molecule. The competitions among drugs, fatty acids, and bile pigments render studies of the pharmacology of albumin binding *in vivo* very complex. As an example, it has been shown that elevations of free fatty acids, by displacing albumin-bound tryptophan, will increase the concentrations of this amino acid entering the brain, and hence may modulate the concentrations of serotonin and other biogenic amines in the nervous system.

A secondary transport function of serum albumin is its ability to provide a reservoir for some metal ions (Ca^{2+}, Cu^{2+}) as well as certain hormones (thyroid hormones, glucocorticoids, sex steroids). Specific binding proteins of the globulin fraction have higher affinities for binding these hormones, but albumin serves as a low-affinity-but-high-capacity buffer pool for the overflow from the more specific binding proteins, and stabilizes the plasma concentrations of the free active forms. Albumin also appears to play a significant part in solubilizing and carrying the poorly soluble disulfides of cysteine and glutathione (cystine or Cy-S-S-Cy, and G-S-S-G respectively). This is accomplished by thiol-disulfide exchange with the free SH group on albumin, one of the rare examples where a ligand forms a covalent link with albumin.

Two additional physiological functions have been attributed to serum albumin, although other proteins of the blood may participate nonspecifically as well. The first function is the use of albumin as a source of protein nutrition for peripheral cells. In fact, any protein that may be taken into cells by the process of pinocytosis (see Chap. 4) may also play a nutritional role. The reason for considering a special contribution from albumin is the evidence for very high rates of turnover of albumin in the peripheral tissues. It is estimated that 15 g of albumin, representing one tenth of the total body synthesis of proteins, is taken up by the tissues and replaced by

the liver each day. The second function of serum albumin is in the maintenance of osmotic balance between the blood and other extracellular fluids
of the body. In the blood capillaries on the arterial side of the circulation
there is a tendency for fluid to escape because of the high hydrostatic
pressure outward exerted by pumping of the heart. The dissolved proteins
of the blood plasma are not so easily extruded through the cells lining the
vessels as are smaller molecules, and therefore the concentrations of these
proteins in the interstitial fluid outside the capillaries is considerably
lower than in the bloodstream. Owing to its abundance, its relatively
small size as a protein, and its highly charged surface, albumin, of all the
proteins in plasma, acts most effectively to draw water back from the
extracapillary space by osmotic action. As the blood passes through the
capillary bed the hydrostatic pressure outward will drop greatly to reach
values below the inward osmotic pressure exerted by plasma proteins, so
that, on the venous side of the circulation, fluid will be induced to return
from the interstitial spaces into the blood. This movement of fluid back
into the venous capillaries counteracts the outward movement to prevent
excessive water loss from the circulation.

In conditions of protein malnutrition, an inadequate supply of amino
acids to the liver results in depression of albumin synthesis by 50% or
more. Continuing losses of albumin from the circulation by leakage and
use in the periphery with inadequate replacement leads to very rapid
depletion of plasma albumin before changes in the globulins are noteworthy. The consequence is failure to maintain normal osmotic retention of
water in the blood with excesses in the interstitial water content (edema).
Treatment of patients with this type of water loss from the circulation is by
nutritional replacement of adequate dietary protein which leads to rapid
albumin replacement. Hypoalbuminemia and consequent edema is a
common result of dietary inadequacy (e.g., in the infant protein deficiency
syndrome, kwashiorkor, and in adult chronic alcoholism). In the latter
situation the nutritional problem is frequently worsened by the toxic effects of ethanol upon the liver cells; the combination of hepatic cirrhosis
and insufficient amino acid supply often leads to drastic curtailment of the
synthesis of albumin by the liver. A lowering of serum albumin is also seen
in conditions where extensive losses occur in the urine (nephrotic syndrome, glomerulonephritis) or in the gastrointestinal tract (protein-losing
enteropathy). In acute fluid loss from the vascular system (burns, hem-
morrhagic shock) it is essential to replace the albumin by intravenous administration immediately. The body can apparently provide some chronic
adaptive mechanisms whereby other plasma proteins regulate the osmotic
balance, since patients with hereditary analbuminemia do not show the
symptoms of edema.

In recent years the chemistry and structure of albumin have been well
elaborated. The molecule consists of a single polypeptide chain of 584
amino acids, with a molecular weight of 66,300. Charged amino acids

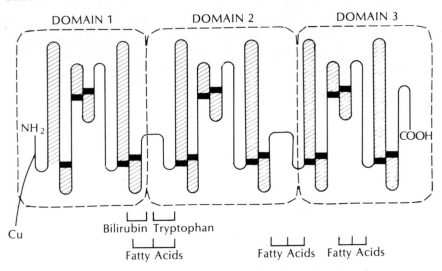

FIG. 1–8. Albumin structure, showing regions of the three domains where hydrophobic ligands bind.

(glutamate, aspartate, lysine) constitute a quarter of the total thus rendering the molecule very polar; the acidic residues outnumber the basic amino acids by two to one, which confers a high net negative charge accounting for the high mobility of albumin towards the anode in electrophoresis at alkaline or neutral pH's. Secondary structure of the protein is over half in the α-helical configuration, 15% as β-sheet and the remainder random coil. The tertiary structure is approximately that of an ovoid 38×150 Å, with physical and chemical evidence for a large globular domain in the middle flanked by two smaller spheres (Fig. 1–8). Considerations of the folding properties of the polypeptide chain, and predictions from the probable localization of the sites for the 17 Cy-S-S-Cy disulfide intrachain bridges have produced a visualization of the three domains at a molecular level. Each domain consists of two subdomains, the latter containing three α-helical rods about 20 residues long looped back to form a trough-like structure. The troughs, and the interfaces between domains have a preponderance of the hydrophobic side-chains, while the polar amino acid residues are arranged to face the exterior surface of the albumin molecule. This is consistent with the high solubility of the molecule in aqueous media due to the exposed hydrophilic residues; binding sites for long-chain fatty alkyl groups, bile pigments, and other hydrophobic ligands lie in the clefts between loops of the polypeptide chains. Two pairs of fatty acid binding sites occur near one end of the molecule; a third pair coincides with the region where bilirubin, aspirin and tryptophan bind. This may explain why the competitive interactions mentioned earlier are seen when fatty acid levels are high enough to saturate all three pairs of binding sites. Binding of certain metal ions such as Cu^{2+} occurs by a tight chelation complex involving a histidine residue near the NH_2 terminus, and is stoichiometric. Calcium ions are bound loosely by ionic interactions at 16 acidic sites.

LIPOPROTEINS

The lipid-bearing proteins of the blood traditionally have been classified into four broad groups on the basis of their sedimentation properties: the high-density lipoproteins (HDL), the low-density lipoproteins (LDL), the very low-density lipoproteins (VLDL), and the chylomicrons. The first two types contain most of the serum cholesterol, and are involved in the movement and modification of circulating free sterol and sterol esters, while the last two act as major vehicles for the triacylglycerols in blood. The *chylomicra*, the lightest and largest particles, have the highest content of lipid relative to protein. They produce the milky turbidity in serum after ingestion of a fat-containing meal and seem to be synthesized exclusively by mucosal cells of the gastrointestinal tract during fat absorption. Their triacylglycerol content therefore represents the exogenous or dietary fat intake; chylomicra will be absent from normal serum of fasting individuals. The *VLDL*, next in size and lipid content, are produced and secreted by the liver, and their main component of triacylglycerol is chiefly of endogenous origin arising from the fats synthesized within the body by hepatic cells. Both the chylomicra and VLDL may be acted on by those extrahepatic tissues, such as the muscles or the adipose tissue, that contain the hydrolytic enzyme, lipoprotein lipase. The latter is localized extracellularly on endothelial surfaces and attacks the triacylglycerols of lipoproteins to release free fatty acids which are then taken up by muscle cells for oxidation or by the adipose cells for storage. The *LDL* in man arise from catabolism and modification of the larger VLDL particles. As VLDLs are progressively depleted of triacylglycerols they also lose some protein and free cholesterol, the latter being replaced by fatty acyl esters of cholesterol. As they circulate through extrahepatic tissues the LDL combine with specific receptors on cell surfaces and are endocytosed into the cells where the ester linkages become hydrolysed to deliver free cholesterol to the peripheral tissues. The *HDL* produced mainly by liver cells are the smallest lipoprotein particles with the lowest lipid content. They facilitate the VLDL–LDL modifications in the circulation by exchanging lipid and protein components with the other lipoprotein classes. Moreover, the conversion of cholesterol to its esters in blood plasma by the enzyme lecithin–cholesterol acyl transferase (LCAT) is associated with HDL particles. There is growing evidence that the HDL may also control the deposition of cholesterol in peripheral cells or facilitate its removal to the liver from sites of peripheral deposition.

In progressing from the largest (lightest) to the smallest (heaviest) lipoproteins there is a progressive decrease in total lipid, mainly by deple-

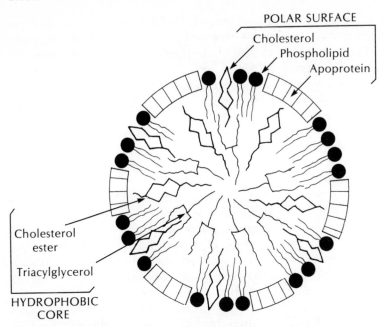

POLAR SURFACE
Cholesterol
Phospholipid
Apoprotein

Cholesterol
ester

Triacylglycerol

HYDROPHOBIC
CORE

FIG. 1–9. Structure of a typical plasma lipoprotein with phospholipid
polar head groups, cholesterol hydroxyls, and apoproteins oriented
toward the surface, and the esterified alkyl residues oriented toward the
hydrophobic core region of non-polar lipids.

tion of triacylglycerols, with an accompanying increase in total protein,
and increased proportions of phospholipids and sterols, particularly the
cholesterol esters. Each density category will consist of a range of particles,
and there will be overlaps at the boundaries between categories. Nonetheless, the families, HDL, LDL, and so forth do have discrete properties and
contain characteristic apoproteins.

The apoproteins have been designated alphabetically, A through E, by
their occurrence as major components within lipoprotein categories: the
main polypeptide of the HDL or α-lipoproteins are termed apo-A; the main
components of the LDL or β-lipoproteins are apo-B; important constituents of the VLDL in addition to apo-B are the apo-C peptides. In several
cases these categories may be further subdivided (*e.g.*, apo-A-I, apo-A-II),
and apo-D and apo-E may accompany these subdivisions as minor components.

The chylomicra are up to 90% triacylglycerol, 4 to 5% each of cholesterol and phospholipids, and only 1% protein. The latter comprises apo-
C > apo-B > apo-A; together with cholesterol and phospholipid it provides a surface coating of these large particles, with the triacylglycerol and
cholesterol esters occupying the hydrophobic core (Fig. 1–9). The VLDL
are about 50% triacylglycerol, 20% each of cholesterol and phospholipids,
and 10% proteins. The latter are equal portions of apo-B and apo-C with

traces of apo-A. Though smaller in diameter, VLDL have a spherical structure similar to that of chylomicra, with a hydrophilic outer layer enclosing
the lipid core. The LDL are up to 50% cholesterol (principally as the ester),
20% phospholipid, 10% triacylglycerol, and 20% protein. The latter is almost entirely in the form of apo-B with traces of apo-C. The particles have
a small highly viscous core of cholesterol ester and triacylglycerol surrounded by the more polar lipids and apoproteins, possibly forming a
trilayer. The apo-B is exposed at the surface, where it may react with cell
surface receptors for the LDL. The HDL is up to 30% phospholipid, 20%
cholesterol ester, with only traces of free cholesterol or triacylglycerol, and
50% protein. The latter is mainly apo-A with traces of apo-C. The phospholipid (chiefly lecithin) and apo-A can associate together to form
bimolecular discs; when cholesterol ester is added it assumes a core location with equal areas of an enveloping sphere of protein and phospholipid
as in circulating HDL particles. Thus, although the apoproteins and the
extent of the interior core may vary, in each of the lipoprotein classes a
central hydrophobic domain exists with polar substituents providing the
outer interface with the aqueous milieu of the blood (Fig. 1–9).

Triacylglycerol transport to peripheral tissues involves the synthesis
of chylomicrons by cells of the intestinal mucosa and the synthesis of
VLDL particles by cells of the liver (Fig. 1–10). This classical view has had
to be modified upon the realization that very large lipoprotein particles in
the range of chylomicra may also be produced in endogenous hypertriglyceridemia induced by high carbohydrate feeding in humans. Moreover,
during alimentary lipemia some smaller particles, termed intestinal
VLDL, containing fatty acids of dietary origin will accompany the
chylomicra. The release of lipid in peripheral tissues requires the action of
lipoprotein lipase. This crucial enzyme is inactive against triacylglycerol
emulsions, and requires a specific protein activator in lipoprotein proteins.
The most effective activator appears to be apo-C-II which is transferred
both to chylomicra and to the VLDL particles in the circulation from the
HDL fraction. The lipoprotein lipase of muscle tissues, such as the heart, is
maximally active in the fasting state, thus allowing the release and uptake
of fatty acids for energy generation by these tissues. On the other hand, the
adipose tissue lipoprotein lipase is depressed by starvation preventing the
uptake of fat for storage purposes (Fig. 1–10). A major positive stimulus to
lipoprotein lipase in fat cells is operative in the fed state, and is the result
of insulin converting the adipose enzyme into a more active state, thus
favoring fatty acid uptake in the depots for lipogenesis. The larger
chylomicron particles are cleared from the circulation most rapidly; their
partial breakdown by tissue lipoprotein lipase leaves a group of smaller
remnant particles enriched in apo proteins and cholesterol esters that are
removed from the circulation by the liver. The VLDL have somewhat
longer survival times; as they are attacked by the lipoprotein lipases the
apo-B becomes enriched, the apo-C content declines, and the residual

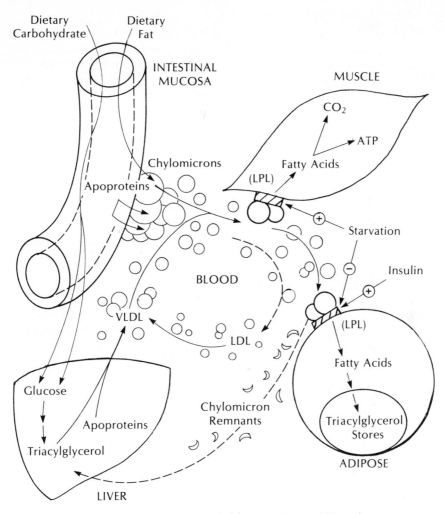

FIG. 1–10. Triacylglycerol transport and delivery to tissues. LPL = tissue lipoprotein lipase. Positive and negative modifiers are shown by + and − signs.

cholesterol becomes more highly esterified as the VLDL are transformed into LDL particles.

Cholesterol transport in blood requires cooperative action of all lipoprotein species (Fig. 1–11). Three sources of circulating cholesterol contribute to the outflow to peripheral tissues: the ingested sterols from the diet, sterols synthesized by cells lining the gastrointestinal tract, or sterols which are synthesized by the liver. In humans the serum cholesterol is only moderately enhanced by high cholesterol ingestion, and the last two sources seem to be the most significant. The dietary cholesterol and that which is synthesized in mucosal cells is conveyed from the intestines to the

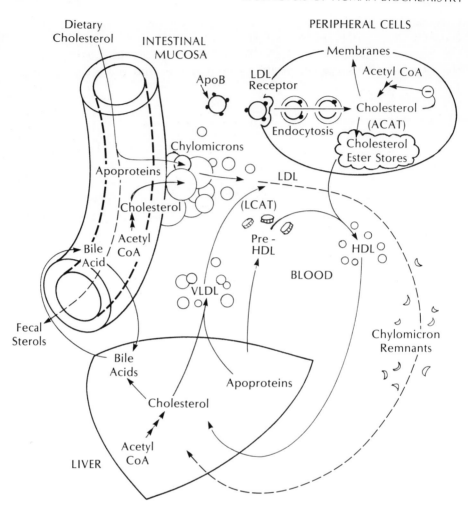

FIG. 1–11. Cholesterol transport and delivery to tissues. LCAT = lecithin–cholesterol acyl transferase; ACAT = acyl-CoA-cholesterol acyl transferase.

liver in the chylomicron remnants, while some enters the circulation directly as VLDLs of intestinal origin. The remnants of chylomicra are taken up by the liver and acted upon by hepatic lipase. The free fatty acids and cholesterol resulting may be utilized by liver for lipogenesis, for bile acid synthesis, or for secretion in the form of VLDLs of hepatic origin. The latter will also contain the cholesterol which is synthesized *de novo* in liver cells, probably the most important single source of endogenous cholesterol in the human. The VLDL particles, as noted above will be converted into LDL particles after they have served their function in carrying the triacylglycerol component to extrahepatic tissues. The LDLs so formed then act to dispatch cholesterol to the periphery.

The transformation of the VLDL to LDL requires the participation of the LCAT enzyme as well as lipoprotein lipase, and the HDL particles. Following removal of the bulk of the triacylglycerol by lipoprotein lipase action, the VLDL transfer their apo-C proteins to the HDL particles. The LCAT enzyme is associated with the HDLs which also contain the specific activator for this enzyme, apo-A-I. Cholesterol molecules at the surface of VLDL particles may transfer to the HDL where they serve as substrates for the LCAT reaction, transesterification of the unsaturated fatty acyl groups from the 2-position of lecithin molecules. The cholesterol esters of the HDL may then transfer back to the gradually diminishing VLDL; eventually the latter, containing apo-B as the sole protein, and cholesterol ester as the major lipid, assume the properties of LDL particles.

A great deal of interest surrounds the peripheral uptake of LDL particles, since the elevation of this lipoprotein fraction in serum and the excessive deposition of its cholesterol ester content in blood vessel walls have been implicated in development of the cardiovascular lesions of atherosclerosis. A number of nonhepatic cell types, including arterial smooth muscle cells and human fibroblasts growing in culture, have high-affinity receptor sites for LDL which seem to bind at the cell surface, through their apo-A component (Fig. 1–11). Once bound the LDL are rapidly internalized in the cells by a temperature-dependent mechanism of endocytosis that delivers the intact LDL particles into the cellular lysosomes. Here the apo-B is broken down by proteases, and the cholesterol esters are broken down by an esterase to free cholesterol. The latter is used by the cells for membrane synthesis, and any excess cholesterol is reesterified for storage. Each of these processes is subject to adaptive regulation in normal cells. Thus the number of LDL receptors declines when cells are exposed continuously to exogenous LDL to prevent the overaccumulation of intracellular cholesterol. Moreover, the rate of synthesis of cholesterol in the cells is reduced in the presence of added LDL. The mechanism of this negative feedback regulation is by the internalized cholesterol acting to repress the synthesis of the key enzyme, β-hydroxy, β-methyl glutaryl CoA (HMGCoA) reductase in the *de novo* pathway for cholesterol biosynthesis. At the same time the internalized cholesterol activates the cytoplasmic acyl CoA cholesterol acyl transferase (ACAT) several hundredfold to esterify and store the excess as cholesterol esters.

Transport of cholesterol from the extrahepatic tissues back to the liver has been postulated to involve the LCAT enzyme system and the HDL particles. The latter facilitate loss of cholesterol from cells, and also reduce the binding, uptake, and breakdown of LDL by cells. Moreover, the size of the extravascular pool of cholesterol is inversely related to the concentration of HDL in plasma suggesting that the rate of removal of tissue cholesterol for transfer to the liver may be dependent upon the HDL particles, in analogous fashion to the role of LDL in transporting cholesterol in the opposite direction. It seems significant that although factors that contribute to greater risk of atherosclerotic cardiovascular disease are associated with elevated LDL concentrations in blood, they show an inverse correlation with levels of HDL particles. Thus preventive measures designed both to lower LDL and raise HDL concentrations are being sought in order to arrest or reverse the atherogenic process.

Lipoprotein Disorders

The familial deficiencies of serum lipoproteins that have been observed in human populations fall into two categories: *abetalipoproteinemia* in which apo-B and LDL are absent; *Tangier disease* in which reduced amounts of HDL with abnormal structure are found. The abetalipoproteinemia is a rare autosomal recessive trait characterized by profound disturbances in fat absorption, severe ataxia, progressive degenerative pigmentation of the retina, and presence of characteristic spiky red cells (acanthocytes) in the blood. The inability to absorb fat is not surprising in view of the fact that chylomicra and VLDL, in addition to LDL, require apo-B for their formation. As expected there is defective secretion of triacylglycerol from both intestinal mucosa and liver, the cells of which show fatty infiltration. Plasma lipids, particularly the triacylglycerol and lecithin fractions, are markedly depressed. All the serum cholesterol is borne by the HDL particles. The retinal lesions may be related to defective absorption of vitamin A, which is normally taken up with the chylomicron fat. The neurologic and hematologic defects are unexplained but could be associated with defective membrane structures arising from deficient absorption of the polyunsaturated fatty acids and vitamin E which protects them from auto-oxidation. Treatment consists of restriction of long-chain dietary fat and substitution with medium-chain (C_8– C_{14}) fat which is absorbed via the portal route rather than in chylomicra. Supplementation of diets with high doses of fat-soluble vitamins (A, E, K, and D) is also important to offset the secondary effects. Tangier disease is also a recessive trait, with enlargements of the spleen, and the tonsils which show unique orange striations. There is deposition of cholesterol esters in cells of these organs and of liver, skin, lymph nodes, intestinal mucosa, and the cornea. Peripheral neuropathy with loss of pain and temperature sensitivity is frequently observed. Serum cholesterol is very low, particularly in the HDL fraction, but triacylglycerol content is normal, thus providing a clear distinction from abetalipoproteinemia. The major proteins of normal HDL, apo A-I and apo-II, are detected in serum of Tangier disease in their usual proportions (3 : 1). The nature of the biochemical defect is unknown, although it is likely that a molecular anomaly of HDL synthesis or assembly is responsible for excessive cholesterol ester uptake by peripheral cells, or defective removal for transport back to liver. Despite the often massive deposits of cholesterol esters there does not seem to be the anticipated tendency toward atheroma formation in the cardiovascular system in patients with Tangier disease.

The inherited hyperlipoproteinemias have been categorized into five major types. *Type 1* is a rare familial syndrome with massive increases of chylomicrons in the blood. There are excessive amounts of circulating triacylglycerols and cholesterol of dietary origin; feeding a fat-free diet for several days leads to normalization of the blood lipoproteins. The condition is readily diagnosed from the characteristic white creamy layer floating on a sample of plasma after it has stood overnight in a refrigerator. The inability to clear chylomicrons from the circulation is attributed to deficient extrahepatic lipoprotein lipase. In typical patients with this condition the activity of this enzyme in blood is not elevated following an injection of heparin as it is in normal individuals. Patients with this syndrome show symptoms of spleen and liver enlargement, acute abdominal pain particularly after a fatty meal, and eruptive xanthomas. The most serious complication is pancreatitis, which may prove fatal. Despite the excessive plasma levels of cholesterol and triacylglycerol from birth, the surviving adult patients have no signs of premature atherosclerosis. Treatment is by dietary management with restriction of total fat intake, and supplementation with medium chain triacylglycerols which do not contribute to the chylomicron output. *Type II* is the most common of the hyperlipoproteinemias. It has a dominant mode of inheritance, and has very severe consequences. The afflicted individuals with homozygous genotype show great enhancement of serum cholesterol, principally in the LDL fraction, from birth, with massive deposition of lipid in the tendons (xanthoma) and the blood vessels (atheroma). The latter generally lead to death from ischemic heart disease in early life. Heterozygotes, although less severely afflicted, show similar symptoms and accelerated atherosclerosis. The biochemical lesion is a failure of peripheral cells to bind the LDL particles and take them up by endocytosis. In heterozygotes the cells possess only one-half of the normal number of LDL receptors, causing intermediate effects to those seen in the homozygotes where the receptors are totally absent. These important findings suggest that internalization of LDL by tissue cells is a crucial rate-limiting step in maintaining normal levels of serum cholesterol, and hence in staving off the atherogenic process. Moreover, the defective cells lack the normal negative feedback by LDL uptake on the HMG–CoA reductase step, and hence synthesize cholesterol in an accelerated, uncontrolled fashion. Treatment directed toward lowering the plasma LDL–cholesterol level is considerably more effective in the less severely afflicted heterozygotes. Low cholesterol intake combined with cholestyramine supplementation to remove sterols from the body by binding, and excretion of bile acids in the gut is currently the most useful therapy; new methods and drugs to lower serum cholesterol are under active investigation, particularly for management of the more severely affected homozygotes. *Type III* hyperlipoproteinemia is a rare syndrome, characterized by elevations of both cholesterol and triacylglycerols in serum, xanthomas in palms of the hand and on the elbows, peripheral vascular disease, and an increased incidence of ischemic heart disease. Biochemically the VLDL fraction is grossly increased, but the lipoproteins appear in an anomalous position as a broad beta band on electrophoresis. This abnormal lipoprotein, which has more cholesterol than normal VLDL, is designated as β-VLDL. The particles are smaller than normal VLDL, and lack the apo-C components. After a fatty meal there is a persistence of chylomicrons, but these are also smaller in size and contain relatively more cholesterol esters. Based upon these and other findings it has been suggested that abnormal lipoproteins in Type III serum are partially processed "remnants" of VLDL and chylomicra, and that there is a defect in conversion of these intermediate forms to normal end-products. Recent studies have indicated that the β-VLDL remnants are a common companion of vascular diseases induced by high cholesterol intake, suggesting that these remnant lipoproteins, if not metabolized normally, may play a major role in atherogenesis. Decline in blood lipids and disappearance of xanthoma are readily affected by dietary control of caloric, fat, and cholesterol intake, supplemented by lipid-lowering drugs such as nicotinic acid, thyroxine, or clofibrate. *Type IV* is a hypertriglyceridemia with excess VLDL in serum that is produced by endogenous lipogenesis. The condition is inherited as an autosomal dominant trait. Patients characteristically are obese with premature propensity to ischemic heart disease, but they generally lack the xanthoma deposits or acute abdominal pain seen in other hyperlipoproteinemias. The cause of the triglyceridemia is controversial. Some evidence suggests that patients have elevated insulin levels in serum, and a hypersensitivity of hepatic lipogenesis to induction by carbohydrates. To this end diets are frequently devised with restriction of both caloric and carbohydrate intake. Long-term improvement of the hypertriglyceridemia following such dietary therapy may be attributable primarily to weight reduction. While some turnover studies support the idea that VLDL production is enhanced, there is also an indication that there is a defect in the removal of triacylglycerols from the circulation. However, this removal defect is not related to defective lipoprotein lipase, which in contrast to Type I hypertriglyceridemia, appears to be normal in individuals with the Type IV lesion. A combination of dietary regulation to correct the obesity problem and administration of the drug, clofibrate, seems to be the most effective method for reducing the serum triacylglycerol levels, and hopefully in offsetting the atherogenic process. The *Type V* syndrome is a very severe hypertriglyceridemia in which the triacylglycerol levels are 10 to 20 times higher than normal, and both chylomicra and VLDL fractions are increased. The clinical severity of the condition is also greater with eruptive xanthoma, acute abdominal pain, and accumulation of fat-laden cells in spleen and other organs of the reticuloendothelial system. There is no evidence for premature vascular disease. The etiology is not clear, but the severe hypertriglyceridemia is alleviated by restriction of caloric intake. In particular the extreme chylomicronemia which is associated with abdominal pains and development of pancreatitis may be dramatically and rapidly relieved by fasting. For maintenance patients require reduced fat and strict caloric balance of the diet.

IMMUNOGLOBULINS

Defense of the body against invasion by foreign organisms or macromolecules is effected by protective leukocytes (cell-mediated), and neutralizing antibodies (chemically-mediated). The latter humoral agents are contained in the gamma globulin fraction of the serum. Even without an active infection the normal serum contains hundreds of different antibody molecules. When a foreign protein molecule enters the body it acts as the trigger (antigen) for the synthesis of the specific globulin (antibody) which can combine selectively with the antigen and lead to its inactivation. The primary and three-dimensional structures of the immunoglobulins, the molecular biology of their biosynthesis and their interactions with antigens, the complement system, and elements of the cell-mediated response are active areas of inquiry.

The immunoglobulin (Ig) proteins of the blood can be divided into three major classes IgG, IgA, IgM, and two minor classes, IgD, IgE, based on their physical and chemical properties. The subunits in all immunoglobulins are of two types: a shorter or light chain (L), and a longer or heavy chain (H). Within each of the chains there are regions or domains which differ little from one molecule to another, and these are known as the constant (C) domains; other regions will have different amino acid sequences depending upon the antigen the molecule reacts with, and these are known as the variable (V) domains. Upon protease digestion immunoglobulins are readily cleaved into fragments containing the constant domains (Fc), and those fragments containing the variable antibody determinants (Fab).

The chemistry and shape of the various classes of immunoglobulins follow a similar pattern, and are epitomized by the structure of the major G class of molecules (IgG) which have molecular weights of about 150,000 (Fig. 1–12). Each molecule consists of four polypeptide chains, two H chains of about 450 amino acids (molecular weights of 50,000), and two L chains of about 220 amino acids (molecular weights of 25,000). The L chains comprise the variable region (V_L) of 110 residues at the NH_2 terminus, joined to the constant region (C_L) of similar size, while the H chains comprise the variable domain (V_H), also of 110 residues at the NH_2 terminus, joined to a sequence of three constant regions (C_{H1}, C_{H2}, C_{H3}). Overall shape of the molecule resembles a Y: C_{H3} and C_{H2} regions of the H chains

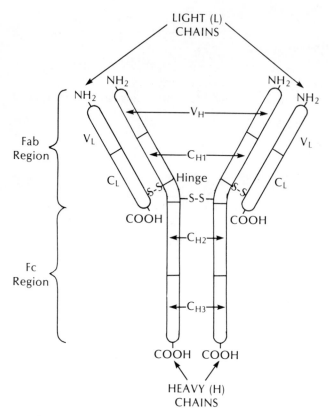

FIG. 1–12. Immunoglobulin structure, showing the variable (V) and constant (C) regions of the heavy (H) and light (L) chains.

intertwine to form the base of the Y; a disulfide bond links the L chains near the COOH terminus of the C_L regions to C_{H1} regions of the H chains to form the arms of the Y. A hinge-region at the juncture of C_{H1} and C_{H2} regions is the point where disulfide bonds hold the two heavy chains together to produce the F_c portion at the base, which also contains several carbohydrate residues. The variable regions (V_L and V_H) bearing the antigen reactive sites are exposed at the tips of the extended arms, thus forming the bifunctional Fab regions.

The classes of immunoglobulins vary somewhat in their sizes, distribution, and functions. The main chemical differences are found in their H chains designated α for IgA, γ for IgG, δ for IgD, and so forth; however, the L chains fall into only two families, κ or λ, found in all of the Ig classes. Thus, the structure of IgG molecules could be $\kappa_2\gamma_2$ or $\lambda_2\gamma_2$, the structure of the IgD molecules could be $\kappa_2\delta_2$ or $\lambda_2\delta_2$, and so forth.

The *IgG class* constitutes 70 to 80% of total serum immunoglobulins, or 0.9 to 2.0 g per 100 ml. It has the lowest carbohydrate content (2–3% by weight), and is the only class of antibodies that is capable of crossing the placental barrier from the maternal to fetal circulation. The *IgA class* accounts for 10–20% of total immunoglobulins in serum, or 0.2–0.35g per 100 ml. The molecular weights range from 180,000–500,000 since the IgAs can exist as multimers of the types $(\kappa_2\alpha_2)_n$ or $(\lambda_2\alpha_2)_n$ where n = 1, 2, 3. The H chains of multimers are joined by a short polypeptide J chain of 15,000 molecular weight. As for all other Igs, except IgG, the carbohydrate content of IgA is quite high (12–13% by weight). In addition to its role as a serum antibody, IgA is the major class present in secretions of mucous membrane cells into external fluids of lung and gastrointestinal tract. Mucosal cells add an extra polypeptide of molecular weight about 70,000 known as SC (for secretory component) joining the H chains of IgA dimers before passage into external secretions. By producing unabsorbable aggregates with harmful antigens in the gut and respiratory tract these excretory IgAs probably prevent the entry of such deleterious substances to the body. The *IgM class* represent 5 to 10% of total serum immunoglobulins, or .075 to .15 g per 100 ml. It is the largest of the immunoglobulins with molecular weights of 900,000, and was at one time known as the γ-macroglobulins for this reason. The IgMs are pentamers of the basic molecule, and may be represented by the formulations $(\kappa_2\mu_2)_5$ or $(\lambda_2\mu_2)_5$. As in the case of the IgA class the multimeric form of IgM contains the linking J chains near the carboxyl ends of the H chains, and the SC component has been detected also in IgM when it occurs in external secretions from mucous membrane cells. The IgM molecules are the first to appear as newly synthesized in infancy; since the fetus lacks the capacity for antibody synthesis the neonate possesses only IgG species derived from transfer across the placenta from maternal blood which confers a temporary passive immunity to infections in the newborn infant until endogenous immunoglobulin production develops. IgM shares many antibody functions with the major IgG class, including the fixation of complement. However, it has a shorter half-life (3 to 5 days) than IgG (3 to 4 weeks). IgM possesses a strong tendency for agglutination, probably owing to its multivalent Fab regions and large molecular size, and it acts as the primary isohemagglutin. The *IgD class* accounts for only 0.2% of total immunoglobulins, or a few mg per 100 ml of serum. The molecular weight is about 180,000 and it may be formulated as $\kappa_2\delta_2$ or $\lambda_2\delta_2$. It does not act as a secretory antibody and few specific functions can be attributed to IgD except for the initiation of complement activation by the alternative pathway. The *IgE class* has the lowest concentration in serum amounting to .002% of the total, or a few 1/100ths of a mg per 100 ml. It has a high molecular weight of about 200,000 since, as in the case of IgM, its H chain has a fourth constant domain (C_{H4}). It may be depicted as $\kappa_2\epsilon_2$ or $\lambda_2\epsilon_2$. The IgE class plays a major role as surface antibodies of cells involved in the anaphylactic response. A membrane receptor of basophilic leukocytes or mast cells binds the Fc portion of IgE molecules leaving the Fab regions exposed on the outer surface. When specific antigens to these membrane-bound antibodies react at the surface they trigger the cells to release histamine and other vasoactive amines. The IgE class is also found as secretory immunoglobulins in the lungs and gastrointestinal mucosal secretions, but the IgEs lack the J chain and SC portions found in IgAs and IgMs.

The nature of antigen-binding regions of the immunoglobulins will depend upon the particular antigen in question, but certain general features have been elaborated. At the Fab arms of the immunoglobulins the V subunits of L and H chains lie apposed to one another. The V_L and V_H subunits each have quite flat surfaces formed by extensive β-pleated sheet structures of the polypeptides, and these planar surfaces form the contacts between the subunits. However, at the tips of the Fab arms the L and H chains are not in contact, producing a solvent-filled cavity. It is in this cleft between the two V domains that the antigens react with immunoglobulins. Walls of the cavity are made up of several loops of the polypeptide chain that connect the β sheets and project beyond their planes of contact. The amino acid composition, the sequence, and lengths of these loops are highly variable depending upon the particular antigen which reacts, and hence these are termed hypervariable regions. The nature of the insertions or deletions of amino acids in these hypervariable regions will govern the depth and width of the cleft to accommodate the antigenic determinant, and will expose different binding sites along the walls of the cavity by altering the dimensions and shapes of the loops. Thus the specificity and the affinity of binding of antigens to the combining sites of the immunoglobulins are determined by the sequences of amino acids in their hypervariable regions of the V domains in the Fab arms.

In contrast, effector functions of the immunoglobulins reside in the C regions of the Fc segment. These functions include facilitating the attachments of antibodies to basophilic leukocytes and mast cells, binding of antibodies to lymphocytes and macrophages, passage of antibodies through the placenta, stimulation of the phagocytic actions of monocytes and neutrophilic leukocytes (opsonization), and the activation of cell lysis or anaphylactic responses that are mediated by the complement system (complement fixation). The latter process is a cascade of regulated proteolytic reactions analogous to the blood clotting system, involving some 19 plasma protein components most of which are produced in their inactive precursor forms by macrophages (Fig. 1–13). Activation of the complement system involves components C_1, C_2, C_3, C_4, with recognition of the F_c portion of antigen-bound IgG or IgM by C_1 as the trigger (classical pathway); alternatively, IgA or IgD may directly activate C_3 following reaction of immunoglobulins with bacterial endotoxin or polysaccharides of bacterial cell walls (alternate pathway). As a result the activated com-

CLASSICAL
PATHWAY

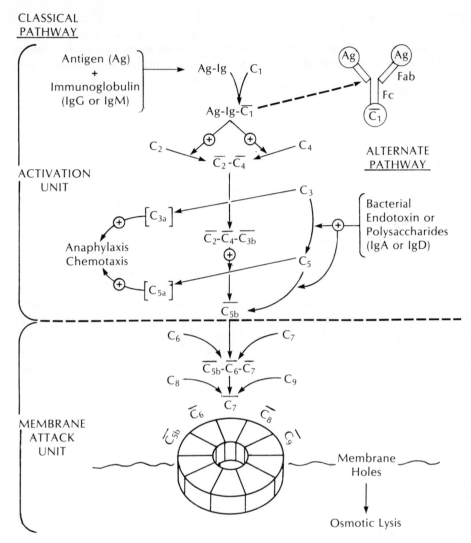

FIG. 1-13. Complement activation. Individual components (C_1–C_9) are shown in their activated forms ($\overline{C}_1$–$\overline{C}_9$). Stimulatory events are represented by + signs.

plex formed with the components C_5, C_6, C_7, C_8, C_9 causes hole-like lesions in membranes of target foreign cells, and produces anaphylatoxins, chemotactic substances, and other by-products to facilitate phagocytosis and to promote the inflammatory responses to infection. This is but one, albeit a very important one, of a number of ways in which the binding of antigen to the Fab portion is transmitted to a biologic effector function through the Fc sector of immunoglobulins. The mechanism of transmission of the signal has not been proven, but from analogy with the intrinsic changes of enzyme or hemoglobin shapes in allosteric modifications, it seems to have many of the characteristics expected for an induced conformational change. Thus crystallographic analysis reveals that immunoglobulin molecules in the absence of their antigens show free mobility of the Fab portion about the hinge region, with extension of the arms in a relaxed configuration. Upon binding of the antigen the molecule adopts a more rigid conformation, presumably because of longitudinal interactions between amino acid side-chains from V_L and V_H to C_L and C_{H1}, from C_{H1} to C_{H2}, and so forth. The decreased intersegmental flexibility of the immunoglobulin may thereby bring about structural changes in the F_c stem to unmask latent binding sites for complement activation, to enhance the affinity for cell surface binding, or to amplify the membrane-mediated signals associated with opsonization and other intracellular responses.

Antibody biosynthesis is a process which has intrigued molecular biologists and immunologists for many years since it provides an important key to the understanding of gene function as well as to the control of infectious diseases. The burgeoning technology in this field has already produced new insights into our basic concepts about genetic expression. It has also provided the methods for the generation of any given human immunoglobulin. The technique referred to as monoclonal antibody production is accomplished in a cell culture system by the hybridization of appropriate plasma cells with cancer cells to form a hybridoma cell-line which combines the capacity for specific antibody synthesis with rapid growth properties of the tumour. In the body's lymphoid tissues of the immune system a category of bone marrow-derived lymphocytes (B cells), along with the regulatory influences of thymus-derived lymphocytes (T cells), are responsible for the production of circulating immunoglobulins. Some of the T cells play a stimulatory role (helper cells), while others play an inhibitory role (suppressor cells). A third category of cells (the A cells) which play an accessory role in immunoglobulin synthesis, are a group of macrophages with the capability to adhere to foreign surfaces and to activate foreign antigens, thus enhancing the production of the specific antibody. The immune system's response of generating a great diversity of such specific antibodies has been explained by the clonal selection hypothesis which postulates that the lymphoid tissue contains a bank of potential immunologically competent cells, each of which may react with a particular antigen. The antigen is proposed to promote development of a clone of such specific immunoglobulin secreting cells. Thus, the antibody diversity must reflect a corresponding genetic diversity of the immune cells present in the germ-line, or it must arise because of mutations during somatic differentiation.

Mechanisms of antibody production are in some respects analogous to mechanisms of enzyme induction; unmasking and accelerated transcription of quiescent regions of DNA in the genome lead to increased mRNA species and translation of the particular polypeptide sequences for H and L chains of the immunoglobulins. The stimulus, however, involves a complex interaction between antigenic molecules, macrophages (A cells), and lymphocytes (B cells and T cells). The macrophages take up antigen by endocytosis and promote its conversion into a more active form, resulting in the proliferation of T lymphocytes and the differentiation of B cells into the clone of antibody-synthesizing plasma cells. The genetic evidence is now clear that each polypeptide chain requires two discrete genes, one for the V region and one for the C region. The question arises where does the fusion occur, at the DNA, mRNA, or protein levels? Much current evidence favours a fusion of DNA. The possibility of fusion of polypeptide chains may be ruled out since a single mRNA species which is responsible for translation of an entire polypeptide chain containing both V and C regions can be isolated from antibody-producing cells. Moreover, both the H and L chains could be shown, through the use of pulse-labelling experiments with radioactive amino acids, to be synthesized in their entirety from a single growing point. Two types of polyribosomes are present in lymphoid cells; they are equipped with the numbers of ribosomes and the length of attached mRNA molecules appropriate for the generation of polypeptide chains with molecular weights of 25,000 (L chains) or 50,000 (H chains). Analogous to the manner of secretion of proteins by pancreatic cells described in Chapters 4 and 5, the L and H chains of immunoglobulins are translated on membrane-attached polyribosomes, and are released vectorially into the lumen of the rough endoplasmic reticulum. The orderly addition of carbohydrate residues occurs, as described for other glycoproteins, during the passage through smooth endoplasmic reticulum and the Golgi complex, prior to release from the cell by exocytosis.

BLOOD CLOTTING

The basic defense mechanism for preventing the escape of blood involves several cellular and plasma components, as well as physiological effects upon the blood vessels involved. This section deals with the biochemical factors which regulate and confine the process of blood coagulation. Instantaneous formation of a blood clot is the critical response to injury of the blood vessels, and genetic or acquired defects in the clotting factors will result in *hemorrhagic diseases*. Much is now known about the chemical pathology of such bleeding disorders, and as a result afflicted patients have greater prospects for survival than heretofore. Of great significance for clinical medicine is the opposite situation of uncontrolled or dislocated blood clots in the circulation, producing the disorders which are collectively termed *thromboses*. The pathology of unwanted thrombosis, particularly when it occurs in vessels nourishing the brain or the heart, is dire

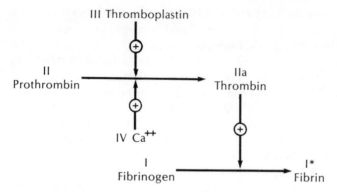

FIG. 1–14. Outline of blood clotting cascade, showing factors I–IV. Stimulatory factors are represented by + signs.

indeed, and affects great numbers of people in the prime of life. Because of the widespread nature of thrombosis, the resulting high rate of mortality, and the relative helplessness of medical science to avert the onset or modify the outcome, the thrombotic diseases of humans constitute one of the most important frontiers of clinical and biochemical study.

In its original formulation the process of blood clotting was thought to depend upon four factors (Fig. 1–14): I—fibrinogen, a plasma protein, which is the soluble precursor of fibrin, the insoluble polymeric form of which produces the basis for the clot; II—prothrombin, also a plasma protein, which is the inactive zymogen form of thrombin, the latter serving to convert fibrinogen to fibrin; III—thromboplastin, a factor released from tissues upon injury which triggers the conversion of prothrombin to its active state; IV—calcium ions, which are required in the ionized state for the activation of prothrombin. Over the years a number of different plasma factors, mostly proteins, have been shown to participate in the trigger mechanism, and the early four factor model has been expanded to include a dozen factors. Several of the protein factors have been shown to depend upon vitamin K for a key post-translational modification that is required for Ca^{2+} binding. Most of the proteins have been shown to possess enzymatic activity, and the mechanism whereby they promote the activation of other clotting factors entails selective, limited hydrolysis of certain peptide linkages, much the same as the mechanism for the conversion of inactive trypsinogen to active trypsin in the intestinal tract. The activation and specificities of these proteases, the characterization of protease-inhibiting agents in blood, the identification of "mopping-up" enzymes which clear the activated clotting factors from the circulation, and the characterization of the system for stabilization or eventual dissolution of the fibrin clot, all are subjects of continuing, active investigation by hematologists and biochemists.

COAGULATION MECHANISMS

Studies of the coagulation of blood from normal people or from patients with various bleeding disorders led to the discovery of additional clotting factors. Normal blood plasma gradually loses its ability to respond to thromboplastin upon storage, but the clotting reaction of the stored plasma is accelerated upon addition of small amounts of fresh plasma. This labile factor was isolated and termed proaccelerin, or factor V. The plasma from vitamin K-deficient patients lacks prothrombin, and, in addition, lacks a factor necessary for conversion of prothrombin to thrombin. Since the new factor, but not prothrombin, could be obtained from normal blood serum, the second vitamin K-dependent factor was readily separated and named proconvertin, or factor VII. The blood from patients with hereditary hemophilia failed to clot upon standing, but a concentrated globulin fraction from normal plasma could restore the coagulation defect. In this way antihemophilic globulin (AHG), or factor VIII, came to be identified and purified. The administration of AHG to hemophiliacs has extended their life expectancy from 16.5 years to approximately that of normal individuals. Further work with hemophiliacs demonstrated that there were genetic and biochemical distinctions among hemophiliac patients. For example, mixing blood from hemophiliac patient A who has a deficiency of factor VIII, with that from hemophiliac patient B, who does not have a deficiency of factor VIII, showed mutual correction of the clotting defect. Such studies established the existence of hemophilia B as distinct from hemophilia A, the former characterized by normal amounts of factor VIII, but the absence of a different factor, factor IX, sometimes termed the Christmas factor from the name of the original afflicted family. Factors X to XIII were elaborated later. Factor VI turned out to be a modified form of V and has been dropped, leaving a gap in the numbered sequence.

The identification, naming system, and postulated modes of action of the many clotting factors described in the medical literature have proven contentious and are often confusing. A standardized nomenclature has now been agreed upon based on the extension of the original roman numeral designation for factors I through IV. Although there is still some argument about the independent existence and action of all the factors, there is presently a fair consensus that the coagulation mechanism operates by sequential activation of clotting factors in a chain of events triggered either by the release of some tissue-damage factor from outside the circulation (extrinsic system), or by some abnormal contact within the blood (intrinsic system) (Fig. 1–15). This waterfall or cascade activation process is analogous to the process observed for the cyclic AMP-mediated activation of glycogen catabolism, and provides a finely regulated rapid response to an alarm stimulus. Typical of the cascade type of process is the large overall amplification resulting from sequential enzymatic activation which is of at

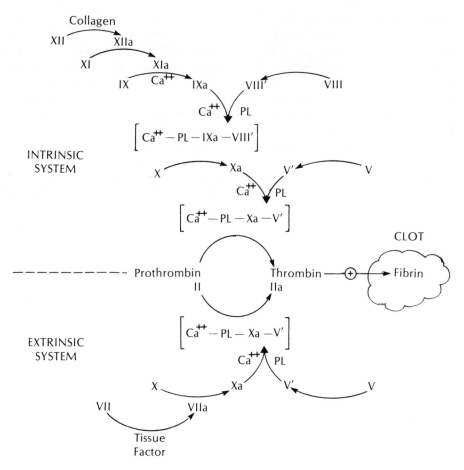

FIG. 1–15. Detail of blood clotting. Activated components are indicated either by the conversion II → IIa or VIII → VIII′.

least an order of magnitude at each stage. Consequently, activation of a single molecule at the first of a series of six linked reactions can give rise to 10^6, or greater activated molecules at the end of the cascade.

The intrinsic system is the more complex of the two clotting processes, involving all of the plasma factors except VII (Fig. 1–15). The initial activation process is apparently prompted *in vivo* by contact of one of the clotting factors, XII, with a suitable surface such as collagen fibrils, or the basement membrane of vascular endothelium. *In vitro* the activation can be accomplished by exposure of blood to glass, asbestos, and several other inorganic surfaces. Formation of the active form, XIIa, is accompanied by appearance of esterase activity toward arginine esters or peptides. Both the proteolytic action of XIIa and its ability to trigger coagulation are blocked by the serine–protease inhibitor DFP (diisopropylfluorophos-

phate), indicating that XIIa initiates the clotting process by partial digestion of the next factor in the cascade, XI. The latter is thus converted to its
active form, XIa, which is another DFP-sensitive protease with selective
endopeptidase activity toward factor IX. The generation of IXa requires
calcium ions, and involves removal of a substantial polypeptide chain
from the interior of the molecule of factor IX. The next reaction is the
activation of factor X by a concerted reaction of the activated factors IXa,
VIII', calcium ions, and phospholipids. Although factor VIII may need
conversion to its active form VIII', it does not seem to possess any enzymatic activity itself, but potentiates the proteolytic action of IXa upon X. The
latter thus becomes converted to its active form, Xa, with release of two
large glycopeptides containing all the bound carbohydrate residues. As
with the previously characterized factors XIIa, XIa, and IXa, the activated
Xa is a protease with a reactive serine at the catalytic site, and hence its
enzymatic and clotting activity are blocked by DFP. It is Xa which finally
carries out the critical conversion of prothrombin to thrombin. The reaction bears a striking resemblance to the activation of X itself: factor Xa, the
activated form of V (or V'), calcium ions, and phospholipids, combine to
promote stepwise cleavage of prothrombin by removal of approximately
one-half of the polypeptide residues; V' like VIII' appears to play a regulatory rather than an enzymatic role in the activation process.

The extrinsic clotting system, so-named because it requires the contribution of an extravascular component "tissue factor" is considerably
simpler (Fig. 1–15). It does not require factors VIII, IX, XI, or XII for the
initiation of coagulation, but it involves an additional factor, VII, as well
as factor X, and the tissue factor. The latter has not been well characterized to date. It seems to be a lipoprotein presumably released by trauma
to the vascular endothelium, and reportedly contains a peptidase activity
that is DFP-insensitive. When combined with factor VII tissue factor promotes the conversion of factor X to Xa, by a reaction analogous to that
conducted by factor IXa. There is some evidence that factor VII is the
enzymatic agent for activation of factor X by a DFP-sensitive reaction, but
it is not clear whether tissue factor is responsible for accelerating this
reaction (by analogy with factor VIII in the intrinsic system), or for an
activation process to produce VIIa (by analogy with factor XI). The remainder of the extrinsic process beyond production of Xa is common to
that for the intrinsic system, and leads to the generation of thrombin.

Thrombin, derived from either an intrinsic or extrinsic system of activation, is a well-characterized serine protease which is readily inhibited by
DFP as well as by several natural protein inhibitors including hirudin
(anticoagulant from leeches), heparin (isolated from liver and present in
most cells), and antithrombin (plasma thrombin antagonist which potentiates the effects of heparin). The structure of fibrinogen, the specific substrate of thrombin, is established, as is the reaction between the two molecules (Fig. 1–16). Thrombin is very highly selective, cleaving only 4 out of

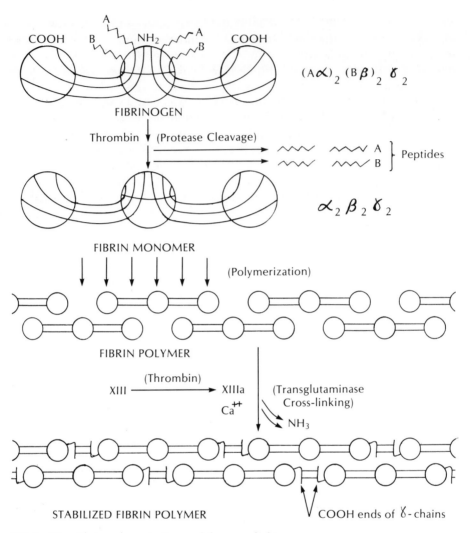

FIG. 1–16. Fibrin polymerization and the cross-linking process.

the 3000 peptide bonds in fibrinogen. The latter consists of three different polypeptide chains (Aα, Bβ, and γ) knotted together in pairs (Aα_2B$\beta_2\gamma_2$) near their NH$_2$-termini by several disulfide linkages. Thrombin attacks the molecule sequentially by releasing the 16-residue peptide segments, A, from each NH$_2$-terminus of the two α chains, followed by release of two additional 14-residue peptides, B, from the β chains to produce fibrin ($\alpha_2\beta_2\gamma_2$). Polymerization of the fibrin molecules appears to result from unmasking of "sticky" ends by removal of the N-terminal blocking groups. The polymer forms first by end-to-end linkages of fibrin monomers associated with removal of A peptides, then by lateral aggregations facilitated

by the later loss of the B peptides. These early stages in fibrin polymerization are reversible; the fibrin polymer is rendered more stable and insoluble by further modifications requiring an additional clotting factor, XIII,
which is converted to its active catalytic form, XIIIa, by thrombin. The
enzyme reaction catalysed by factor XIIIa is termed transglutaminase, and
results in a covalent cross-link between two γ chains of adjacent molecules; the process requires calcium ions, and results in elimination of NH_2
from a glutamine residue of one chain and replacement of the amide bond
at the γ-carboxyl group with an ϵ-NH_2 group from a lysine residue on the
adjacent chain. There is no structural alteration associated with this modification. It appears that it serves simply to stabilize the existing fibrin gel
by irreversibly stitching together those monomers which had come together spontaneously.

VITAMIN K AND ANTIMETABOLITES

Synthesis of the clotting factors II (prothrombin), VII, IX, and X by the
liver requires vitamin K. Consequently, deficiency of vitamin K will lead
to an acquired bleeding disorder which is distinct from the genetically
determined hemophilias which are inborn. The vitamin K requirement is
satisfied by synthetic reactions of the bacteria that normally inhabit the
intestinal tract, and absorption of vitamin K from the gut, a lipid-soluble
vitamin, is dependent upon normal mechanisms of fat absorption involving the bile. In cases where the synthesis or absorption of vitamin K is
impaired, uncontrolled bleeding may result, as in the hemorrhagic disease of newborn infants who have not yet developed the normal population
of intestinal flora, or as in adults with defective fat absorption symptomatic of intestinal diseases or obstructive jaundice affecting the secretion of
bile salts. Intravenous supplementation with vitamin K_1 will restore the
deficiency of clotting factors after a lag period of several hours. More rapid
restoration of normal clotting ability requires transfusion with fresh whole
blood or plasma, both of which contain prothrombin and the other clotting
factors. The hemorrhagic diathesis associated with depressed levels of
prothrombin is also seen after the ingestion of structural analogues of
vitamin K, the coumarin compounds (*e.g.*, Warfarin, dicumarol), which
antagonize the action of vitamin K in promoting synthesis of the clotting
factors. These antimetabolites of vitamin K have been used as anticoagulants at low concentrations for therapy of thrombotic diseases in medicine,
or as very potent pesticides at high concentrations to induce fatal bleeding
in rodent eradication.

In recent years the mode of action of vitamin K has been established to
involve post-translational modifications of the clotting factors. The plasma
of vitamin K-deficient subjects, or of patients receiving anticoagulant
therapy with dicumarol, contains an abnormal prothrombin that is biologically inactive. Unlike the normal prothrombin, the molecule is incapable

of forming the complex with Ca^{2+} and phospholipids that is necessary for the conversion to thrombin by factor Xa. A detailed study of the abnormal prothrombin revealed that it differed from the normal molecule with respect to ten glutamic acid residues near the NH_2 terminus of the protein. In the normal prothrombin molecule these residues carried an extra carboxyl group attached to the γ-carbon atom, providing a high affinity binding site for calcium; these carboxyl groups were absent from the K-deficient prothrombin. The vitamin K-dependent process is an enzymatic γ-carboxylation of the glutamic acid residues. In isolated liver microsomes the γ-carboxylase enzyme converts the abnormal prothrombin into biologically active prothrombin at the same time as CO_2 is incorporated into the protein. The reaction is inhibited by Warfarin, and the inhibition can be overcome by high concentrations of vitamin K.

PLATELET FUNCTIONS

The blood platelets, or thrombocytes, are tiny membrane-bound cellular fragments that are released into the circulation by budding off from the large megakaryocyte cells of the bone marrow. They play a role in several stages of the process of hemostasis whereby bleeding is arrested. In the first place platelets will bind tenaciously to exposed areas of the basement membrane underlying endothelial cells of the blood vessels; consequently, when the vessel lining is damaged, a layer of adhering platelets will rapidly cover the site. This adhesion process is probably triggered by a response of the platelet membrane to primary contact with collagen fibrils, but a variety of chemical agents, including thrombin (produced in the clotting reaction), ADP (released from damaged cells), or catecholamines (formed in response to stress) will potentiate the adhesiveness of platelets. The formation of a layer of platelets covering the damaged area of the vessel wall is followed quickly by the aggregation of additional platelets on top of one another to generate a hemostatic plug that is colorless, and is referred to as the white thrombus. In the usual response to injury, platelet adhesion and aggregation provide a focal point for coagulation, while a meshwork of fibrin filaments with entrapped platelets and erythrocytes will form the typical red thrombus. The clumping of platelets together seems to be activated by similar stimuli as for their adhesion to other surfaces. In addition, aggregation is strongly influenced by certain prostaglandin compounds that are produced in platelets from polyunsaturated fatty acids such as arachidonic acid. Recent studies indicate that it is the complex endoperoxide metabolites of the prostaglandins, termed thromboxanes, which induce aggregation; the drugs, indomethacin and aspirin, block the synthesis of thromboxanes and inhibit platelet aggregation. As noted later, much interest is focused on clinical investigations of such agents in the management of thrombotic diseases.

Adhesion and aggregation of platelets are associated with morphologic and permeability changes of the membranes. Firstly, the change in shape is probably related to movements of microtubules and microfilaments producing extrusions of cytoplasm as elongated pseudopods, and increasing the effective area for contacts with other surfaces. Secondly, there is a release of intracellular ADP (to accelerate aggregation in the immediate vicinity), of serotonin (to induce vasoconstriction of injured blood vessels), and of several stimulators of coagulation, including the phospholipids (to provide the focus for activations in the cascade process of coagulation). Finally, analogous with muscle contraction, the microfilament proteins within the pseudopods eventually undergo an ATP-dependent shortening, to condense or retract the material of the tangled blood clot. Clot retraction may contribute to hemostasis by pulling attached surfaces of the vascular lumen together; later in the healing process, contraction of the mass of the clot may allow reopening of the vessel to the flow of blood.

FIBRINOLYSIS

Under normal circumstances blood clots do not persist in the vascular system indefinitely, and there is a hydrolytic enzyme system, specific for fibrin, which is responsible for removal of the clot once the danger of bleeding is past. There is evidence for the existence of a dynamic equilibrium between formation of fibrin and fibrinolysis, with disturbances of either process serving as major contributing factors in hemorrhagic disorders or thrombotic disease. The fibrin-dissolving enzyme is termed plasmin; it is present in blood plasma chiefly as its inactive zymogen, plasminogen, which may be activated by a number of stimuli, including heavy exercise, stress, or trauma. The mechanism by which activation is triggered is unknown, although it is clear that the activators of plasminogen from the blood or from damaged tissue cells are proteolytic enzymes which release a small peptide from the zymogen together with active plasmin.

Plasmin is capable of hydrolysing both fibrin and fibrinogen, but in the presence of inhibitory substances in blood plasma the enzyme preferentially digests fibrin. The fibrin degradation products (FDPs) are quite characteristic small peptides that are readily detected in the blood, and whose presence provides a handy index of the activity of the fibrinolytic system. Inhibitory substances against plasmin (antiplasmins) have been detected both in the α-globulins of the plasma and in the platelets. The means for controlling the antiplasmin levels, as well as the plasminogen activators, may play a very significant role in the understanding and treatment of several types of vascular disease. Clinical investigations have demonstrated interference with normal fibrinolytic mechanisms for dissolving clots in patients with intermittent claudication, pulmonary embolism, and deep vein thrombosis; there are also reports that hyperlipidemia, a common concomitant of atherosclerotic plaque formation possibly involving white thrombus deposition on vessel walls, is also associated with an impaired fibrinolytic response.

HEMOPHILIAS

Congenital deficiencies of the clotting factors are relatively rare. The most common classical picture of hemophilia is associated with defective synthesis of factor VIII, the A type of hemophilia. Depending upon the population surveyed, the incidence is one in every 5,000 to 20,000 live births. The occurrence of hemophilia B, with deficiency of factor IX, is reportedly one-fifth as common. Both types of hemophilia are transmitted as sex-linked recessive traits, that is the gene is on the X chromosome, and hence is passed to the next generation only by the mother who is an unaffected carrier; only male offspring who are hemizygotes are affected. Severity of the bleeding varies from patient to patient, but generally involves spontaneous hemorrhaging into the muscles and joints leading to crippling deformities with age, as well as loss of blood in the urine and gastrointestinal tract. After an injury, bleeding may continue for prolonged periods (days or even weeks). It is essential to distinguish between the A and B forms of the disease by *in vitro* corrections of the coagulation defect with either factor VIII or IX before attempting replacement therapy in the patient. Various concentrated fractions of normal plasma obtained by selective precipitation (*e.g.*, by freezing or cryoprecipitation for factor VIII), or by adsorption (*e.g.*, by complexing to inorganic and ion exchange adsorbents for factor IX) have been used to treat hemophiliacs. The management of these conditions obviously depends upon the degree of severity of reduction in concentration of the factors, the exposure of patients to risk of trauma, surgery, and so forth.

Bleeding disorders of mild to marked severity, equally affecting both sexes, have been described in association with lowering of plasma fibrinogen, prothrombin, or factors V, VII, X, XI, XII, or XIII. Transmission in each case seems to be as an autosomal recessive trait. Another hemorrhagic disease inherited in an autosomal dominant manner is von Willebrand's disease. These patients characteristically have prolonged bleeding from superficial wounds which is related to abnormal adhesion and clumping of blood platelets. In some of the patients there is also a tendency, associated with a lowering of the factor VIII concentrations, to bleed into the deep muscle tissues. The rationale and treatment of this disorder is perplexing because of the resulting multiple defects, and the fact that replacement of factor VIII causes an overshoot phenomenon. It appears that normal plasma contains substances which stimulate factor VIII production in the patients with von Willebrand's syndrome.

THROMBOSIS

The presence of unwanted blood clots and their derivatives in major blood vessels is a life-threatening occurrence which has become one of the main causes of morbidity and mortality. It will be apparent from the foregoing discussion that inappropriate activation of the clotting factors, inappropriate aggregation of the platelets, and possibly failure of the fibrinolytic system, could all contribute to accumulation of thromboembolisms in the vasculature. Moreover, normal blood flow will remove platelet aggregates and clotting factors for clearance by the reticuloendothelial system, so that stasis of the blood, particularly within the veins, will predispose to thrombus formation. An important trigger of thrombosis in the arteries is the excessive platelet–platelet interaction in regions of turbulent flow and at sites of damage to vessel walls. A complete catalogue and rationale for all of the causative agents in thrombosis, atherosclerosis, and other occlusive vascular diseases has not yet been determined, but sufficient evidence is available regarding the risk factors involved in thrombosis so that the problem can be approached from the stand-point of preventive medicine. Specifically, the association of stress, hyperlipidemia, cigarette smoking, and family history of coronary disease with abnormal platelet function in humans indicates the increased likelihood of the appearance of thrombotic disease at some point. For individuals who fit into this high risk group a change in habits and life style may be more significant than treatment after the fact.

Therapy of thrombotic lesions generally falls into two categories: the use of anticoagulants to deter production of new thrombi; the activation of fibrinolysins to remove pre-existing clots. In practice it may be advisable to combine both approaches. Among the anticoagulants the most common agents in clinical use are warfarin and heparin. Warfarin is a synthetic, water-soluble vitamin K antagonist that is effectively absorbed from the gastrointestinal tract; hence it is administered orally. Since the effect is on the synthesis of clotting factors II, VII, IX, and X in their effective forms, the anticoagulant action of warfarin is delayed and sustained (1–3 days). Overdoses may be corrected in the event of excessive bleeding either instantaneously, in critical situations, by infusion of stored blood bank plasma which contains the stable factors II, VII, IX, and X, or more slowly by ingestion of vitamin K, which, as expected, corrects the defect in the endogenous factors by resynthesis in 1 to 2 days. Provided that careful monitoring of blood clotting tests is continued, warfarin is the most convenient agent for long-term prevention of thrombus formation (*e.g.*, in patients with a tendency for arterial embolization in mitral valve disease, or following the implantation of prosthetic heart valves). Heparin is a natural mucopolysaccharide of sufficiently large size (MW 12000) that it must be administered intravenously or subcutaneously. In this case the agent potentiates the blood proteinase-inhibitor, antithrombin III, and thus produces an immediate, but brief, blockage of thrombin–fibrinogen and thrombin–platelet interactions. Because of its immediate short action, heparin is favored in therapy of acute thrombotic lesions (*e.g.*, in venous thromboembolism, particularly those cases affecting the pulmonary circulation, and in pregnancy). In the case of pregnancy, oral anticoagulants could cross the placenta and induce fetal hemorrhage, while heparin is too large to pass the placental barrier. The most dramatic effects are reported in initial treatment of pulmonary embolism where rapid reduction of clotting and platelet aggregation by large doses of heparin reportedly allows close to 100% survival; up to one quarter of untreated patients die, and a further one quarter suffer from nonfatal emboli reoccurring. In the long-term management of such patients heparin is generally supplanted by warfarin through a process of gradual conversion. Despite some 30 years of clinical trials the use of the anticoagulants in myocardial infarction is controversial. Early reports were encouraging, but subsequent extended control studies have raised serious doubts that anticoagulants affect the thrombotic lesions within coronary vessels, or mechanical and physiological anomalies of myocardium leading to death. Anticoagulation during the acute period when patients are immobilized with high risk of stasis clots is frequently advised, but long-term use of anticoagulants in ambulatory patients does not seem to provide benefits outweighing the risks from hemorrhage. In current clinical studies the chronic use of antiplatelet agents such as aspirin is showing evidence of greater benefits than the usual anticoagulants, and may be of particular importance in the prevention of strokes resulting from cerebrovascular embolism.

DISSEMINATED INTRAVASCULAR COAGULATION

The pathological process termed disseminated intravascular coagulation (DIC) is characterized by the inappropriate general formation of clots throughout smaller vessels of the circulation. The consequence, in addition to possible damage to organs such as the kidney from blockage of the microcirculation, is depletion of the clotting factors of the blood owing to their continuous consumption, with attendant danger of hemorrhagic crises. The disorder is sometimes called "consumption coagulopathy" for this reason.

Many causative factors of DIC have been identified. The most common and serious contributing factor is infection by micro-organisms which may trigger the continuous activation of the coagulation cascade through reaction of their toxins or immune complexes with leucocytes, platelets, or clotting factors. DIC may also be observed in certain malignancies, complications of pregnancy (*i.e.*, retention of placenta, or dead fetus), shock, heatstroke, and a variety of other traumatic injuries. Circulating levels of platelets or fibrinogen are generally depressed, while amounts of peptide degradation products from fibrinolysis are elevated. Since these fibrin-

fibrinogen degradation products (FDPs) may inhibit the conversion of prothrombin to thrombin and platelet aggregation, the combination of lowered clotting factors and raised FDP concentrations will exacerbate the tendency of hemorrhagic diathesis.

The management of this condition is difficult and controversial, with the conservative course being treatment of the infection or other underlying causative condition. In cases of continuing DIC the cautious infusion of heparin has been used together with antibiotics (*e.g.*, to cure the underlying infection). Heparin treatment may offset the deleterious effects from thrombi in the microcirculation in end-organs and allow buildup of platelets and clotting factors, but it can also lead to increased bleeding. The administration of platelets or fibrinogen to correct hemorrhagic emergencies with continuing DIC could accentuate the microcirculatory blockages. Thus this type of replacement therapy is usually instituted only after the episode that triggered the coagulation cascade has been remedied (*e.g.*, by delivery of the fetus and placenta or removal of sources of endotoxins). Judicious alternation of the two opposing types of therapeutic intervention may be required, with constant monitoring of laboratory indices of clotting factors and clinical signs.

SUGGESTED READINGS

Adamson JW, Finch CA: Hemoglobin function, oxygen affinity, and erythropoietin. Annu Rev Physiol 37:351–369, 1975

Amzel LM, Poljak RJ: Three-dimensional structure of immunoglobulins. Annu Rev Biochem 48:961–998, 1979

Biggs R (ed): Human Blood Coagulation, Haemostasis and Thrombosis, 2nd Ed, Oxford, Blackwell Scientific Publications, 1976

Bunn HF, Forget BG, Ranney HM: Human Hemoglobins, Philadelphia, WB Saunders, 1977

Bunn HF, Forget BG, Ranney HM: Hemoglobinopathies, Philadelphia, WB Saunders, 1977

Capra JD, Edmundson AB: The antibody combining site. Sci Am 236(1):50–59, 1977

Cash JD: Disseminated intravascular coagulation. In Poller L (ed): Recent Advances in Blood Coagulation, Number Two. Edinburgh, Churchill Livingstone, 1977

Cerami A, Peterson CM: Cyanate and sickle-cell disease. Sci Am 232(4):44–50, 1975

Donoso E, Haft JI (eds): Current Cardiovascular Topics, Vol II, Thrombosis, Platelets, Anticoagulation, and Acetyl-salicylic Acid, New York, Stratton Intercontinental Medical Book Corporation, 1976

Doolittle RF: Fibrinogen and fibrin. Sci Am 245(6):126–135, 1981

Friedman, MJ, Trager W: The biochemistry of resistance to malaria. Sci Am 244(3):154–164, 1981

Goldstein JL, Brown MS: The low-density lipoprotein pathway and its relation to atherosclerosis. Annu Rev Biochem 46:897–930, 1977

Goldstein JL, Brown MS: The LDL receptor locus and the genetics of familial hypercholesterolemia. Annu Rev Genet 13:258–289, 1979

Jackson CM, Nemerson Y: Blood coagulation. Annu Rev Biochem 49:767–812, 1980

Lewis B: The Hyperlipidaemias Clinical and Laboratory Practice, Oxford, Blackwell Scientific Publications, 1976

Miller GJ: High-density lipoproteins and atherosclerosis. Annu Rev Med 31:97–108, 1980

Mustard JF, Kinlough-Rathbone RL, Packham MA: Prostaglandins and platelets. Annu Rev Med 31:89–96, 1980

Ogston D, Bennett B (eds): Haemostasis: Biochemistry, Physiology, and Pathology, London, John Wiley & Sons, 1977

Perutz MF: Regulation of oxygen affinity of hemoglobin: Influence of structure of the globin on heme iron. Annu Rev Biochem 48:327–386, 1979

Peters T, Sjöholm I (eds): Albumin—Structure, Biosynthesis, Function, Oxford, Pergamon Press, 1977

Porter RR: Structure and activation of the early components of complement. Federation Proc 36:2191–2196, 1977

Putnam FW (ed): The Plasma Proteins: Structure, Function, and Genetic Control, 2nd Ed, Vols I, II, III. New York, Academic Press, 1975

Reid KBM, Porter RR: The proteolytic activation systems of complement. Ann Rev Biochem 50:433–464, 1981

Rifkind BM, Levy RI (eds): Hyperlipidemia: Diagnosis and Therapy, New York, Grune & Stratton, 1977

Smith LC, Pownall HJ, Gotto AM Jr: The plasma lipoproteins: Structure and metabolism. Annu Rev Biochem 47:751–779, 1978

Tall AR, Small DM: Body cholesterol removal: Role of plasma high-density lipoproteins. Adv Lipid Res 17:2–53, 1980

Weatherall DJ, Clegg JB: Molecular genetics of human hemoglobin. Annu Rev Genet 10:157–178, 1976

Zucker MB: The functioning of blood platelets. Sci Am 242(6):86–103, 1980

Chapter Two **Nutrition**

The study of nutrition cuts across many disciplinary boundaries. The foods we choose to eat and their relevance to the needs of our bodies are subjects for cultural–psychological considerations as much as they are subjects for medical–biochemical considerations. Moreover, the field is generously endowed with self-styled experts both within and outside the health-related professions. Recommendations by nutrition experts are frequently unsupported by reliable evidence, and dietary advice reaching the public from different sources is often at variance. Mass marketing of foods requires a certain amount of processing for storage and preservation purposes; this only adds to the confusion of the consumer since he must weigh the preservative value of food processing against the destruction of nutrients which can also be caused by food processing. Our approach to nutrition is also profoundly affected by social fads and taboos, parental and peer group preferences and promotional conditioning. In many cases we have become willing to compromise good nutrition with palatability, appetite satisfaction, or convenience. In responding to these factors we may not obtain equally excellent nutrition but we will each get our just desserts!

Medical interest in nutrition dates back to the discovery of vitamins and the role of such micronutrients in prevention of deficiency diseases. Some food producers, having removed many vitamins and minerals by milling and extraction processes, are now attempting to reconstitute the natural state by supplementation. Currently the nutritional emphasis has switched to pharmacologic effects of supplements to the diet far in excess of natural content in foods, or the so-called megavitamin concept in diet therapy. Epidemiologic and experimental work has brought to light associations of certain macronutrients, particularly fats and refined carbohydrates, in the development of major diseases such as cancer, diabetes, and cardiovascular disease. Similar studies have led to an awareness of the elimination of various types of fibre from our highly processed supermarket foods, and the possible consequences this poses for the retention of potentially dangerous by-products of digestion in the gastrointestinal tract. In each of these areas conflicting positions are common among representatives of food industries, government regulatory agencies, the medical establishment, and nutrition counselors with varying backgrounds and credentials. It seems fair to conclude that although much data is in evidence on the scientific basis of nutrition, the application of these findings to prevention of illness in populations, or to treatment, management, and recuperation of diseased persons remains an important subject for continued study.

MAJOR NUTRIENTS

Among the factors to be considered in relation to the three main dietary components, carbohydrates, lipids, and proteins, are quantity or relative proportions of each, quality with respect to essential nutrient content, digestibility or utilization of these components by the body. From the standpoint of quantity each of these classes of nutrients can contribute to energy metabolism, and thus play a role in the balance between obesity at one extreme and starvation at the other. The qualitative factor is most acute for the protein classes which may vary greatly with respect to provision of the essential amino acids. The digestibility factor is probably of greatest significance with differing protein sources and for some lipid substituents, although certain food incompatibilities may relate to problems with intestinal utilization of common natural carbohydrates. Moreover, in many pathological conditions problems of digestion, absorption, and retention of foodstuffs may be grossly exacerbated, and the minimum daily nutritional requirements may be met only by the direct administration of nutrients into the bloodstream.

CARBOHYDRATES

Glucose is the most prevalent natural sugar, and it is found generally in combination with other sugar molecules. The most abundant of these combined forms are the polysaccharides such as starch, glycogen, and cellulose. In starch or glycogen the connecting linkages are susceptible to cleavage by digestive enzymes (amylases), but cellulose and several related structural polysaccharides found in plants are not broken down and utilized in the body. Such indigestible carbohydrates are important constituents of dietary fiber and will be discussed in a subsequent section. Prior to absorption and metabolism the combined sugars must be liberated in their free or monosaccharide forms. In most cases (e.g., for the polysaccharides such as starch, or the common disaccharide, sucrose) this presents no problems. However, a very significant proportion of the world's population lacks the intestinal enzyme, lactase, which is necessary to cleave the milk disaccharide, lactose, into its constituents, galactose and glucose. As a consequence the ingestion of milk products by such individuals can lead to passage of substantial amounts of undigested lactose right through the small intestine where sugars are normally absorbed. Upon reaching the large intestine the undigested lactose may be utilized as an effective growth and energy substrate for intestinal bacteria with resulting flatulence, diarrhea, and cramps owing to distension of the gut. The condition is most frequently seen in adult blacks who cease to produce lactase beyond the weaning period. Intolerances to most other disaccharides are uncommon, although there are certain oligosaccharides found in legumes (*i.e.*, raffinose and stachyose) which cannot be digested, and lead to postprandial flatulence. Efforts are being made by plant geneticists to breed varieties with lowered content of the offending saccharides with the goal in mind of selecting a "silent bean."

It is apparent that prior to utilization by body tissues all polysaccharides must be digested, and that monosaccharides of differing form are mainly converted to glucose in the body. Thus it would seem that the total amount of carbohydrate rather than the type of carbohydrate in the diet would be the most important consideration. However, recent studies have suggested associations of various disorders with the increased consumption of refined sugars rather than starch in modern diets. The clearest cause–effect relationship is seen between the high consumption of sugars, such as sucrose, and the proliferation of dental caries. Sucrose is an excellent substrate for oral bacteria which hydrolyse the sucrose to its components glucose and fructose. Both monosaccharides may provide energy for bacterial metabolism and growth, and lead to organic acid end-products, chiefly lactic acid, which lower the pH in the mouth and demineralize the tooth. The bacteria also utilize the dietary sugars to produce extracellular insoluble polysaccharides (dextrans, levans) which deposit on the tooth surface and provide the adherent gelatinous layer of plaque which harbors the microorganisms and serves as a focal point for the decay process. High sucrose intake has also been implicated in the etiology of cardiovascular diseases and diabetes in adults. The evidence is inconclusive, though suggestive, that replacement of free sugar in the diet with starch as the carbohydrate source may be more healthful at all ages. In part this may be related to the more gradual rise in blood sugar from starch digestion, in contrast with surges of hyperglycemia and consequent hyperinsulinism upon ingestion of free sugars.

LIPIDS

Fat in the diet is composed mainly of glycerol with esterified long-chain fatty acids. The differences between various food fats relates to their fatty acid components. In general the animal fats have a higher content of saturated chains, while vegetable fats, particularly those from seeds, have more of the unsaturated chains with two or more double-bonds. Although they act rather similarly as energy sources, the polyunsaturated fatty acids have special functions in cell membranes and frequently must be supplied in the diet since they cannot all be synthesized within the body. Cholesterol is a minor component of dietary lipids, and since it can be formed from small molecular precursors in virtually all cells it is not a required nutrient. The lipids must be acted upon by emulsifiers and digestive enzymes to aid their dispersion into small absorbable particles in the intestinal tract because of their low water solubility. Generally, the dietary fat is broken down and absorbed in its entirety. However, in some instances of malabsorption there is incomplete uptake of lipid-soluble materials including the fats and fat-soluble substances such as vitamins A, D, E, and K. In these circumstances the escape of unabsorbed fats into the feces may cause a steatorrhea as well as symptoms from deficiencies of the fat-soluble vitamins over a long period of time. Substitution of the long-chain triacylglycerols by more soluble fats containing fatty acids of medium-chain lengths has been used to alleviate absorption defects of this type.

A great deal of public debate has centered around the merits of regulating the lipid composition of the diet. Apart from the provision of a modicum of the essential polyunsaturated fatty acids, such as linoleic and arachidonic acids, the body has little requirement for dietary lipids. However, the latter provide an excellent energy source and contribute to the feeling of satiety after a meal. Thus, although it is prudent to reduce the current over-burden of fat in modern diets (chiefly because of the association with increased incidence of certain forms of cancer such as mammary carcinoma), rigourous exclusion of lipid from the diet is neither practicable nor desirable. A reasonable course, as recommended by the American Heart Association, would seem to be a restriction to less than 35% of total diet energy intake in the form of lipids. Since there is evidence that saturated fat contributes to elevation of serum cholesterol, decreased intake of animal fats and replacement with vegetable fats are recommended as general dietary guides. Such a regimen, particularly when enriched in high linoleic acid-containing fats, will provide a desirable balance of polyunsaturated to saturated fats, and a lower cholesterol intake. As for the total intake of fats in the diet, it does not seem reasonable to advocate the rigourous exclusion of cholesterol from diets for the general population in hopes of preventing atherosclerosis. Firstly, ingested cholesterol is poorly absorbed from the intestinal tract in humans, and is probably a minor contributor to hypercholesterolemia, which arises chiefly from endogenous sources. Secondly, benefits to the population as a whole arising from restrictions of high cholesterol foods such as milk products and eggs are more than offset by eliminating such excellent nutrient sources from the diet. It seems more reasonable to identify by serum analysis those people with a genetic predisposition to the various forms of hyperlipidemia, and to prescribe appropriate restrictions of cholesterol intake for these high-risk individuals rather than for the general population.

Caloric Balance

A majority of the body's energy needs are met by the carbohydrates and lipids ingested as part of a normal diet. As noted below, it is more appropriate to consider protein intake in relation to nitrogen balance rather than to caloric balance. A reasonable rule of thumb for a balanced diet is the provision of roughly 10% of energy needs in the form of proteins. To stay within the limits recommended for lipid intake it follows that 50 to 60% of caloric needs should be provided by carbohydrate, mainly in the form of starch.

Carbohydrate foods and proteins yield roughly the same amount of energy (4 kcal/g) to serve the body's needs. This is not surprising since a majority of the amino acids in protein are interconvertible with glucose and follow similar pathways of catabolism. Production of glucose from amino acids (gluconeogenesis) occurs mainly during starvation and involves the utilization of tissue proteins. Dietary protein may provide a source of glucose directly, but only when very low carbohydrate–high protein diets are ingested. Fats produce the greatest caloric yield (9 kcal/g), and hence they serve as concentrated energy sources. However, apart from the consequences of long-term ingestion as discussed earlier, high fat–low carbohydrate diets may produce an excess of ketone bodies (ketosis) because of inefficient oxidation of fatty acids by the liver without carbohydrates to prime the citric acid cycle. Glucose and other sugars are readily converted to fat in the body as noted below; however, the converse is not true, and the fatty acids cannot serve as precursors for gluconeogenesis. A significant caloric component in some diets is ethyl alcohol, which has an energy yield (7 kcal/g) intermediate between that of fat and carbohydrate. Hepatic cells preferentially utilize alcohol with a result that fat may accumulate abnormally and result in cirrhotic degeneration of the liver with prolonged excess consumption. Even in moderate drinkers alcohol may contribute significantly to the caloric balance.

The body has a remarkable capacity for energy storage. Most of this is in the form of fat depots, or adipose tissue. Circulating albumin and tissue structural proteins may contribute a reserve of carbon for glucose synthesis by gluconeogenesis, and storage glycogen of liver, muscle, and other organs may provide short-term energy needs through carbohydrate catabolism. However, these stores would become depleted rapidly during the course of starvation, while the adipose storage reserves would persist. Since an average-well-nourished adult may carry upwards of 10 kg of fat in the depots, the total energy storage in this form will exceed 90,000 kcal or over one month's requirements at moderate physical activity.

The energy reserves in fat depots play a critical role in survival of man in a primeval setting, with food intake dependent upon periodic successes in the hunt and cycles of heavy physical activity, exposure to cold environment, and so forth. In modern society the combination of easy accessibility to foods rich in fat and refined sugars, and the tendency towards a less active, sheltered life-style has removed the survival value of fat deposition. A large segment of the population has reached a state of physical excess, otherwise called overweight or obesity, in which a steady imbalance between caloric intake and energy expenditure over some years has resulted in fat accumulation beyond any reasonable demands of the body for reserve needs. Among the consequences of excess weight are stresses upon the feet and joints which further reduces physical activity, stresses upon the cardiovascular system promoting hypertension and risks of ischemic heart disease, and stresses upon the endocrine system leading to increased incidence of diabetes in later life. The normal values for fat content are 10 to 15% of body weight for adult men, and 5% higher content for adult women. Values in excess of 25% for males, 30% for females, roughly a doubling of the fat stores or 20% above the desirable body weight, are considered to be indicators of obesity. A simple calculation based upon the subcutaneous fat layer measured by the skin fold thickness with the "pinch-test" allows determination of % fat and lean body mass, which can be compared with ideal weights adjusted for age, sex, and body frame.

The reversal of obesity is considerably more difficult and more complex than its acquisition, and a multimillion dollar, quasi-scientific industry has grown "fat" from the attempted reductions of its clients. Although new diets and reduction aids spring up continually most of them are variations on the same theme,—rapid and painless weight loss. A common factor in most regimens is low carbohydrate intake. Ketonuria is a planned accompaniment of such unbalanced diets, and is promoted as a means of eliminating fat metabolites. Fast weight loss is usually obtained in the first week owing to loss of body sodium and accompanying water excretion; slower weight loss is obtained thereafter as a result of the reduced caloric intake. Several untoward results argue against the crash-diet plans. The early water loss is regained upon resumption of normal carbohydrate intake which promotes sodium retention. Apart from the toxic effects of keto-acidosis the high fat diets have been shown to produce an increase in serum cholesterol. The enhanced gluconeogenesis induced by low carbohydrate intake may cause wasting of functional tissue proteins and place additional stress upon liver and kidneys to eliminate the excess nitrogenous wastes from protein catabolism. It is difficult to maintain appropriate levels of minerals, vitamins, and other essential nutrients with the unbalanced diets.

Although it may be necessary to introduce drastic measures (controlled fasting, surgical intervention) in management of patients with overt obesity (patients who are 2–3 times their ideal weight), for most moderately overweight individuals it is preferable to institute a slow weight loss of 1 to 2 lbs per week for several weeks. Since 1 lb of fat is roughly equivalent to 3500 kcal in energy terms the elimination of food, or the increase of physical activity, to the tune of 500 kcal/day will result in a 1 lb loss per week. Moderate activities such as swimming or jogging consume 8 to 10 Kcal/min so that a 30 minute exercise period will provide half the reduction, while the elimination of a piece of rich cake or a couple of martinis will account for the other half. The gradual mobilization of depot fat by mild caloric deficits is preferable to the physiologic and psychologic ups and downs of the feast or famine approach encouraged by crash diets. Moderate reduction entails an ongoing self-evaluation of food and activity habits, with even small adjustments attaining significance when viewed as part of a long-term project. These could involve restriction of candy and other high sugar foods, replacement of whole milk or cheese with low-fat counterparts such as skim milk or cottage cheese, and replacement of passive forms of transport with walking or cycling. A reduced mixed diet allows for a reasonable variety and balance of all foodstuffs and does not leave the individual feeling deprived. On the contrary, it encourages a positive attitude change and modification of life-style that may help to arrest, or at least slow down, the ravages of time on the body.

PROTEINS

Dietary proteins contain some twenty amino acids, only one-half of which (arginine, histidine, isoleucine, leucine, lysine, methionine, phenylalanine, threonine, tryptophan, valine) must be ingested to maintain normal growth and health in humans. Before they can be utilized, the polypeptide linkages of the food proteins must be cleaved by action of proteolytic enzymes of the gastrointestinal tract to yield the free amino acids. Thus, in addition to the total protein content of a given food, the composition of the protein with respect to the essential amino acids and the susceptibility of the protein to intestinal digestion must be considered in determining nutritional value. The composite index which measures both the biologic value of absorbed amino acids and digestibility of food proteins is the net protein utilization (NPU). Topping the scale are the animal proteins such as fish, milk, and eggs with values of 80 to 90%, while vegetable protein sources, such as grains and beans, frequently give NPU figures in the 50 to 60% range. The key principle, as in many aspects of nutrition, is to provide a balanced mixture of varied food sources. This is particularly critical with vegetarian diets where the lysine, which is deficient in cereal proteins, may be supplemented with that in legumes, and the low tryptophan content of most plants may be augmented by that which is in milk or cheese. Obviously the ovo-lacto-vegetarian who restricts only meat from his diet will have fewer difficulties achieving adequate protein nutrition than the strict "vegan", (vegetarian) who also abstains from eggs and dairy products. In order to be effective the complementation of inadequate foods must occur at the same time since there is no storage of the unbalanced amino acid mixtures until the next meal. For efficient synthesis of cellular proteins in the body all of the amino acids must be present simultaneously.

Requirements for dietary protein are obviously greatest during periods of growth, but there are also situations of repair, such as recovery from surgery, hemorrhage, burns or wasting diseases, where extra protein intake may be needed. For the average adult the maintenance of protein nutrition as determined by nitrogen balance measures is met by a minimum requirement of about 30 g of protein per day. To allow for individual variability and NPU values of 75% on a mixed diet, it is recommended that the daily protein allowance should be 0.8 g/kg body weight. Babies require almost three times the adult protein intake on a body weight basis at birth, preferably in the form of mothers' milk which is not only sterile and conveniently packaged but is also much more efficiently utilized than other protein sources. Since the protein of cows' milk is less suitable for infant needs, the amount in formulas should be increased somewhat upon weaning. During the first year of life adequate growth and health will require 2 to 2.5 g protein/kg daily. High protein intake should be maintained for growing youngsters through the adolescent period, gradually tapering off to adult levels in the later teenage years. During pregnancy 30 g of extra protein should be ingested daily by the mother, continuing with an extra 20 g per day during lactation. In the nutritional therapy of patients under prolonged stress such as sepsis, major burns, or restoration of body cell mass, protein intakes of 1.5–2 g/kg are advocated. On occasion oral administration of protein hydrolysates or intravenous hyperalimentation with amino acid mixtures may be necessary to circumvent digestive problems (*i.e.*, in cancer patients), and to replete the debilitated state in a rapid fashion.

In many underdeveloped areas of the world the synergistic effects of protein deficiency and reduced resistance to infection contribute to the high mortality rates among children. When total food intake is inadequate (protein–calorie malnutrition) the infant usually develops emaciation in the first year of life resulting from complete mobilization of fat stores; this condition is termed marasmus. When a child is weaned to a diet with low protein content but one which is adequate in energy terms (protein malnutrition) he becomes chubby in appearance with abundant subcutaneous fat and edema developing generally after year one; this syndrome is called kwashiorkor. Marasmus and kwashiorkor may occur in combination and these patients, in addition to being underweight and short of stature, show only moderate signs of edema but much greater tissue wasting than if they suffered from only kwashiorkor alone. Treatment of these severely malnourished children requires appropriate diagnosis and therapy for accompanying infections, and replacement of the fluid and electrolytes (chiefly potassium and magnesium) lost from diarrhea. Owing to the serious gastrointestinal problems from deficiencies of digestive enzyme secretion, and the anorexia that accompanies chronic protein deficiency, patients often require parenteral nutrition at first, only gradually replaced with feeding by mouth. Since many of these children may be lactase-deficient, introduction of milk products into the diets must be undertaken cautiously and removed if diarrhea or other reactions develop. The caloric intake should be increased depending on the degree of marasmus, but even in the cases of pure protein deficit, diet therapy should include more calories to provide the energy for protein synthesis as well as to spare the dietary protein from catabolism. Supplementation of indigenous vegetable diets with high protein additions to provide up to 4 g/kg daily, particularly from animal sources, is essential in order to reverse the retardation of development. The degree of irreversible impairment depends upon the severity of the nutritional insult and the stage at which it is detected and treated. In many instances protein malnutrition commences *in utero*, and the best preventive measures are directed toward adequate nutrition for the mother during pregnancy and lactation.

The protein intake of the general population in Western industrialized nations seems to be more than adequate. In fact, as noted in the following section, the high incidence of diet-related cancer in such populations may be correlated with excessive intake of meat. While some of the problem is no doubt related to the high content of animal fats in meats, high animal protein ingestion may also contribute to production of carcinogenic nitrosamines in the intestinal tract. Thus a reduction by half from present consumption figures would approach the more moderate recommended allowances of 50 to 60 g of protein per day. Along with this reduction, a replacement of meat proteins with mixed vegetable and dairy sources of protein would probably be healthier for the general population.

Despite the apparent surfeit of dietary proteins in the majority of adults the health professions should resist becoming complacent regarding protein malnutrition. Within the last five years there has been a growing awareness that an alarming proportion of general hospitalized patients are seriously malnourished, and that among medical–surgical patients with hospital stays over two weeks, between one-quarter and one-half have been found to suffer from some degree of protein–calorie malnutrition. Because of the associated susceptibility to infections and delay in wound healing resulting from insufficient protein intake, it is imperative that such patients be identified and given the appropriate nutritional therapy. A number of simple parameters concerning a patient's physical state, dietary habits, intestinal problems, and so forth may be determined at admission by physicians or paramedical staff to assess the patient's nutritional status. Ongoing assessment during hospitalization can only be provided if there is an aware, co-operative nutrition service to coordinate the special skills of doctors, nurses, dietitians, and pharmacists. Among the signals of high nutritional risk are evidence of underweight or overweight, excess use of alcohol or fad diets prior to admission, dermatologic anomalies, edema, friability or discoloration of hair and nails, muscle wasting, low levels of circulating albumin or lymphocytes. The adult equivalents of kwashiorkor, marasmus, or the combined syndrome, can be identified in hospital patients, and the proper corrective feeding programs instituted with appropriate vigor and timing. Since the stress induced by acute infection or trauma will trigger release of the catecholamines and glucocorticoids which promote protein catabolism and gluconeogenesis while reducing the anabolic response to insulin, it may not be beneficial to institute protein supplements in this early phase. Moreover, in patients with liver or renal insufficiency large increments of protein may tax the already marginal capacity to deal with end-products of nitrogen catabolism. Obviously the restoration of protein balance in recuperating patients requires a sophisticated appreciation of the hormonal adaptations, with monitoring of blood levels of glucose, electrolytes, urea nitrogen, and tests of liver and kidney function. Prevention of large protein deficits from the body, as in the case of severe burns, requires hyperalimentation with both protein and calories to achieve positive nitrogen and energy balance in the anabolic phase of recovery. In the past conventional intravenous nutrition with glucose– saline has been used inappropriately for patients undergoing stress-induced negative nitrogen balance. The modern development of total parenteral nutrition with provision of adequate nitrogen, vitamins, and minerals, as well as calories has provided an invaluable mechanism for proper nutritional rehabilitation of malnourished patients, and for enhancement of immunogenic resistance to infections, wound healing, and ability to withstand further surgery or chemotherapy.

FIBER

One of the most neglected constituents of modern diets has been the indigestible material which is termed fiber, or roughage in layman's terms. The term embraces a group of heterogeneous compounds with varying properties depending upon their chemical makeup and their natural source. Fiber includes the polysaccharides, cellulose, the hemicelluloses such as agar and pectin which escape digestion by intestinal amylases, and the phenylpropane derivatives termed lignin. All of these compounds act as structural reinforcements for various plant cell walls, and hence are found only in vegetable sources. Dietary fiber promotes water uptake by contents of the large bowel thereby increasing the bulk and fluidity of the feces. In addition some of the polysaccharide components are broken down by bacterial action to provide substrates for and enhance the growth of intestinal flora. Thus a high fiber diet results in bulkier, looser stools, and increases the speed of movement of contents through the bowel.

These effects of fiber components upon intestinal function may explain some of the epidemiologic data relating various disease states to diet. For example, cancer of the colon and diverticulosis are much more common among Western peoples who ingest low-residue diets than in primitive populations whose staple foods contain a high proportion of fiber; when the latter move to urban areas and adopt the civilized diet the risk factors for colon disease rise correspondingly. Although such findings are not conclusive there is supporting evidence of an inverse cause and effect relationship between dietary fiber intake and diseases of the colon. High fiber regimens have been fed to patients with spastic colon diverticulosis, and in a high proportion of cases such a regimen brought about reversal of pain and other symptoms presumably by distending the wall of the colon and relieving the pressure. It is speculated that colon cancer may develop from exposure to agents that are either present in foods or generated in the intestinal tract: in the first category are the mutagenic hydrocarbons which are produced by high temperature cooking (broiling or frying) of meats; in the second group are the endogenous carcinogens, such as nitrosamines, which originate in the stomach from action of nitrates upon proteins, and which probably play a role in the etiology of gastric cancers; also in the second group are derivatives of the bile acids, such as deoxycholate, which arise from bacterial metabolism in the lower bowel, and which promote the tumor-forming capacity of the carcinogens. Obviously the dietary contents, their mode of cooking, and other factors can interact in a highly complex way to provide conditions for a long-term transformation that leads to tumor production at one or another site in the gastrointestinal tract. It is also believed that there are correlations between cancer and the absorption of nitrosamines and other dietary carcinogens in the other organs of the body, and between certain types of cancer and high meat intake, or high fat intake (leading to increased bile acid excretion). It is postulated that the larger volume of the stools and the more rapid transit of contents through the intestinal tract resulting from high fiber ingestion may explain the associated decrease in colon cancer since there is dilution and reduced contact with the cancer-initiating and -promoting substances in the bowel. Until all the evidence is in it is probably not justified to attempt to duplicate the diet of the aboriginal Bantu (25 g of fiber per day). Nonetheless, a shift in that direction by ingestion of relatively more fruit and vegetables, and particularly unrefined whole grains and bran, seems to be a prudent movement.

VITAMINS

Vitamins have played a prominent part in the scientific study of nutrition since their characterization around the turn of the century. In spite of the fact that much is known about their functions in the body and the molecular bases of their effects upon metabolism, there remain several important unanswered questions concerning vitamins and their actions, as well as areas of some controversy concerning the amounts that are optimal for health.

Vitamins are accessory food factors that are organic and, unlike the major nutrients, are required only in small amounts in the range of mg per day. There are 13 well-defined vitamins, not all of which must be provided per se in the diet. Vitamin D, for example, may be produced in the body, although not always rapidly enough to sustain growth. Vitamin A is produced from food precursors termed provitamins. Vitamin K is synthesized to a large extent by bacterial metabolism in the intestinal tract. Chemically, the vitamins are quite diverse and they are generally broken down into the four fat-soluble vitamins, A, D, E, and K, and the water-soluble vitamins, B complex and C. The division is important since the water-soluble vitamins are generally more readily absorbed, while the fat-soluble compounds are taken up by mechanisms similar to those for dietary lipids, and hence may be grossly deficient in fat malabsorption states. Moreover, there is virtually no storage for many of the water-soluble substances and the ingestion of surplus amounts above body requirements leads to excretion of the excess in the urine. In addition to providing a useful test indicator for the state of an individual with respect to a particular water-soluble vitamin, the efficient urinary excretion of these compounds allows for a fair margin of safety in terms of overdosages and explains why excessive amounts are rarely toxic. On the other hand, the fat-soluble vitamins can be stored in very large amounts, chiefly in the liver. Thus they need not be eaten daily, as in the case of water-soluble vitamins, and a single large dose may serve to counteract deficiency states. However, too great an overdose, or repeated administration of excessive amounts of fat-soluble vitamins has been shown to lead to severe toxic effects because of excess deposition. This pathologic condition is referred to as hypervitaminosis.

Deficiency of a vitamin in the body is referred to as avitaminosis, and may be caused by a primary lack in the diet or by secondary problems in eating or digestion. As in other types of malnutrition vitamin deficiencies are most frequently observed in association with poverty, but may also be commonly found in people who choose fad diets, go on prolonged alcoholic binges, or exclude certain groups of foods from their meals. Other common concomitants of avitaminosis are problems of dentition, malabsorption defects, anorexia, or continued use of oral, broad-spectrum antibiotics that knock out vitamin-producing intestinal bacteria. Vitamin deficiencies are seen most frequently in those regions of the world where single crop farming and subsistence upon a limited selection of staple foods is the rule. In our own society the situation is most serious among the elderly who may have been on marginal diets for a long time. Such a state of chronic undernutrition may produce subclinical deficiencies in which the frank symptoms of avitaminosis are not evident but the body has reduced resistance to stress, trauma, or disease.

Many laboratory techniques have been devised to evaluate the status of individuals with respect to vitamin nutrition. However, even in the absence of such sophisticated tests, careful observations of various clinical signs are of prime value in the assessment. These will include body measurements, growth patterns of children, rapid weight changes of adults, alterations in texture or pigmentation of the skin, anomalies of the eyes and oral cavity, skeletal malformation, mental confusion, irritability, and other neurologic signs. Other clinical markers associated with vitamin deficiencies are various types of anemia, lowering of certain serum or red cell enzymes that require the vitamin as cofactor, or decreases in circulating levels of the vitamins themselves or of their excretory end-products in urine. A saturation loading test may be used in follow-up investigations; a well-nourished individual will excrete the excess of an overload of a water-soluble vitamin as noted earlier, while a deficient individual may excrete little or none since the entire dose would be retained in the body to replete the exhausted tissues. Individual assays will be described for the various vitamins in the following discussions.

VITAMIN B₁—THIAMIN

Thiamin was the first of the B-complex vitamins to be purified and characterized. Some 75 years ago Funk isolated the substance from rice polishings and showed that it counteracted the disease known as beriberi. Since it was vital for survival and had chemical properties of an amine Funk termed the compound a "vitamine," and the shortened form without the "e" has persisted as the generic name for such accessory food factors despite the fact that a majority are not amines in nature. Thiamin occurs abundantly in whole grains, in beans and peas, and in beef or pork. Between 1 to 1.5 mg of thiamin is required per day for adults, with somewhat higher allowances during growth, pregnancy, lactation, or during high carbohydrate intake.

Sufferers of beriberi were the first casualties in man's attempts to improve upon nature by the process that is euphemistically referred to as food refinement. In the Orient the disease was endemic among those people who consumed a staple diet of rice which had been polished to remove the brown outer casing. Up until the 1940s white flour marketed in Western countries had 80% of the natural thiamin removed by the milling process, but this is now replaced in so-called enriched flours. Prolonged ingestion of thiamin-depleted foods has led to progressively more serious, irreversible effects involving the digestive tract, the musculature, the nervous system, and finally the heart. Early symptoms are anorexia and nausea, followed by fatigue, weakness, and muscle wasting. Literal trans-

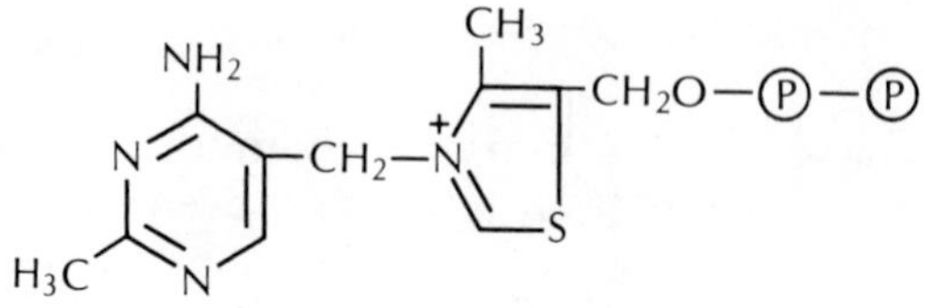

Thiamine Pyrophosphate (TPP)

FIG. 2–1. Thiamine coenzyme structure.

lation of the Singhalese term, beriberi, is "I can't, I can't." Development of
the neurologic signs of the deficiency is referred to as dry beriberi and is
characterized by a peripheral neuritis with numbness and tingling sensations in the lower legs and feet, ataxic gait, and various degrees of
paralysis. In most cases 5 to 10 mg thiamin supplementation per day
improves the condition. Wet beriberi refers to the edema found in severe
cases which accompanies cardiac disturbances. Cyanosis, enlargement of
the heart, palpitation, and chest pain are common concomitants. Patients
may die of sudden heart failure unless treated with large amounts (50– 100
mg) of thiamin. In infants who receive milk from thiamin-deficient mothers the beriberi is particularly acute and pernicious, leading rapidly
through episodes of gastrointestinal dysfunction and abdominal pain,
dyspnea, convulsions, and culminating in death from cardiac failure. Injections of 2 to 5 mg of thiamin may lead to dramatic recovery within
hours.

　　One of the more common causes of thiamin deficiency is chronic alcoholism. In time many alcoholics develop peripheral and central neurologic
defects known collectively as the Wernicke– Korsakoff syndrome. Typical
symptoms include nystagmus and paralysis of eye muscles, ataxia, confusion, memory loss, and delusion. A high percentage of patients will die of
heart failure unless thiamin is administered in large doses (10– 20 mg
several times daily), preferably by injection since alcoholics absorb the
vitamin poorly. In many instances thiamin is not the only deficient vitamin, and it is wise to administer a mixture of the B complex.

　　The molecular basis of thiamin action is well understood, and the
rationale for the pathologic changes in beriberi is clear in light of the key
role the vitamin plays in carbohydrate metabolism and oxidative energy
production. Two decarboxylating enzymes of the citric acid cycle responsible for conversion of pyruvate to acetyl-CoA and of α-ketoglutarate to
succinyl-CoA employ thiamin in its pyrophosphate form, TPP, as the obligatory coenzyme (Fig. 2– 1). Thiamin possesses two heterocyclic ring
systems, pyrimidine and thiazole. It is the latter which plays the functional
part in the coenzyme by formation of an activated intermediate through
the carbon α to the carboxyl group of the substrate being attacked (Fig.
2– 2). After elimination of CO_2 from the latter the remainder of the sub-

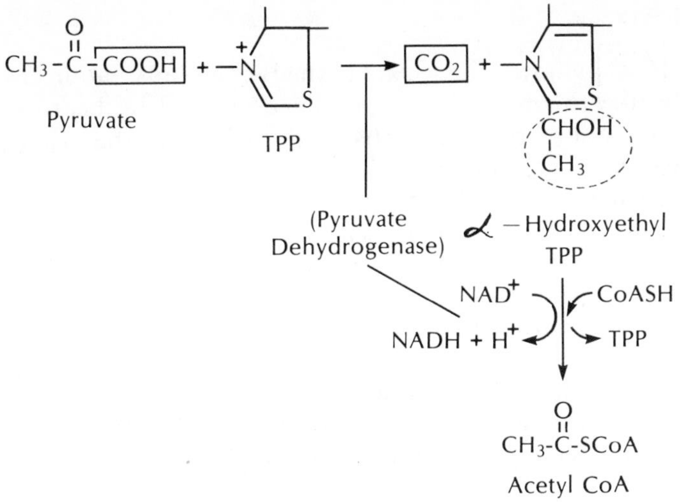

FIG. 2–2. Coenzyme function of thiamine pyrophosphate in pyruvate oxidation.

strate, α-hydroxyethyl in the case of pyruvate, becomes further oxidized to acetyl-CoA (succinyl-CoA in α-ketoglutarate oxidative decarboxylation) to generate great sources of energy for the cell in mitochondrial respiration. Consequently lack of thiamin in the tissues prevents the mitochondrial oxidations from carbohydrate by preventing entry of pyruvate to the citric acid cycle, and since it also impedes an intermediate reaction within the cycle, it interferes with oxidative catabolism of the end-products from fatty acids (acetyl-CoA) and amino acids (oxaloacetate, α-ketoglutarate). The aerobic production of ATP by mitochondrial oxidative phosphorylation will thereby become grossly impaired and cells will have to rely upon the meagre ATP generation from anaerobic glycolysis. The latter is quite inadequate for the vast energy requirements in nerves and heart muscle with the resultant malfunctions noted in the discussions of dry and wet beriberi respectively.

Blood levels of pyruvate and α-ketoglutarate are elevated in beriberi, and these measurements may be used as indicators of thiamin deficiency. The enzyme, transketolase, of the pentose phosphate pathway is also dependent upon TPP as its coenzyme, and the activity of this enzyme in the patient's erythrocytes and the stimulation by added TPP provide a very sensitive indicator of thiamin concentrations in tissues. An additional test is the measurement of the amount of thiamin excreted in urine after administration of a standard load; a decreased elimination from the body by comparison with values from normal humans is indicative of low tissue levels of thiamin and a need for the retention of the test dose to replete the exhausted stock.

VITAMIN B_2—RIBOFLAVIN

As its name implies riboflavin was the second of the B-complex vitamins to be characterized. Unlike thiamin which is easily destroyed by heating, riboflavin is quite heat-stable, but it absorbs light strongly in the visible range and is inactivated upon exposure to illumination. This may present problems in sources such as milk which should not be left out in glass containers; some of the riboflavin content will also be lost when milk is treated with ultraviolet irradiation to raise its vitamin D potency. Milk, and milk products such as yogurt, are the most important sources of ribo-flavin. The vitamin is stored to a significant extent in visceral organs, so that liver and kidney are good dietary sources. Grains are generally low in riboflavin, but reinforced breakfast cereals have substantial amounts added as noted on the package contents. The average requirements are estimated at 1.3 to 1.8 mg per day for average adults with some increments for pregnancy or lactation.

In humans there is no life-threatening syndrome of riboflavin deficiency analogous to beriberi. Nonetheless there are many people who seem to be in a marginal position with respect to riboflavin ingestion. The deficit is greatest in strict vegetarians, such as those on the Zen macrobiotic diet, and particularly in their infant children. Low riboflavin intake is also common among alcoholics. In test situations ariboflavinosis produces a number of dermatologic anomalies including characteristic crusted fissures of the lips (cheilosis), painful purple-coloured tongue (glossitis), scaly dermatitis of the face and groin area, and various eye lesions such as irritation and vascularization of the cornea, iritis, and severe photophobia. Although none of these symptoms are disabling, the lesions may form foci for infections, and the deficiency of riboflavin may lower an individual's resistance to stress and his ability to recover from trauma. In experimental animals subjected to long-term B_2-deficient diets not only are the flavin enzymes (discussed below) decreased in activity, but the tissue mitochondria apparently fail to divide normally thus producing gigantic organelles (megamitochondria) in the cells. With prolonged ariboflavinosis hypoplasia of the bone marrow develops resulting in macrocytic anemia, and various neurologic anomalies may ensue. Ability of tissues to store riboflavin may account for the failure to observe any very serious deficiency disease in humans on marginal intakes. Oral administration of riboflavin (10–15 mg/day) usually clears up the dermatologic lesions within a week or so.

The mechanism of action of riboflavin is explained in terms of the coenzymatic role of the vitamin as prosthetic group of several flavoprotein oxidation–reduction enzymes. Before it functions as a coenzyme, riboflavin must be phosphorylated to produce flavin mononucleotide (FMN), (Fig. 2–3a), or reacted with an additional molecule of ATP to generate flavin

a)

Riboflavin Phosphate (FMN)

b)

FMN or FAD
(Oxidized)

$+2e^-$
$+2H^+$

$-2H^+$
$-2e^-$

FMNH$_2$ or FADH$_2$
(Reduced)

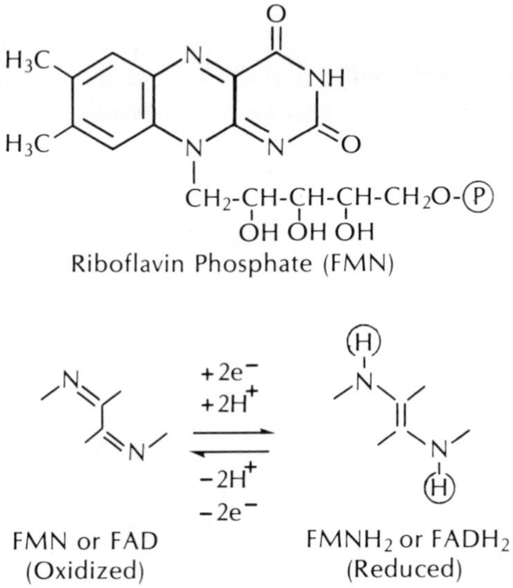

FIG. 2–3. Riboflavin and its coenzyme forms: (A) flavin mononucleotide structure; (B) redox functions of riboflavin coenzymes.

adenine dinucleotide (FAD). Coupled with the appropriate oxidoreductase enzyme, the heterocyclic isoalloxazine ring of the flavin coenzymes can undergo reversible reduction by transfer of two hydrogen atoms to produce the reduced forms, FMNH$_2$ or FADH$_2$ (Fig. 2–3b). The flavin nucleotides generally become tightly bound, sometimes covalently, to their respective apoproteins so that the coenzymes are non diffusible. Oxidations catalysed by flavoproteins often serve to extract hydrogens from $-CH_2-CH_2-$ structures, as in succinate dehydrogenase or fatty acyl CoA dehydrogenase, both of which utilize FAD as coenzyme. In the respiratory chain cytochrome c reductase utilizes FMN as the intermediate to abstract H from NADH and relay it to the cytochromes. Flavoprotein oxidoreductases also play important roles in amino acid and purine catabolism, in the α-keto acid oxidative decarboxylation mentioned earlier in connection with thiamin's role, in metabolism of thyroxine, in metabolism of biogenic amines and certain drugs, and in interconversions of folic acid coenzymes; the latter may explain the occurrence of macrocytic anemia in pronounced ariboflavinosis.

The activity of glutathione reductase in erythrocytes is particularly sensitive to riboflavin depletion, and response of the enzyme *in vitro* to added FAD is a good indicator of the level of riboflavin nutrition. As for thiamin, the determination of urinary excretion of the vitamin following a test load may also be used to estimate tissue stores. It is a common experience for normal adults who have taken a vitamin B supplement to notice the presence of intense yellow-colored fluorescent products of riboflavin in their urine within hours.

VITAMIN B₃—NIACIN

Niacin, or nicotinic acid as it is properly called, was identified as the B vitamin which was required to cure or prevent the disease known as pellagra. It is a derivative of pyridine and it may also originate within the body from metabolism of the amino acid, tryptophan. As part of a normal diet it is estimated that roughly one-half of the daily niacin requirement comes from tryptophan conversion. Many of the foods that are low in niacin are also low in tryptophan. Good sources of both are lean meats, legumes, peanuts, several green vegetables, whole grains, enriched flour, and breakfast cereals; milk and eggs are low in niacin but rich in tryptophan, and therefore act as supplementary sources. For normal adults 15 to 20 mg per day seem to meet the requirements with extra allocations for pregnancy and lactation. It is estimated that 60 mg of tryptophan give rise to 1 mg of niacin, and since 500 to 600 mg of tryptophan are ingested as part of a good minimal protein-containing diet up to 10 mg niacin may derive from this source.

The syndrome of pellagra has been endemic in certain populations who feed exclusively upon niacin-deficient cereals, such as millet in Asia or corn in the Americas. Sharecroppers in the Southern United States in the last century were particularly prone to the deficiency of niacin since they subsisted on corn, and pork fat which are not good sources of the vitamin or tryptophan. Pellagra is characterized by the so-called 3 Ds—dermatitis, diarrhea, and dementia—and if untreated, is followed by a final D, death. In chronic cases the condition may occur in cyclic fashion with most severe exacerbations occurring in the spring possibly as a result of increased physical activity and exposure to sunlight. The acute dermatitis is light-sensitive and resembles sunburn at first, and then progresses through darkening, blistering, and scaling of the exposed skin areas to produce the rough surface appearance that led to the term for the syndrome (pele = skin, agra = rough). The diarrhea is symptomatic of the severe inflammation of mucous tissue of the intestinal tract. As a result eating becomes painful and absorption inefficient with further deterioration of the patients' nutritional status. Anxiety, depression, confusion, and delirium are hallmarks of the demented state of pellagra sufferers who may eventually become comatose in the terminal stages. Up to 0.5 g per day of the vitamin in its amide form should be administered to patients with acute pellagra, preferably in several intramuscular injections if serious gastrointestinal disorder is evident. It is important to use nicotinamide rather than nicotinic acid since the former does not produce the vasodilation reaction that the latter does upon injection. As the symptoms clear up smaller doses (50– 100 mg per day) may be taken orally. A diet with adequate high quality protein and other B-vitamins, especially pyridoxine (see below), should be continued as the intestinal problems become alleviated.

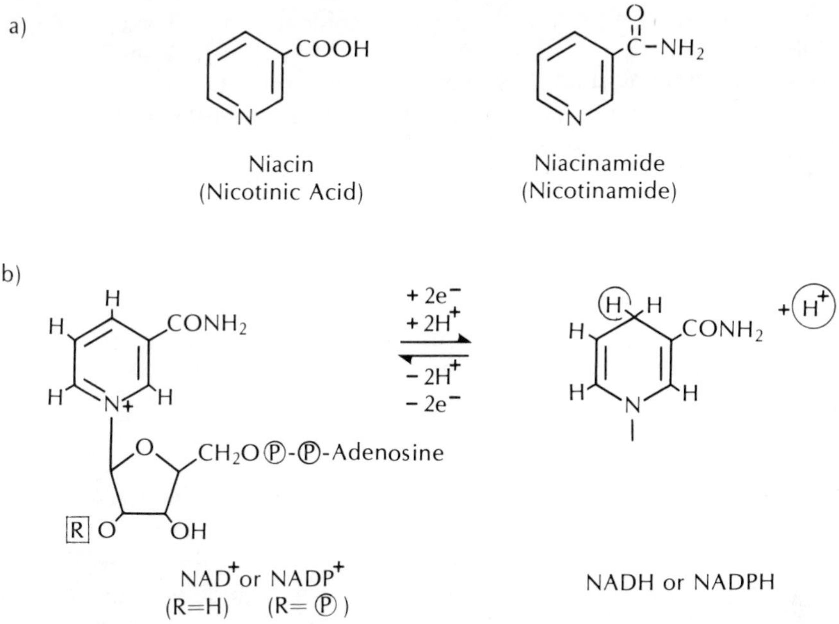

a)

Niacin
(Nicotinic Acid)

Niacinamide
(Nicotinamide)

b)

$$+ 2e^-$$
$$+ 2H^+$$
$$- 2H^+$$
$$- 2e^-$$

CH$_2$O Ⓟ-Ⓟ-Adenosine

R O   OH

NAD$^+$or NADP$^+$
(R=H)   (R= Ⓟ)

NADH or NADPH

FIG. 2–4.   Niacin and its coenzyme forms: (A) niacin and niacinamide
structures; (B) redox functions of nicotinamide coenzymes.

The amide form of nicotinic acid (Fig. 2– 4a) has its functional role in
the body as the active portion of the pyridine–nucleotide coenzymes,
nicotinamide adenine dinucleotide (NAD) and nicotinamide adenine dinucleotide phosphate (NADP). As noted earlier for the riboflavin coenzymes, the nicotinamide nucleotides serve as hydrogen carriers in biologic
oxidation–reductions. In this case only one hydrogen atom from the substrate is attached to the pyridine ring of the appropriate coenzyme, and
reduced forms are therefore represented as NADH and NADPH respectively (Fig. 2– 4b). Certain enzymes are quite specific for one or the other of
the pyridine nucleotides and only rarely may they be used interchange-
ably. As a general rule NAD is used for the oxidative catabolic reactions of
glycolysis, the citric acid cycle, fatty acid oxidation, and amino acid
catabolism, and generates the NADH that is used principally for ATP generation in mitochondrial oxidative phosphorylation. On the other hand,
NADPH is produced mainly from oxidative reactions of the pentose phosphate shunt pathway, and is utilized as the reductive energy to drive the
biosynthesis of fatty acids, steroids, and glutamic acid in tissue anabolism.
Within cells both NAD, NADP, and their reduced forms diffuse from one
enzyme to another and therefore serve as mobile carriers of reductive
energy. However, since they are not attached to cellular proteins the
nicotinamide coenzymes are not stored to the extent that the flavin coen-

zymes are, which may explain the greater incidence and severity of symptoms of deficiency from niacin than from riboflavin in humans.

Apart from the characteristic clinical symptoms niacin deficiency may be detected from the decrease in urinary excretion of its metabolites, N^1 methyl–nicotinamide and 2 pyridone. The most definitive proof however is response of the dermatitis and other symptoms to therapeutic doses of niacin. It is important to be aware that pellagralike symptoms can be secondary to disturbances of tryptophan metabolism as in the hereditary condition known as Hartnup disease, and to consider the possibility that pyridoxine deficiency could be a cause of this disturbance. As noted in other instances (*i.e.*, chronic alcoholics with pellagra symptoms), it is wise to use a B-complex mixture in therapy.

VITAMIN B₆—PYRIDOXINE

Numbering in the B-complex series continued with the isolation of various natural factors that were required for the growth of bacteria, but a majority failed to meet the criteria as true vitamins in animal and human nutrition. Survivors of the numbered compounds include only pyridoxine (B_6) and cobalamin (B_{12}). References may be seen in the popular press but not in refereed scientific publications to vitamins B_{13-15}; nutritionists have found no evidence that such compounds (pangamic acid, laetrile) have any vitamin functions in humans. Pyridoxine is quite heat-stable, but decomposes in the light or in alkaline solutions. It is almost completely eliminated during the milling process. Major sources of pyridoxine in the diet include whole grains, meat (especially organ meats), eggs, and legumes. The estimated adult requirement is 2 mg daily with an extra 0.5 mg during lactation. The same level of additional supplementation has been suggested in pregnancy, but as noted below there is much current evidence that such an allowance may be inadequate.

The symptoms of pyridoxine deficiency include glossitis of the tongue, pigmented scaly dermatitis similar to that seen in pellagra, a characteristic hypochromic microcytic anemia with high serum and tissue iron levels, and variable effects upon the nervous system including numbness and tingling sensations in the extremities, irritability, and depression. The neurologic symptoms are most severe in pyridoxine-deficient infants who may develop convulsive seizures. In some instances the convulsions are prevented by normal dietary supplements of pyridoxine, but in other cases the children have inherited a defect, pyridoxine dependency, in utilization of the vitamin and require ten times the normal amounts. A similar situation seems to apply to men and women who are exposed to high levels of estrogenic hormones. There is serious cause for concern that pregnant women, or women taking oral contraceptives which contain large amounts of estrogens seem to have a greatly increased need for pyridoxine. In several studies it has been found that a high proportion of such individuals

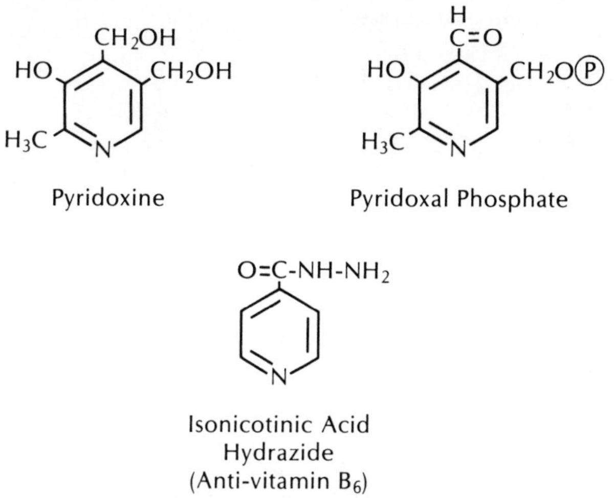

FIG. 2–5. Pyridoxine, its coenzyme form—pyridoxal phosphate and the antivitamin—isonicotinic acid hydrazide.

show laboratory signs (*e.g.,* anomalies of tryptophan metabolism) of vitamin B_6 depletion, and that high doses (10–30 mg daily) are required to normalize the situation. Therapeutic intervention with high doses of pyridoxine seems to be most useful in curing the mental depression that occurs in some women taking oral contraceptives. It is difficult to assess the true nutritional requirements of vitamin B_6 during pregnancy, but the recommended allowance of 2.5 mg per day seems to be too low and should probably be doubled, particularly in patients exhibiting evidence of anemia or depression. Another situation which may require therapy with supplementary pyridoxine is the long-term treatment of patients with anti-tuberculosis drugs such as isonicotinic acid hydrazide (Fig. 2–5). These agents antagonize the effects of vitamin B_6 and in time can produce symptoms such as peripheral neuritis. The latter may be cleared up by treatment with large doses of pyridoxine (50–100 mg per day). Large doses of pyridoxine along with thiamin are also frequently advocated in the treatment of chronic alcoholism.

Coenzymatic functions of pyridoxine have been demonstrated for a wide variety of metabolic reactions chiefly involving amino acid substrates. Typical of such reactions and best understood from the standpoint of mechanism of action of pyridoxine coenzymes is the process of transamination or aminotransfer. A number of different aminotransferases with different specificities for the amino acid substrates have been found, but all interact with the same pyridoxine coenzyme, pyridoxal phosphate (Fig. 2–5). This derivative of the vitamin is produced in the body, first by a

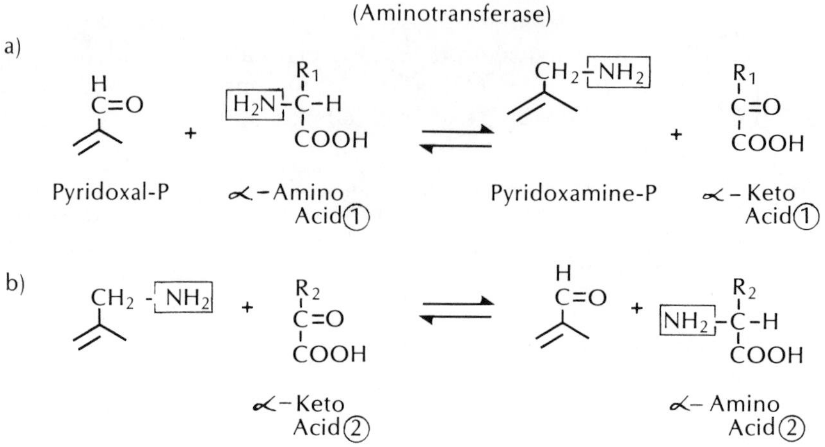

(Aminotransferase)

a)

Pyridoxal-P α-Amino   Pyridoxamine-P α-Keto
                 Acid①   Acid①

b)

α-Keto α-Amino
Acid②   Acid②

FIG. 2–6. Coenzymic function of pyridoxal phosphate in aminotransfer reactions: (A) the coenzyme accepts the NH_2–group from an amino acid; (B) the coenzyme donates the NH_2–group to an alpha keto acid.

specific kinase which phosphorylates one of the outer CH_2OH substituents on the pyridine ring, and then by a specific flavoprotein dehydrogenase which oxidizes the second CH_2OH at the position para to the ring N to the aldehyde form. The pyridoxal phosphate becomes firmly bound to the appropriate apoproteins functioning as a prosthetic group of the aminotransferases. The amino acid substrate binds to the aldehyde function of the prosthetic group through its α-amino group, first forming the intermediate Schiff base, then hydrolysing from the enzyme surface as the α-keto acid product and leaving the NH_2 behind on the coenzyme as pyridoxamine phosphate (Fig. 2–6a). The latter may then transfer the NH_2 to a second α-keto acid as the other end-product of the reaction (Fig. 2–6b). In this way the aminotransferase is able to transfer amino groups reversibly from one amino acid to produce another, with the pyridoxal phosphate acting as intermediate in the process. The system is very important in providing a means for synthesis of a balanced mixture of amino acids in tissues for efficient protein synthesis, and a means to mobilize the carbon chains of amino acids for gluconeogenesis or for energy production from the α-keto acids in caloric deprivation. Other important enzyme systems that require the pyridoxine coenzymes include the amino acid decarboxylases that generate histamine, serotonin, norepinephrine, γ-aminobutyric acid, and other important amines. Many of the latter compounds are vital to the functions of nervous tissues, and mental depression and infantile convulsions may be related to altered balance of such powerful modulating agents in pyridoxine deficiency. The vitamin is also needed for amino acid dehydration, desulfhydration, transsulfuration, as well as amino acid transport across cell membranes. The conversion of porphyrin precursors, glycine, and succinate to produce the heme ring system by the enzyme, delta aminolevulinic acid synthetase, is also critically dependent upon pyridoxal phosphate, which explains the occurrence of hypochromic anemia in B_6 deficiency despite the presence of adequate iron stores. Pyridoxal phosphate is also essential for the normal metabolism of tryptophan, and the production of nicotinamide. In the absence of pyridoxine, tryptophan is incompletely metabolized, and the accumulated byproducts, xanthurenic acid and kynurenic acid, are eliminated in the urine. It is noteworthy that excess amounts of these compounds in urine reflect not only a pyridoxine deficiency but also a possible shortage of niacin.

Tests for the adequacy of B_6 in the diet have been based upon the measurement of xanthurenic acid excretion following a standard load of tryptophan. However, it should be noted that other factors, for example the presence of large amounts of leucine in the diet, may also produce interference with tryptophan utilization. The induction of the liver enzyme, tryptophan oxygenase, by glucocorticoids, which are in turn stimulated by estrogens, may distort the tryptophan utilization pathway and create a false impression of pyridoxine requirements to correct the B_6 imbalance during pregnancy or as a result of chronic use of oral contraceptives. Other tests that may be applied in the determination of vitamin B_6 status include measurements of erythrocyte glutamate–aspartate aminotransferase and its response to excess added pyridoxal phosphate, blood levels of the vitamin, or the urinary excretion of its major metabolite, 4-pyridoxic acid.

VITAMIN B_{12}—COBALAMIN

Cobalamin was identified as the component of raw liver extracts which was able to reverse the symptoms of the condition known as pernicious anemia. This syndrome was recognized to be no ordinary vitamin deficiency, however, and it was demonstrated that in addition to an extrinsic diet factor (vitamin B_{12}) the body required an intrinsic factor synthesized in the gastric mucosa in order to absorb and utilize the vitamin. Cobalamin, as its name suggests, contains cobalt complexed within a ring system resembling the porphyrin structure of heme (Fig. 2–7). It is quite a stable compound, and survives ordinary cooking since it is bound in a protein complex in natural foods. Vitamin B_{12} is restricted to animal foods with liver, meat, fish, and eggs providing the major sources; vegetable foods are completely devoid of the vitamin, and even yeast, which is an excellent provider of other B-vitamins, is not a source of B_{12}. Only miniscule amounts are required by the body (1 μg/day) but since only one-third of the amount ingested is absorbed the recommended allowance is 3 μg daily, with an extra 1 μg during pregnancy or lactation.

Vitamin B_{12} deficiency of strictly dietary origin is quite rare, except among vegetarians who eat no meat products or eggs. The classic syn-

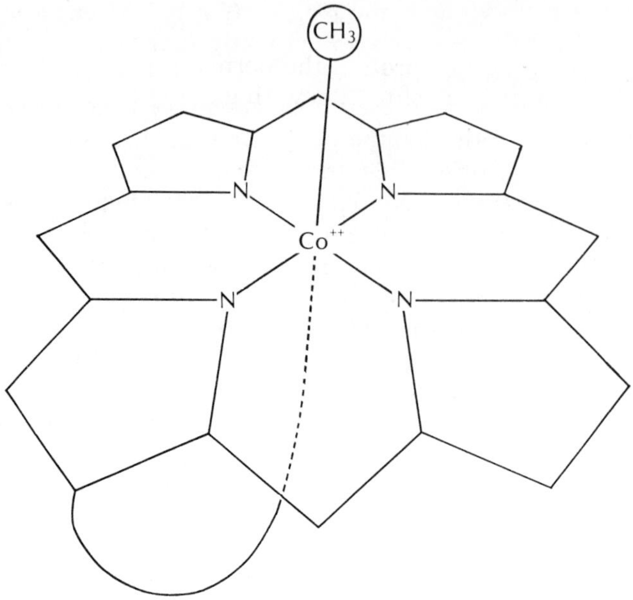

Vitamin B_{12} - Methyl Cobalamin
Coenzyme

FIG. 2–7.   Vitamin B_{12} coenzyme structure—methyl cobalamin.

drome of pernicious anemia is the result of a defective secretion in the stomach of an obligatory glycoprotein (the intrinsic factor) which normally complexes with vitamin B_{12} (the extrinsic factor) and facilitates its absorption from the small intestine. A deficiency of B_{12} in the body may also result after surgical gastrectomy, resection of the ileum, or in malabsorption syndromes that interfere with uptake from the intestinal tract. The depletion of tissue stores, present mainly in the liver, generally takes some years since the body has a great reserve of B_{12}. Consequently, the disease may develop quite slowly and late in life; however, children who have been raised on strict vegetarian diets may show early signs of pernicious anemia since they have had no opportunity to build up their liver stores. Thus it is important to urge the parents of growing children and adolescents to include some animal products, such as eggs or a cobalamin-containing vitamin preparation, in their vegan diets.

Pernicious anemia is characterized by disorders of the blood-forming tissue, the gastrointestinal tract, and the nervous system. In the bone marrow the maturation of erythrocytes is arrested with the accumulation of abnormal, gigantic cells (megablastosis), and the release into the circulation of few red cells that are enlarged (macrocytes) but which contain

normal amounts of hemoglobin (normochromic). Leukocytes and platelets
are also generally reduced in numbers in the blood. In addition to the
reduced production of intrinsic factor protein the pernicious anemia patient shows other signs of intestinal dysfunction with gastric atrophy and
decreased secretion of hydrochloric acid and pepsin in the gastric juice.
The neurologic symptoms include a progressive peripheral neuritis with
tingling and numbness of the extremities, loss of coordination, and muscular weakness of the limbs resulting from combined degeneration of tracts
in the spinal cord. Psychiatric problems of varying severity occur including irritability, apathy, depression, confusion, memory disturbances, and
paranoia. Unless treated the condition is fatal. Treatment requires injections of large amounts (50–100 μg) of the vitamin, several times weekly at
first until the hematologic anomalies are corrected, and monthly thereafter as a maintenance regimen. Because of the body's storage capacity for
B_{12} patients may be maintained on a massive dose (1000 μg intramuscu-
larly) every three months. Reversal of the peripheral neuropathy and mental changes may occur relatively more slowly than changes in the blood
cells depending upon the duration of the deficit and the severity of the
effects upon the nervous system.

Cobalamin has been implicated in a coenzymatic role for several varied metabolic reactions in cells. The coenzyme form of the vitamin has a
5-deoxyadenosine group attached to the free sixth coordination position of
the cobalt atom. In one of its major actions the cobalamin coenzyme has
this position filled instead by a $-CH_3$ group (Fig. 2–7); the carbon–metal
bond is unique in living systems and has led to reference to this derivative
as nature's Grignard reagent. The methyl B_{12} form serves as an intermediary in the generation of methylated compounds such as methionine
and thymine. Requirement for the latter in DNA synthesis provides a likely
explanation for the arrested development in pernicious anemia of the
erythrocytes at the stage of the rapidly dividing cells that are normally
undergoing maturation in the bone marrow. Coenzyme forms of B_{12} also
appear to be essential for reduction of ribonucleotides to their respective 2-
deoxy derivatives that are also required for DNA synthesis and hence cell
division. Other coenzyme functions of B_{12} involve its participation in
mutase reactions of the type $-CX-CH- \rightarrow -CH-CX-$, where the group X
is transferred to a vicinal carbon atom. An example is the reaction whereby
methyl-malonyl-CoA, a metabolite derived from odd-carbon fatty acids
and certain amino acids such as valine, isoleucine, and methionine, is
transformed into succinyl-CoA for catabolism in the citric acid cycle. In
pernicious anemia the reaction is defective and methylmalonic acid which
accumulates is excreted in excessive amounts in the urine, an indicator of
B_{12} deficiency.

Another useful test of vitamin B_{12} status and its absorption is the
Schilling test whereby radioactive B_{12} is administered orally with or without intrinsic factor. In normal individuals 30 to 40% of the ingested

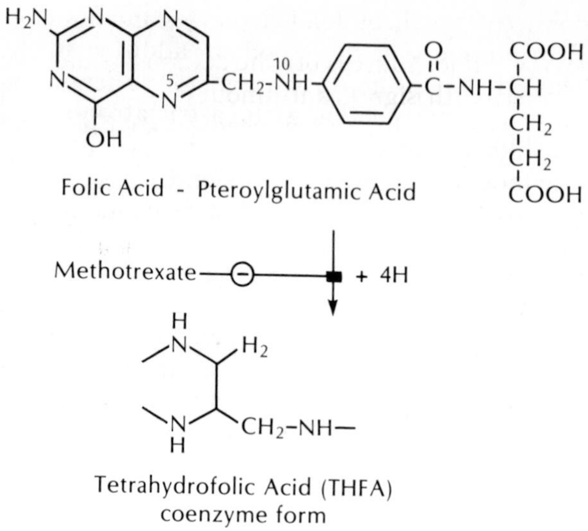

Folic Acid - Pteroylglutamic Acid

Tetrahydrofolic Acid (THFA)
coenzyme form

FIG. 2–8. Folic acid and its coenzyme form—
tetrahydrofolic acid. Reduction to the active coenzyme
(THFA) is blocked by the antifolate drug—methotrexate.

cobalamin is absorbed in either condition, while less than 2% will be
absorbed in persons suffering from pernicious anemia without intrinsic
factor. Patients with classical deficiency of the gastric factor will show
enhanced absorption when it is administered, and they can be differentiated from sufferers of other malabsorption defects who show no improvement of B_{12} uptake with added intrinsic factor.

FOLIC ACID — PTEROYLGLUTAMIC ACID

The vitamin, pteroylglutamic acid, was first isolated from the foliage of
the spinach plant and hence received the trivial name of folic acid. This
factor was found to be capable of curing a type of megaloblastic anemia
that is quite common among pregnant women, and it turned out to be
distinct from the factor in liver extracts which cured pernicious anemia.
The compound contains the heterocyclic pteridine ring joined to the N of p
amino-benzoic acid—this moiety is called pteroic acid; the carboxylic
group is joined to the N of glutamic acid by an amide linkage—this compound is termed pteroylglutamic acid (Fig. 2–8). In its natural state additional molecules of glutamic acid (up to seven) are attached in a chain by
similar amide links to the –COOH of glutamic acid; an enzyme in the
small intestine, conjugase, breaks these extra links to yield free folate
which is absorbed. Folic acid is quite stable in acidic solutions, but decomposes slowly at room temperature and rapidly when heated at neutral pH

or above. Consequently, there are severe losses of this vitamin upon storage and in most methods of cooking. Major sources of folic acid in the diet are fresh green vegetables and liver, with significant amounts also found in whole grain cereals, meats, and legumes. Individual needs for folic acid are difficult to assess. In most cases 50 μg per day seems adequate to prevent or correct the symptoms of macrocytic anemia. However, there are wide differences in the availability and utilization of folate derivatives from different foods, and recommendations for adult daily allowances range from 200 μg in Canada to 400 μg in the United States. Demands increase markedly in pregnant women and an additional 50 to 200 μg should be ingested during pregnancy or lactation.

Folic acid deficiency appears to be a very widespread phenomenon, particularly in infancy and during pregnancy when increased demands from growth may be superimposed upon low intake; it is also commonly observed in patients suffering from sprue and other malabsorption syndromes. Estrogen in high concentrations appears to interfere with folate absorption, and signs of deficiency may appear in women taking oral contraceptives or in postmenopausal women on long-term estrogen therapy. Symptoms are similar to those of B_{12} deficiency with megaloblastosis of the marrow, leukopenia, thrombocytopenia, macrocytic anemia, and gastrointestinal disorders; however, there are none of the neuropathic changes seen in pernicious anemia among people with folate deficiency. If untreated, the patients develop sequelae of the anemia including severe skin pallor, tissue anoxia, weakness, and cardiac enlargement associated with congestive heart failure. Oral administration of up to 5 mg folate can be prescribed in severe deficiency states. If the condition is associated with a malabsorption syndrome the intravenous route is advocated. Some supplementation of the diet may be provided after the blood changes are normalized to prevent recurrences, but amounts in excess of 100 μg are rarely necessary. It has been reported that the psychological performance of mentally defective children with Down's syndrome (mongolism) may be improved by large doses (50 mg daily) of folic acid. Several studies are currently in progress to assess the efficacy of this and other vitamin administration in larger numbers of patients with mental retardation and chromosomal anomalies such as Down's children. Although high dosages of folic acid do not appear to be toxic there is some danger in their indiscriminate use. The anemia due to vitamin B_{12} deficiency is corrected with high folate intake, which does not unfortunately improve the neuropathic effects of pernicious anemia. Thus the ingestion of large folate supplements could mask the potentially lethal deficiency of cobalamin. It is for this reason that food and drug laws in North America prohibit more than 100 μg of folic acid in over-the-counter, one-a-day vitamin preparations.

The actions of folic acid coenzymes are highly complex since they involve the transfer of one-carbon units at oxidation levels ranging from

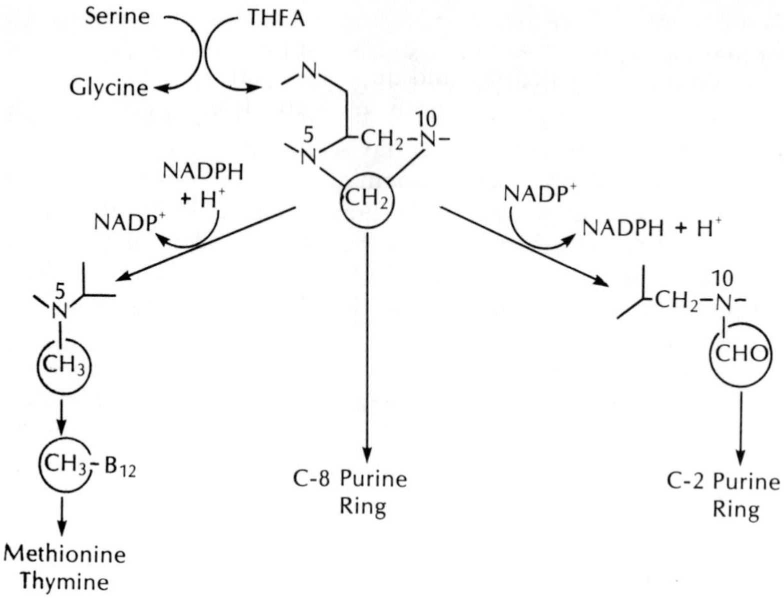

FIG. 2–9. Coenzyme functions of tetrahydrofolic acid and methyl cobalamin in one-carbon transfer reactions.

$-CH_3$ (methyl) through $-CH_2OH$ (hydroxymethyl), $-CHO$ (formaldehyde), to $-COOH$ (formic acid). These processes are of great biologic and clinical interest since they are crucial to DNA synthesis and thus cell division, which explains their key role in the maturation of erythrocytes and lymphocytes in the bone marrow. The active coenzyme form of the vitamin is produced by a two-step reduction of the pteridine ring utilizing two moles of NADPH with generation of tetrahydrofolic acid (Fig. 2–8). It is in this form that the coenzyme can accept a one-carbon unit either at the nitrogen of the para-aminobenzoic acid residue (N^{10}), or at the adjacent nitrogen in the pteridine ring (N^5); in some intermediate forms the one-carbon unit may form a bridge between the two nitrogens (N^5, N^{10}) (Fig. 2–9). Among the reactions in which these one-carbon shuttles participate are the reversible addition of a $-CH_2OH$ group to glycine to produce serine, conversion of histidine to glutamic acid, and fixation of formic acid from glycine or tryptophan metabolism. In each of these reactions of amino acid metabolism an activated one-carbon derivative of tetrahydrofolic acid is generated which may be reduced or oxidized to the appropriate level for further utilization by the body. Complete reduction at the expense of NADH and NADPH results in the synthesis of the N^5-CH_3 form which acts to furnish methyl groups for methionine or thymine with the help of vitamin B_{12} as an intermediary. In the biosynthesis of the purine rings of adenine and guanine by hematopoietic and other rapidly growing cells two of the carbon atoms are contributed by folic acid coenzymes: C 2 from the N^{10}-

formyl form and C 8 from the N^5, N^{10}-methenyl form. Thus the dividing cells are dependent upon folic acid coenzymes (and vitamin B_{12}) for the thymine required in DNA synthesis, and upon the folates for *de novo* production of purine bases required for both DNA and RNA synthesis. This double-dependence of blood-forming tissues provides the rationale for the application of the so-called antifolates in chemotherapy of leukemia. These anticancer agents are prime examples of a group of competitive analogues with such structural similarity to the natural vitamin that they are capable of acting as antivitamins by displacing the natural coenzyme. Methotrexate is one of the antifolates that has found considerable use as an antileukemic agent. It blocks the conversion of folic acid to the tetrahydro form (by acting as competitive inhibitor of the reductase enzyme) and hence prevents the vitamin from acting as a coenzyme (Fig. 2– 8). As a result hematopoiesis is stopped including synthesis of the leukemic cells. Of course, other dividing cells, such as those in the gastrointestinal tract, and other essential cells of bone marrow will also be arrested. For this reason a regimen of antifolate therapy must be countered by rescue of the normal cells, for example with tetrahydrofolic acid administration.

Tests of folic acid status are usually based upon very sensitive bacterial growth assays of the content of the vitamin in blood, particularly in the circulating erythrocytes where the folate concentration is many-fold greater than in serum. Another useful assessment is based upon the requirement for folate coenzymes in the breakdown of histidine to glutamic acid. In the absence of tetrahydrofolic acid this process goes only part way to the intermediate, formiminoglutamate, or FIGLU as it is called in medical literature. An increase of FIGLU excretion in urine beyond normal limits after the ingestion of a histidine test load is indicative of folic acid deficiency.

BIOTIN

A need for this vitamin in nutrition has been seen only under the highly unnatural circumstance when a large proportion of dietary protein is furnished by egg white. This is because the whites of eggs contain a glycoprotein which has been found to have a very high binding affinity for biotin, and hence was named avidin. In the gastrointestinal tract the vitamin may therefore become tied up in a high molecular weight form as the avidin–biotin complex which cannot be absorbed. The deficiency syndrome was originally known as egg white injury. Biotin was first isolated from the yolk of eggs, and was called the anti-egg white injury factor because of its ability to counter the effects produced by avidin. The vitamin consists of two five-membered heterocycles, the imidazole and thiophane rings fused together, with a valeric acid side-chain attached to the latter (Fig. 2– 10). Biotin is quite heat-stable, but like thiamin, it contains a sulphur atom and is susceptible to oxidation and to alkaline or acid decomposition. Good sources of biotin besides egg yolk are organ meats (*i.e.*, liver and kidney),

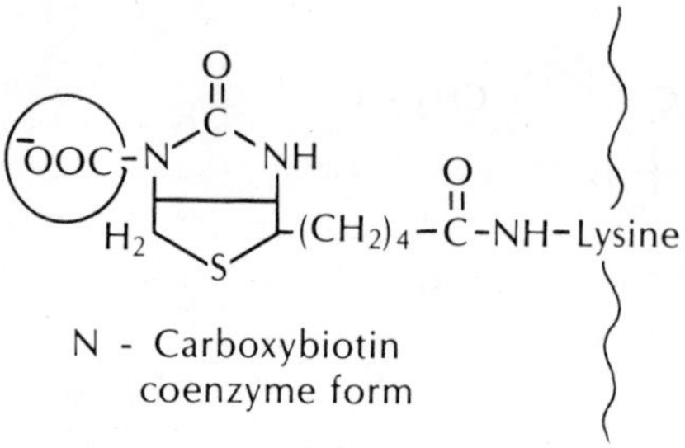

N - Carboxybiotin
   coenzyme form

FIG. 2–10. Biotin coenzyme structure.

legumes, and nuts. The requirement for adults is estimated as between 0.15 to 0.3 mg daily.

Dietary deficiency of biotin is an extremely improbable occurrence, and has been documented only on rare occasions in individuals on fad diets that contain large numbers of raw eggs. Avidin is destroyed by mild heating, so that cooked eggs or pasteurized commercial eggnogs present no problems. The symptoms of biotin deficiency in experimental subjects include anorexia, depression, insomnia, muscle pain and dermatitis, all of which were reversed rapidly within a few days of biotin administration in amounts similar to daily needs. In addition to food sources a large amount of biotin is generated within the gastrointestinal tract by the microflora. In fact, the amount of biotin that is excreted in the urine exceeds the amount ingested by a large margin in normal individuals, indicating that humans probably satisfy their biotin needs from biosynthesis by the intestinal microorganisms.

As in the case of folic acid the coenzymic function of biotin has to do with one-carbon metabolism, namely in certain enzyme reactions involving carbon dioxide fixation into organic compounds. Some of the important enzymes which require biotin include the key priming reaction in fatty acid biosynthesis, acetylCoA carboxylase, and the important gluconeogenic enzyme, pyruvate carboxylase. In both cases the biotin is attached by a covalent linkage between the side-chain carboxyl and the ε-amino group from a lysine residue in the enzyme protein. The biotin prosthetic group acts as a CO_2 carrier; in the presence of ATP and CO_2 the intermediate N-carboxybiotin (Fig. 2–10), is formed at the enzyme's catalytic site, and then on the addition of substrate (*e.g.*, pyruvate) transfers the $-COO^-$ group to form the carboxylated product (*e.g.*, oxaloacetate). There are also suggestions that the CO_2 fixing reactions in purine and pyrimidine biosynthesis may be dependent upon biotin coenzymes.

PANTOTHENIC ACID

Despite clear evidence for the participation of this vitamin in essential coenzymatic roles, there is little indication that pantothenic acid plays a significant part in human nutritional disease. This is because the vitamin is

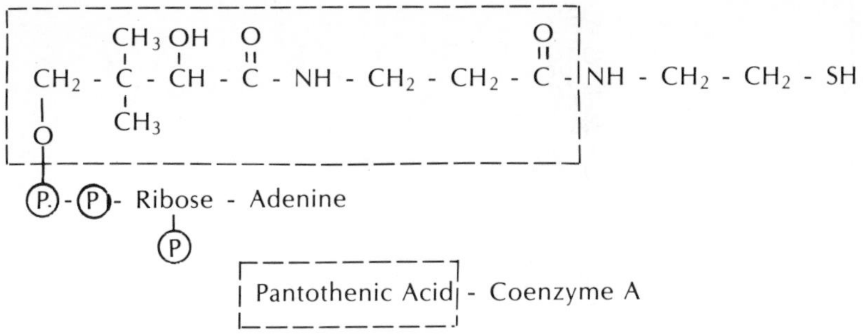

$$CH_2 - \underset{\underset{O}{|}}{\overset{\overset{CH_3}{|}}{C}} - \overset{\overset{OH}{|}}{\underset{\underset{CH_3}{|}}{CH}} - \overset{\overset{O}{||}}{C} - NH - CH_2 - CH_2 - \overset{\overset{O}{||}}{C} - NH - CH_2 - CH_2 - SH$$

(P.) - (P) - Ribose - Adenine
        |
       (P)

Pantothenic Acid - Coenzyme A

FIG. 2–11. Pantothenic acid coenzyme structure—coenzyme A.

ubiquitous and abundant in natural foods, and, as in the case of biotin, is probably produced in substantial amounts by the flora of the intestine. Moreover, it is quite stable to heat and undergoes little loss from foods during usual cooking procedures. Eggs, liver, meat, and a variety of vegetables and grains are rich sources of pantothenic acid. A daily intake of 5 to 10 mg is thought to be adequate.

A state of deficiency in pantothenic acid seems to be quite rare among human populations. Administration of the antivitamin, ω-methylpantothenic acid to a group of volunteers evoked a deficiency syndrome characterized by fatigue, sleep disorders, irritability, weakness, abdominal cramps and a burning sensation of the feet, all of which were reversed promptly upon administration of the vitamin. Blood levels of pantothenic acid are subnormal in some pregnant women and in chronic alcoholics who should receive supplements of the vitamin. Capital has been made of the fact that black rats turn grey when fed a pantothenate-deficient diet. Unfortunately there seems to be no substance to the claims of some advocates of pantothenic acid supplementation that such a regimen could either stem or reverse the ravages of time upon the human scalp.

In its coenzyme form the hydroxyl group at one end of the pantothenic acid is joined to the pyrophosphate end of an adenine nucleotide, while the carboxylic acid group forms an amide link with mercaptoethylamine (Fig. 2– 11). The free SH group of the coenzyme is responsible for activating and transferring a variety of acyl compounds, hence the molecule was named coenzyme A. CoA, as it is abbreviated, carries the activated acetyl group that is generated in pyruvate oxidation and β-oxidation of the fatty acids. Acetyl-CoA is a mobile two carbon donor for a number of synthetic processes including synthesis of the neurotransmitter, acetylcholine, of cholesterol and other steroids, of ketone bodies, and of long-chain acyl compounds for lipogenesis. In catabolism, acetyl-CoA provides substrate carbon for the major energy-generating reactions of the citric acid cycle. CoA derivatives of other acyl compounds, such as long-chain fatty acids and succinic acid, play important roles in synthesis of various lipids and of

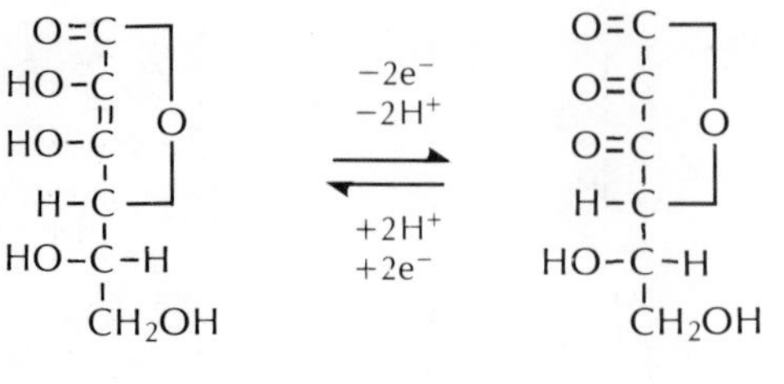

```
   O=C ┐                              O=C ┐
  HO-C │          -2e⁻              O=C  │
  HO-C │ O        -2H⁺              O=C  │ O
   H-C ┘         ⇌                  H-C ┘
  HO-C-H          +2H⁺             HO-C-H
   CH₂OH          +2e⁻              CH₂OH

   Vitamin C                        Dehydro-
  Ascorbic Acid                   Ascorbic Acid
```

FIG. 2–12.   Ascorbic acid redox reaction.

heme, respectively. Pantothenic acid also provides the prosthetic group of the factor termed acyl-carrier protein which is required for the process of fatty acid synthesis in the liver and adipose tissue.

VITAMIN C—ASCORBIC ACID

Despite the fact that it has been over 400 years since the disclosure that a factor in plants is capable of curing the disease known as scurvy, and over 50 years since the isolation and characterization of this antiscorbutic factor from citrus fruit, the substance ascorbic acid, or vitamin C, remains enveloped in a cloud of mystery and mystique. The mystery surrounds its true biochemical roles in cell maintenance; the mystique concerns claims that doses some thousandfold greater than required to offset scurvy may cure or prevent a number of disease states from cancer to the common cold. The vitamin is a simple sugar derivative which is related structurally to and derived metabolically from L-glucose. It is an ene-diol lactone of a hexuronic acid (Fig. 2– 12) it is extremely water-soluble, and it is easily destroyed by oxidation, particularly in alkaline solutions and upon exposure to heat and metals such as copper. Vitamin C content of foods decreases substantially on storage, and much is lost by leaching out or by decomposition during cooking. Best sources, other than juices of citrus fruits, include tomatoes, strawberries, cabbage, cauliflower, broccoli, and because of the large amounts consumed, potatoes. Animal foods such as milk, meat, or eggs are not adequate sources. The optimal intake values are difficult to establish. A daily intake of 30 mg will provide 2 to 3 times the amount required to prevent scurvy and has been recommended by British and Canadian authorities; an additional 100% increase to 60 mg has been advocated to provide for individual variations among most normal persons. Needs for growing children are probably almost as great as for adults, and 20 to 40 mg above the normal intake is suggested during pregnancy or lactation. An upper limit of 100 mg per day might be considered as a prudent allowance for normal maintenance, while larger doses (*e.g.*, in the g/day range) might be considered as therapeutic under special circumstances as discussed below.

Although scurvy had been known since antiquity it first came into prominence during long naval voyages and expeditions between the 16th and 18th centuries when the greater part of a ship's company might become disabled by this dreaded affliction. Nowadays the disease is rare except among the poor, the elderly, food faddists, and alcoholics; infants who are weaned to cow's milk, which has a low ascorbic acid level that is reduced further upon pasteurization, may become scorbutic unless their diet is supplemented with orange juice. Pathologic changes in scurvy chiefly involve bones, teeth, blood vessels, and fibrous tissue of mesenchymal origin. It is a disease of connective tissues with many extracellular manifestations. Fatigue, weakness, aching joints, and aching muscles are gradually noted in the first few months of a vitamin C-deficient diet, during which time blood levels of the vitamin may drop from the normal values of 1 mg to less than 0.1 mg per 100 ml of plasma. Overt signs of the disease usually do not appear until levels in the circulating white cells drop from the normal 15 to 20 mg to 2 mg per 100 ml or less. By 3 to 4 months the symptoms intensify with characteristic papillary lesions on the buttocks and numerous small hemorrhages or petechiae under the skin. As the condition progresses there is bleeding into the gums resulting in a spongy gingivitis around the teeth, into the skin producing large discolored spots over the body, and into the bones, muscles and joints causing severe pain and tenderness of these areas. There follows disruption of the process of wound healing to the point where old fractures may break down and scars of wounds which have been healed are forced open again. Two types of anemia have been observed in advanced scurvy: a hypochromic iron-deficiency type secondary to blood loss from repeated hemorrhaging, and a macrocytic type seen more commonly in infantile scurvy which responds to folic acid. Unless treated the scorbutic patient inevitably dies. For the first week or so fairly large doses (100–200 mg) are administered several times per day to restore the normal body pool of 1 to 2 g. Once the tissue stores have been replenished the amounts should be gradually reduced toward normal levels when signs of wound healing are normalized. In patients undergoing surgery it is advisable to boost the ascorbic acid intake to facilitate the process of repair and recuperation from tissue damage. Chronic ingestion of aspirin or oral contraceptives containing estrogens lower the concentration of ascorbic acid in blood, and extra supplements of vitamin C would seem to be advisable in such instances also.

Several molecular processes have been attributed to ascorbic acid to explain its functions at a cellular level. Some of these may be nonspecific, related to the wellknown property of the vitamin as a water-soluble, reducing agent and free-radical scavenger. Protective actions of ascorbic acid on tissues may complement its specific cellular actions, much in the same way that it is added to certain foods prior to storage as a preservative factor as well as a nutrient. The chief action of vitamin C that may be rationalized with its antiscorbutic activity is in the enzyme reactions of collagen formation. In fact, scurvy may be viewed primarily as a disorder of collagen synthesis. Normal collagen is characterized by its high content of the amino acids, glycine, proline, and 4-hydroxyproline, which are present in a repeating unit that is critical for the three-dimensional coiled structure of the final molecule. It is this long fibrous collagen chain that forms the reinforcing rods of the extracellular ground substance. Vitamin C plays a key role by activating the enzyme, proline hydroxylase, which adds hydroxyl groups on to proline residues of the newly synthesized precursor, or procollagen, molecules in connective tissue fibroblasts. In scurvy, the resulting procollagen molecules pass into the extracellular spaces, but because they are deficient in hydroxyproline they are unable to form the cross-linkages necessary to produce stable, mature collagen molecules. This results in a weakening of the intercellular cement joining endothelial cells accompanied by capillary fragility and hemorrhage, defects in the extracellular matrix of bone and scar tissue accompanied by breakdown of old fractures and wounds, and failure to repair new injuries. Ascorbic acid is also required for metabolism of the amino acid tyrosine. In vitamin C deficiency, particularly in infants, the enzyme p-hydroxyphenylpyruvate oxidase of liver has subnormal activity with the result that tyrosine and its metabolite p-hydroxyphenylpyruvate accumulate and are excreted in the urine. The latter compound may give a false positive test for phenylpyruvate which is diagnostic of phenylketonuria. The defect and the abnormal excretion products are corrected by administration of ascorbic acid. Ascorbic acid also seems to be required for maintenance of the folic acid coenzymes in their reduced tetrahydrofolate form. The macrocytic anemia of infantile scurvy is accompanied by urinary excretion of the abnormal oxidized N^{10} formylfolate derivative; additional folic acid, especially as its reduced form, folinic acid, and vitamin C can correct the hematologic anomaly. Other ancillary roles of ascorbic acid include its ability to facilitate the absorption of iron probably by maintaining it in the reduced ferrous state in the intestinal tract, and the production of bile acids from cholesterol in the liver. Unexplained problems in vitamin C metabolism include its high concentration in adrenal glands and a possible role in steroid production in response to stress. High concentrations of the vitamin also occur in nervous tissue and the eye, raising questions about its role in these locations.

In recent years a lively controversy has developed about the ability of ascorbic acid supplements to counteract a number of common disease states. The argument of the protagonists, led by the Nobel laureate in chemistry, Linus Pauling, was based upon comparative studies with other species. For example, nonprimate mammals who have retained the ability to synthesize ascorbic acid from glucose, produce gram quantities each day rather than milligrams. Primates, such as the great apes in their native state, also ingest large amounts, up to 100-fold greater than antiscorbutic levels. By extrapolation, man in a natural condition, would have provided saturating amounts of ascorbic acid in cells to produce a body pool several times greater than it is today. Many of mankind's illnesses thus could be viewed as the consequence of abandonment of the ancestral fruitarian–vegetarian diet with gram quantities of ascorbate as a macronutrient, to the present intake of the milligram amounts necessary to offset scurvy. The minimal vitamin levels presumably resulted in a type of subclinical hypovitaminosis C with reduced resistance to diseases and stress in the general population. It was claimed that daily ingestion of several grams of ascorbate would reduce the incidence of the common cold, and considerable anecdotal evidence was raised in support of this claim. Several clinical trials involving thousands of subjects with varying doses of ascorbic acid have been conducted. The major conclusion is that high doses of the vitamin do not have a significant prophylactic effect; if one wants to reduce the chances of getting colds the best strategy is to avoid exposure to people who have already got them. However, there have been several studies which established that the severity and duration of some of the toxic symptoms of colds (fatigue, headache, malaise) may be reduced considerably by ingestion of extra vitamin C during the period of infection. There did not seem to be much difference between the effects of 250 milligrams and several grams of supplemental ascorbic acid. This finding is not surprising in the light of studies with radioactively tagged ascorbic acid, which revealed that a daily intake of 150 to 200 mg of vitamin C increased the body stores to their maximum level of 4 g total, and that further increments of dietary ascorbate were effectively eliminated by excretion in the urine.

Claims for the effective treatment of heart disease or cancer with massive doses of vitamin C have appeared, but remain to be proved in proper clinical trials. In some studies serum cholesterol levels have been reported to fall in ascorbic acid-supplemented patients, but in others the change has been in the opposite direction, an increase in serum cholesterol levels. There have also been reports of direct effects of vitamin C upon heart tissue, upon connective tissue of the arteries, or upon the formation of thrombi in the circulation, but the clinical importance of these findings in relation to cardiovascular disease is not certain. A number of terminal cancer cases have been treated with large amounts (10–30 g daily) vitamin C, and some evidence of prolonged survival and relief from suffering have been suggested. Although established cancer may not be curable with ascorbic acid therapy, there are suggestions that some diet-induced malignancies may be preventable. It is now clear that amino compounds from dietary proteins react with nitrites generated in the intestinal tract to produce the highly carcinogenic compounds known as nitrosamines. The simultaneous ingestion of a water-soluble antioxidant (ascorbic acid) or fat-soluble antioxidant (tocopherol) greatly reduces the quantities of ni-

trosamine manufactured endogenously. Fortunately, the amounts of vitamin C needed to provide this protective function fall within the "prudent"
range of about 100 mg daily, and can be obtained from a normal diet rich
in vegetables and fruit without pharmaceutical supplements.

Because of the large numbers of people who have taken supplemental
doses of ascorbate over long time periods it is important to determine
whether acute or chronic toxicity effects may occur. Acute effects are not
observed until the intake exceeds 10 to 15 g, and then they seem to result
from the acidotic effect of the compound. Chronic effects have been noted
in certain individuals with a tendency to produce abnormally large
amounts of oxalic acid; the latter is a metabolite of ascorbate, and may
cause renal problems as a consequence of the insolubility of its calcium
salt in the urinary tract. An important factor in any treatment with high
vitamin C intake is the ability of the body to adapt to the elevated concentrations. This ability seems to depend on the increased activity of enzymes
which metabolize ascorbate, and which are induced as an adaptive response to the high substrate levels. The consequence is that a person who
switches suddenly from high to low vitamin C intake will still possess the
capacity to eliminate ascorbate at the increased rate, and hence the normal
low amounts of vitamin C in an ordinary diet may be insufficient. The
phenomenon was first observed in seasonal orange pickers, who exhibited
signs of scurvy after leaving their high citrus diet. Offspring of women who
took high doses of vitamin C during the pregnancy have also been reported
to develop scorbutic signs readily in the neonatal period. In all cases where
amounts in excess of 1 g per day have been consumed for any length of
time it seems a prudent measure to reduce the intake gradually over several weeks to prevent the abrupt withdrawal decline in ascorbate levels
which occurs when the body has become accustomed to the elevated
intake.

VITAMIN A—RETINOL

This substance was characterized as the growth factor present in butter-fat
or egg yolk that was essential for maintenance of epithelial tissues and for
the process of vision. In order to differentiate it from water-soluble B vitamins it was designed as fat-soluble A. Chemically vitamin A is a 20-carbon
compound produced by the condensation of 4-isoprene units. At one end it
possesses an unsaturated 6-carbon ring (the β-ionone ring), with an attached side-chain bearing a terminal alcohol group (Fig. 2–13). The vitamin is found in nature only in animal tissues, but the plants contain a
series of 40-carbon compounds (the carotenes) which may be split in the
middle by intestinal enzymes to yield two molecules of vitamin A; the
precursor carotene forms are referred to as provitamin A (Fig. 2–13). Because the efficiency of the conversion is low it requires 6 times the amount
of β-carotene to provide the same vitamin A activity as retinol. Vitamin A

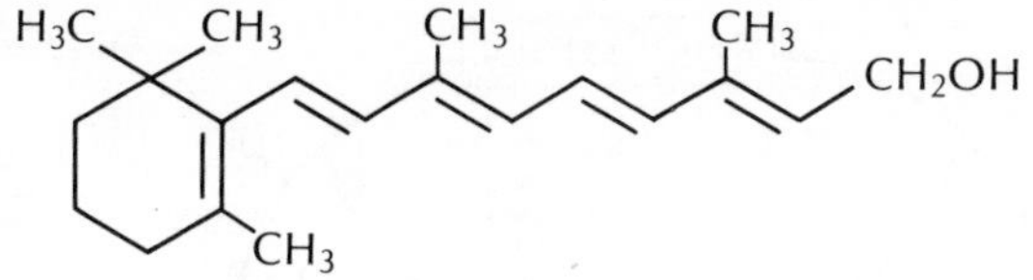

FIG. 2–13. Vitamin A and its provitamin—β-carotene.

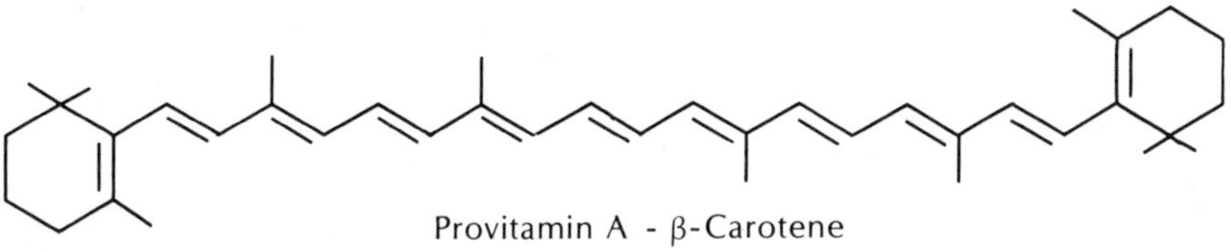

and the carotenes are fat-soluble compounds which are quite stable in normal cooking processes, but can be lost because of oxidation on storage or exposure to ultraviolet light. Although present in adequate amounts in the diet, deficiency of vitamin A can result from conditions where the uptake of fat-soluble substances is impaired as in the malabsorption syndrome. Best sources of preformed vitamin A are to be found in liver where it is stored in large quantities; important plant sources include vegetables such as tomatoes, carrots, spinach, broccoli, squash, and certain fruits, especially apricots, peaches, and melons. Requirements for adults are estimated as between 800 to 1000 μg of retinol, or equivalent quantities of the appropriate carotenes to deliver that amount, with extra allowances of 200 μg and 400 μg during pregnancy and lactation respectively.

The syndrome of vitamin A deficiency, termed xerophthalmia, continues to be a significant problem in underdeveloped areas of the world where it accompanies protein–calorie malnutrition. It is particularly severe among young children, and is estimated to cause 80,000 new cases of preventable blindness each year, with a fatal outcome in one-half of that number. Early symptoms of hypovitaminosis A are inadequate vision in dim light (night blindness), and a dryness of the conjunctiva (xerosis). The skin becomes dry, crackled, and scaly owing to excessive keratin formation on epithelial surfaces, particularly around the eyes. The condition progresses steadily with involvement of the cornea, and if untreated in early stages of xerophthalmia will lead to ulceration and scarring of the cornea (keratomalacia) with irreversible blindness. Growth of bone and formation of tooth enamel are also defective. Deterioration of epithelial surfaces is frequently accompanied by increased incidence of infectious diseases. The prevention and treatment of xerophthalmia is generally accomplished by large oral doses (10,000– 30,000 μg retinol equivalent), administered every six months to children where the deficiency is endemic. Since the hypovitaminosis is complicated by inadequate protein intake resulting in

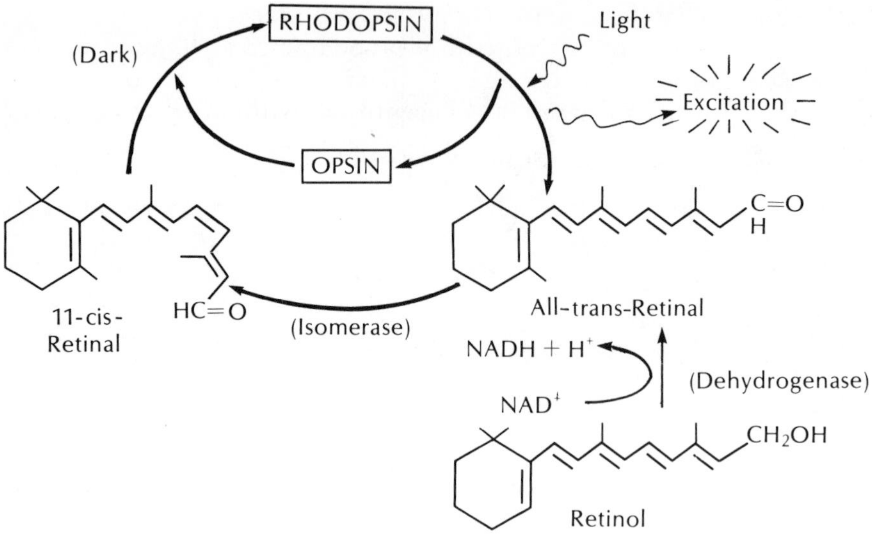

FIG. 2–14. Function of retinal in the visual cycle.

lowered production of the retinol-binding protein that is necessary for vitamin A transport in blood plasma, the protein–calorie malnutrition should be treated simultaneously. Several programs under the auspices of the World Health Organization and other agencies are underway in areas of Latin America and the Orient for the protection of children at risk, and for the education of the adult population concerning the merit of supplementation of the diet with the plentiful local vegetable sources of carotenes.

The ability of the liver to concentrate and store large quantities of vitamin A has several consequences: first, it is not essential to ingest the vitamin every day as in the case of water-soluble vitamins; second, the deficiency syndrome is rare in adults because of the large endogenous reservoir, but occurs readily in children who have had no opportunity to build up their stores; finally, excessive intake of vitamin A will have a cumulative effect since it is retained rather than excreted, and the resulting hypervitaminosis A may promote severe toxic reactions including dermatitis, headache, pain in the extremities, loss of hair, and enlargement of the liver.

Biochemical mechanisms of vitamin A actions at the cellular level are imperfectly understood, except for the role in vision where it functions as the prosthetic group for the visual purple pigment of the retina (Fig. 2–14). The vitamin circulates in the blood associated with the retinol–binding protein of the plasma. Following uptake by the eye the retinol is oxidized by alcohol dehydrogenase at the end of the side-chain from the alcohol ($-CH_2OH$) to the aldehyde ($-CHO$) form, generating retinal. The latter may then be acted upon by an isomerase to change the configuration of one of the double bonds near the middle of the side-chain from *trans* to *cis*. It is in this form, 11-*cis*-retinal, that the vitamin attaches to the protein, opsin, to produce the colored complex, rhodopsin, or visual purple. The visual pigment molecules are concentrated within the retinal photoreceptor cells, the rods and the cones. Upon exposure to light of appropriate wavelength the 11-*cis* double bond reverts to the *trans* form, the pigment loses its color, and the opsin dissociates from the *trans* form of the vitamin. As a consequence, the ion permeability of the membrane of the photoreceptor cell becomes altered, triggering a nerve impulse at the synaptic base of the cell to produce a signal to the visual cortex of the brain. In order to reconstitute the visual pigment the *trans* retinal must be converted back to the 11-*cis* form by the isomerase, and combined with the opsin molecule. In the absence of adequate quantities of circulating retinol the numbers of active pigment molecules will decline so that the retina will be less sensitive to low levels of light, and night-blindness will be the logical outcome.

Effects of vitamin A derivatives upon epithelial cells, and upon bone and tooth development are poorly understood. In some instances it appears that the carboxylic acid derivative, retinoic acid, is the growth-promoting agent. Some effects of vitamin A appear to be mediated by membrane stabilization. In the liver retinol phosphate can act as an intermediary to transfer sugar residues from GDP mannose to membrane glycoprotein by mannosylretinol phosphate. Since glycoproteins play significant roles in recognition sites of membranes, and in the ground substance of connective tissues such as the corneal stroma, a defect of glycoprotein assembly may be an important factor in vitamin A deficiency.

Vitamin A and its derivatives (retinoids) play a primary role in cellular differentiation and the maintenance of proper epithelial cell functions. Since the transformation of normal to cancerous tissue involves a loss of cellular differentiation, the potential of retinoids to retard the malignant transformation is a logical and intriguing approach to chemoprevention of cancer. Experimental studies have confirmed that natural and synthetic retinoids can suppress the conversion of normal into neoplastic cells induced by radiation or chemical carcinogens. The anti-carcinogenic actions of retinoids occur mainly during the later stages of neoplastic progression by counteracting tumor promotion and proliferation. Whether dietary vitamin A or β-carotene can materially influence human cancer risks is still an open question. Since hypervitaminosis A and inappropriate use of retinoids could cause more harm than good, studies in humans must be conducted with great care and constant biochemical monitoring.

VITAMIN D—CALCIFEROL

This growth factor was shown to be a stable component of fat-soluble vitamin concentrates which survived destruction of vitamin A by oxidation. It was termed vitamin D, or the antirachitic factor in the diet, after its

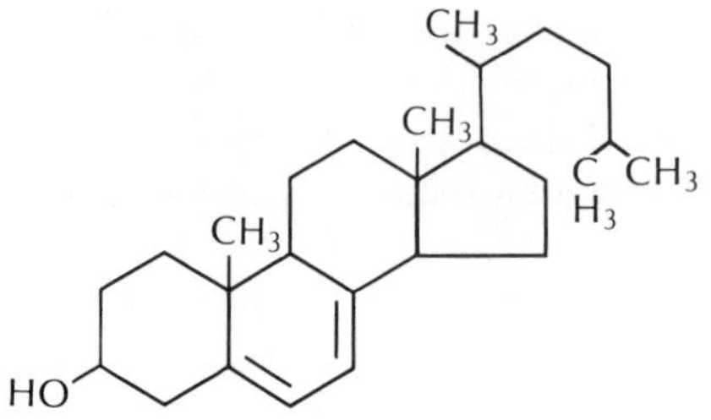

7-Dehydrocholesterol

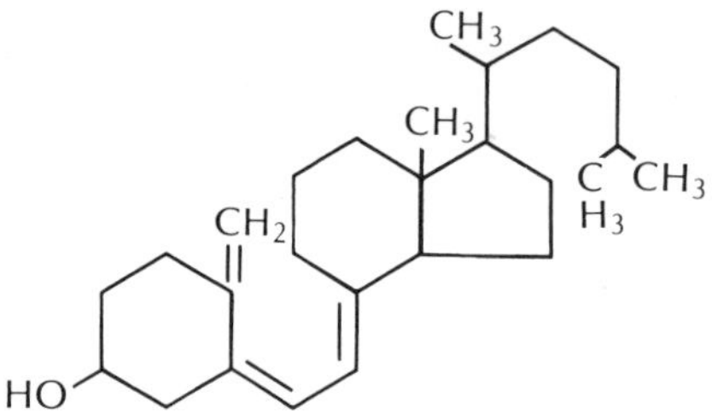

Vitamin D$_3$
(Cholecalciferol)

FIG. 2–15. Structure of vitamin D and its provitamin—7–dehydrocholesterol.

ability to cure the bone disorder of infants known as rickets. Vitamin D and its derivatives, the calciferols, as implied by their name exert crucial regulatory effects upon the metabolism of calcium principally by activating its absorption from the diet in the gastrointestinal tract and by promoting normal deposition of the mineral in the bones. One interesting feature that was evident from early dietary studies was the ability to generate antirachitic activity either by ultraviolet irradiation of various foods (exogenous source), or of the skin of the individual (endogenous source). This phenomenon led to popular description of the factor as "the sunshine vitamin." Recently, some ambiguity has arisen concerning the role of endogenous metabolic reactions in various tissues to generate the active form, and despite retention of the designation as vitamin D many endocrinologists maintain that it belongs more properly among the steroid hormones than in the family of nutritional factors. Chemically, vitamin D is divided into two natural groups: D$_2$ which can be derived artificially from the plant sterol, ergosterol, and D$_3$ which is generated naturally in the pathway of production of the characteristic animal counterpart, cholesterol (Fig. 2–15). These components are rarely present in ordinary food sources, with exception of animal depot fat and liver, particularly fish liver oils; foods such as milk fat may be readily supplemented with the vitamin by exposure to ultraviolet rays. The requirements are difficult to assess since they will depend upon the extent to which individuals have been exposed to the sun. In children it is recommended that 10 μg per day of vitamin D_3 be consumed; for adults 5 μg daily seems sufficient, with a doubling of the intake during pregnancy or lactation.

Rickets is typically a disease of young children, characterized by reduction of serum calcium ions, inadequate calcification of developing bones and teeth, and retardation of growth. Hypocalcemia is the earliest finding in vitamin D deficiency, followed by a progression of deformities in the bones owing to faulty mineralization. Growing ends of long bones become enlarged, while the shafts fail to strengthen sufficiently to bear the weight and stresses imposed as the child stands upright. The result is an abnormal bowlegged or knock-kneed stance, and waddling gait. There are characteristic beadlike protuberances at the costochondral junctions of the ribs termed the "rachitic rosary," and with advanced rickets there is deformation of the skull and scoliosis of the spinal column. The disease is rare today except in areas where general malnutrition is endemic, and among infants fed a strict vegan diet. When encountered in adults vitamin D deficiency causes the syndrome known as osteomalacia. This disorder may be brought on in women with inadequate diets and low sunlight exposure as the result of repeated pregnancies and extended lactation, or in the elderly who are suffering from kidney malfunction or from intestinal malabsorption. Deformities and severe bone pain result from softening and microfractures in vertebrae or weight-bearing bony structures. Spasms and convulsions may ensue when serum calcium levels fall to the extent where they provoke the hyperexcitable state of the nerves known as tetany. In treatment, combined administration of vitamin D and calcium is usually advocated. For infantile rickets from 30 to 100 μg of vitamin D may be administered daily in the first weeks, followed by maintenance doses of 10 μg, or regular sunbaths. Osteomalacia may require larger supplements (up to 500 μg daily), particularly where there are disorders of absorption, or anomalies in the metabolism of the vitamin by the liver or kidneys. The vitamin is toxic in doses above 50 μg per kg body weight daily. Symptoms include weakness, nausea, headache, and digestive disturbances resulting from elevated serum calcium. Severe, even fatal, irreversible damage may occur as the result of calcium deposition in soft tissues and blood vessels of key organs such as the kidneys as a result of prolonged hypervitaminosis D.

Recently a great deal of excitement has been aroused among biochemists, nutritionists, endocrinologists, and clinicians over disclosures about the roles of different tissues in interconversions of vitamin D and its metabolites, and the regulatory interactions with parathyroid hormone (see also Chapter 3). 7-dehydrocholesterol (Fig. 2–15) plays the role of precursor or provitamin in the nonenzymatic formation of cholecalciferol, vitamin D_3, in the skin exposed to ultraviolet rays. From the diet, preformed cholecalciferol and ergocalciferol (vitamin D_2) produced by the irradiation of ergosterol are absorbed in the presence of bile, and together with vitamin D_3 of cutaneous origin are carried to fat depots and the liver for storage. Circulating vitamin D, transported as a complex with an α-globulin protein of serum, may be viewed as a preprohormone which must be processed twice before it can exert its endocrine effects upon target tissues.

The first modification is produced by a hydroxylase enzyme of the liver which converts cholecalciferol to its 25-hydroxy derivative (Fig. 2–16). The enzyme is a mixed function mono-oxygenase of the endoplasmic reticulum which requires oxygen and NADPH. The product, 25-hydroxycholecalciferol, is released from liver and is the main form of circulating vitamin D as the complex with the serum transport protein. The second hydroxylation which results in formation of 1, 25-dihydroxy cholecalciferol is carried out by a different hydroxylase which is confined to kidney mitochondria. In addition to requirements for oxygen and NADPH the specific 1-α-hydroxylase has three other components: a flavoprotein, an iron-sulfur protein, and cytochrome P450. The active 1,25-dihydroxy derivatives of vitamin D_3 or D_2, which are produced in the same manner and have similar biologic activity in humans, are then transported from the kidneys by the carrier protein in blood plasma to the target tissues in the intestinal tract, the bone, and the renal tubules where they exert their hormonal functions on the movement of calcium and phosphate ions.

Molecular mechanisms of the effects of 1,25-dihydroxycholecalciferol appear to be analogous to those of the steroid hormones. In brief there is a latent period of several hours between injection of 1,25-dihydroxy vitamin D_3 and the physiological response of increased uptake of calcium and phosphate ions by the intestinal mucosa of the duodenum. During this latent phase there is evidence of enhanced RNA and protein synthesis in the mucosal cells suggesting that the active D_3 species, like the steroid hormones, may interact with cellular DNA to promote RNA polymerase activation, and thus induction of specific proteins. Following injection of the radiolabelled vitamin metabolite, 1,25-dihydroxy D_3, autoradiography reveals a specific nuclear localization in villus cells of the duodenum prior to initiation of intestinal calcium uptake in rachitic animals. It is estimated that some 2400 molecules of the active metabolite may be bound per intestinal cell, a figure that is comparable to that reported for classical steroid hormone binding in target cells. Moreover, there is also evidence for highly selective and high affinity receptor binding sites for 1,25-dihydroxy vitamin D_2 and D_3 both in cytosol and chromatin of the intestinal cells, with a time-and-temperature-dependent movement of the active vitamin metabolite from the cytoplasm to the nucleus. Subsequent events include an increase of RNA synthesis, and the appearance of certain calcium-binding proteins in the brush-border region of villus cells that correlates with the enhanced uptake of calcium in the intestine. Since both the production of the calcium-binding proteins and the calcium uptake are prevented by inhibitors of RNA synthesis, such as actinomycin D, it is

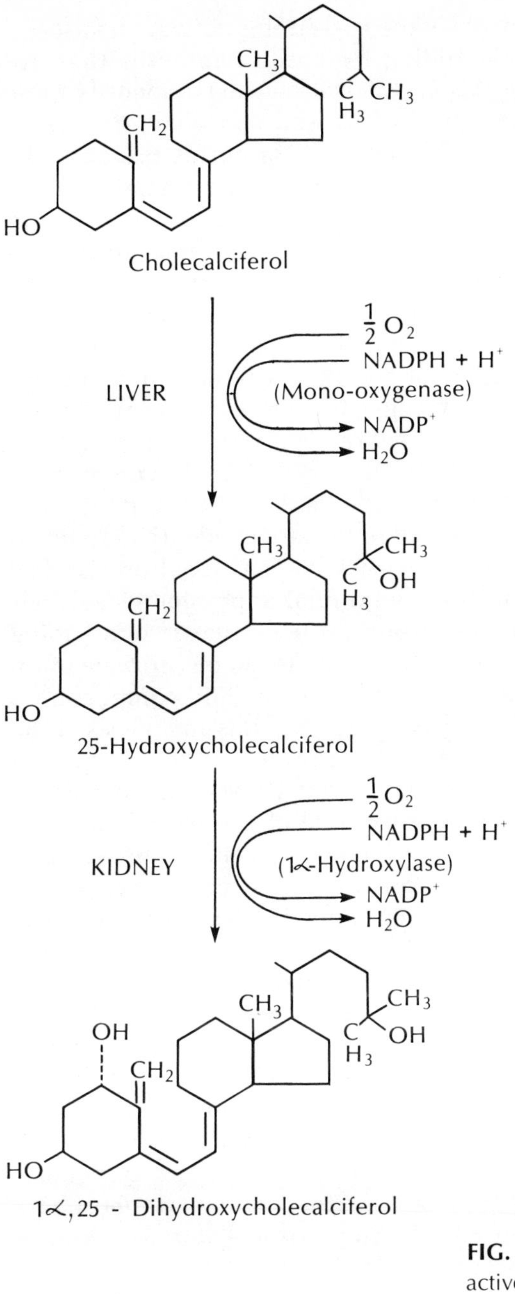

FIG. 2–16. Conversion of vitamin D to its active dihydroxy form.

likely that 1,25-dihydroxy derivatives of the D vitamins activate transcription from a region of the genome coding for specific proteins that are involved in calcium ion transport events. Efforts continue to identify these proteins further and their functions in calcium and phosphate uptake.

Events in the bone and kidney are not so clear-cut at present. The situation is presumably more complicated than in the intestinal cells where calciferol independently exerts physiological effects; in bone and kidney the effects of vitamin D are contingent upon the simultaneous actions exerted by parathyroid hormone. Vitamin D does seem to control the flux of calcium between the bone mineral and fluid compartments, and may play an important role in stimulating reabsorption of calcium by the renal tubules. There is some evidence for specific 1,25-dihydroxy vitamin D_3 receptors in both bone and kidney cells, and for an associated effect upon RNA polymerase. Vitamin D-mediated resorption of calcium in bone is inhibited by actinomycin, while administration of 1,25-dihydroxy vitamin D_3 rapidly stimulates renal RNA biosynthesis. A vitamin-D dependent calcium-binding protein has been detected in the kidney identical to intestinal vitamin-D dependent protein, but no similar functional protein has been described for bone cells. In the bone, vitamin D at physiological levels acts in two ways: to transfer calcium from old bone that is undergoing resorption to regions of new bone deposition and also to mobilize calcium from the bone fluid compartment to the extracellular fluid compartment that is in communication with blood plasma. Several missing pieces of the jigsaw puzzle are needed before a unifying picture of vitamin D actions may be completed. At present it is thought that its actions on various target tissue processes—stimulation of absorption of dietary calcium, enhanced equilibration of calcium between blood and bone, improved calcium retention in renal reabsorption—provide a concerted mechanism whereby the vitamin maintains the circulating levels of calcium and phosphate ions that are required to permit normal skeletal mineralization and other physiological functions of these minerals.

Regulation of vitamin D effects is imposed at several levels and requires an integration with actions of parathyroid hormone. The first control site is in the liver where the activity of 25-hydroxylase may be shut off by feedback inhibition by its product, and by the administration of glucocorticoids. However, the major regulatory reaction is the 1-α-hydroxylase of the kidney, the rate-limiting step of vitamin D metabolism, which responds to the concentrations of either calcium or phosphate ions in the serum. The hydroxylation of 25-hydroxycalciferol by kidney cells is strongly dependent upon serum calcium levels. At calcium concentrations below the normal value of 10 mg per 100 ml, formation of the 1,25-dihydroxy compound is activated, and the combined responses of calcium uptake from the intestinal contents, from bone stores, and from renal reabsorption will then act to raise the calcium level back up to the normal level, at which point the 1-α-hydroxylase is shut off. This au-

toregulating system has been shown to involve an indirect effect of calcium on the kidney enzyme, and requires mediation by the parathyroid
gland. Low calcium in serum triggers release of the parathyroid hormone,
and interaction of the latter with the kidney switches on the production of
1,25-dihydroxycalciferol. Low phosphate in serum seems to have a direct
effect upon renal cells to activate the 1-α-hydroxylation, independent of
the presence of parathyroid hormone. Additional controls by parathyroid
hormone at the level of ion excretion allow for separate modulation of
calcium and phosphate: with *hypocalcemia* the increased parathyroid
hormone will exert a phosphaturic effect upon renal tubules so that vitamin D can produce increased calcium levels without concomitant net increases of the phosphate that is also mobilized; with *hypophosphatemia*
since the direct response is not accompanied by parathyroid hormone elevation and since the reabsorption of calcium by the kidney requires both
parathyroid hormone and vitamin D, some of the extra calcium that is
mobilized will leak into the urine to allow phosphate to rise without a rise
in serum calcium. One additional control is exerted by high calcium levels
in serum (above 10 mg/100 ml). When the 1-α-hydroxylation shuts down
there is activation of a different hydroxylase that adds the – OH group at
the 24 position to produce 24,25-dihydroxy or 1,24,25-trihydroxy calciferol
derivatives. The 24-hydroxylase is present in kidney and in extrarenal
tissues such as the intestine. Activity of the enzyme is suppressed by
parathyroid hormone but activated by elevated calcium levels. The 24-
hydroxylated derivatives have some stimulatory activity upon calcium
uptake in the intestinal tract but are ineffectual in mobilizing calcium
from the bone; their role in calcium homeostasis and mineralization is not
clear at present.

The kidney is the key endocrine organ for production of the active
hormone of vitamin D, and the application of basic information about the
hydroxylation process to treatment and management of mineralization
defects in renal diseases has proliferated in the past few years. Apparent
vitamin D deficiency, characterized by intestinal malabsorption of calcium and dystrophy of the skeleton, has been recognized to accompany
disease of the kidneys for some time. In several instances massive doses of
calciferol (100– 1000 times the usual requirements) were necessary to
normalize calcium metabolism, leading to the designation "vitamin
D-dependent" for such disorders. This situation is epitomized by the genetic disease, hereditary vitamin D-dependent rickets, which is inherited
in an autosomal recessive manner. The condition has all the clinical attributes of childhood rickets because of simple vitamin D deficiency, and it is
accompanied by hypocalcemia, severe rickets, and retardation of growth
in the first two years. The disorder is curable by vitamin D, but only with
pharmacologic doses in the range of 250 to 2500 μg daily, and such large
doses must continue throughout life for normal growth and mineralization
to occur. However, quite small amounts (1– 4 μg per day), within the

physiologic range of 1,25–dihydroxycholecalciferol can correct the clinical and biochemical anomalies. The 25-hydroxy derivative is more effective than vitamin D itself, but is still required at a pharmacologic level (100–900 μg per day) for management of vitamin D dependency. Following therapy with high levels of vitamin D the concentration of 25-hydroxycholecalciferol in blood plasma is increased, but the 1,25-dihydroxy form is not detectable. Thus the liver 25-hydroxylase seems to operate normally, while the kidney 1-α-hydroxylase apparently is defective. This conclusion is validated by the finding that the synthetic vitamin D metabolite, 1-α-hydroxycholecalciferol, when administered in quite low doses (2–8 μg per day) is an effective curative agent, presumably through conversion by the hepatic enzyme to the 1,25-dihydroxy derivative. Similar findings have been reported for adult patients with advanced renal disease who show associated calcium malabsorption and osteomalacia. Small doses (in the μg range daily) of either the 1-α-hydroxy analogue or 1,25-dihydroxycholecalciferol will effect a cure to the same extent as massive amounts (in the mg range daily) of vitamin D. In such patients the situation is complicated by the accompanying defect in renal reabsorption of ions and the effects of dialysis. Prevention and management of renal osteodystrophy requires supplementation with oral calcium and dialysis against high calcium in addition to provision of the 1-α-hydroxylated calciferol derivatives.

VITAMIN E—TOCOPHEROL

If ascorbic acid may be represented as vitamin C for controversy and confusion in the popular nutritional press, the tocopherols may be labelled appropriately as vitamin E, E standing for exasperation and extravagance. The exasperation refers to those involved in medicine and human nutrition who must view with dismay the vast body of experimental work upon the tocopherols in animals. The extravagance relates to the manner in which some clinicians and entrepreneurs of nutribusiness have extrapolated findings from animal experimentation to promote claims for vitamin E as the panacea for a multitude of diseases in humans. In fact, few of the claims for benefits from pharmacologic or megavitamin doses of tocopherols can be substantiated by proper clinical trials, and evidence is accumulating for the occurrence of toxic manifestations in hypervitaminosis E. From the original studies with rats the vitamin was first known as the fat-soluble, anti-sterility factor. In chemical terms the tocopherols consist of a series of substituted aromatic compounds containing the chroman-condensed ring structure with an attached hydrocarbon chain (Fig. 2–17). They are heat-stable but readily oxidized. The most active derivative in nature is α-tocopherol, and other tocopherols (β, γ, δ) are reported in terms of equivalent amounts of the α form. The best natural sources of vitamin E are found in wheat germ, vegetable oils, liver, and green leafy vegetables.

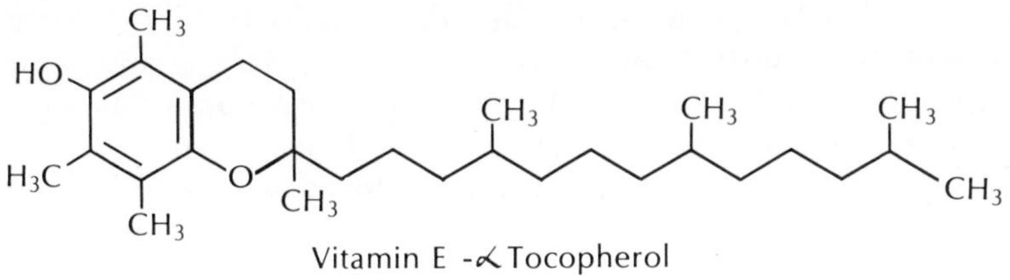

Vitamin E -∝ Tocopherol

FIG. 2–17.   Vitamin E structure.

Fish liver oils, which are rich in the other fat-soluble vitamins, A and D, are deficient in vitamin E. Many cereal products are supplemented with vitamin E. The minimum daily requirements are hard to assess because of extensive storage of the tocopherols in liver and fat depots, and the interrelations with other dietary constituents (vitamin A, polyunsaturated fatty acids, selenium) and environmental pollutants (ozone, nitrosamines). The recommended allowances vary from 10 mg daily for normal, well-fed adults to as much as 30 mg for those living in highly polluted urban areas or eating diets rich in unsaturated fats. In the latter case a rule of thumb of 0.6 mg vitamin E for each g of polyunsaturated fatty acid has been suggested. Pregnant and lactating females should also add a small supplement of α-tocopherol to their diets.

Vitamin E deficiency in humans is an uncommon occurrence because of the wide distribution of tocopherols in foods and the large storage capacity for the vitamin in the body. Individuals at risk of becoming deficient in vitamin E include newborn infants who have low body stores because of poor placental transfer, and patients with malabsorption syndromes and steatorrhea. Early findings are a drop in serum vitamin E levels from the norm of about 1 mg per 100 ml to 0.5 mg or less, signalling a loss of tissue stores; in the neonatal period, for example, values commonly range from 0.2 to 0.4 mg per 100 ml of serum. The first clinical sign is increased susceptibility of erythrocytes to hemolysis, followed by increased excretion of creatine in the urine which is indicative of lesions in the muscles. As depletion of the vitamin progresses hemolytic anemia may occur as well as more severe damage to the muscles. Cow's milk has a lower tocopherol content than human milk so that children who are weaned to simulated formulae, particularly those with high iron or polyunsaturated fat content which raise the vitamin E needs, may develop the deficiency syndrome. The symptoms of vitamin E deficiency are seen also in patients with cystic fibrosis, biliary obstruction, sprue, and other disorders of intestinal absorption of fats. Supplements of up to 100 mg of vitamin E as α-tocopherol daily may be required until blood levels are restored to normal.

Molecular mechanisms of many of the effects exerted by vitamin E upon cells have been rationalized on the basis of anti-oxidant properties of the tocopherols. In this view vitamin E is the lipid analogue of vitamin C which presumably plays a protective role for water-soluble components that are susceptible to oxidative degradation. Thus the tocopherols serve to protect auto-oxidizable lipids of the diet such as vitamin A, the carotenes, and the polyunsaturated fatty acids. In the presence of oxidizing agents, such as hydrogen peroxide, or high concentrations of oxygen the double bonds in such lipids may undergo oxidative cleavage reactions which not only destroy the lipids but which may produce toxic end-products that will damage adjacent molecules within cell membranes to cause lysis of the cells. The tocopherols prevent the peroxidative destruction of unsaturated lipids by acting as nonenzymatic antioxidants and hence protect cells from toxic levels of oxygen. This is the rationale for supplementation of diets with vitamin E in premature infants who were exposed to elevated oxygen environments in incubators. The hemolytic reactions observed in certain astronauts of the United States Mercury program who were exposed for prolonged periods to pure oxygen were found to be preventable with tocopherol supplements. The other consideration that arises from this interrelation is the fact that excess consumption of oxidizable lipids, as promoted by some advocates of highly unsaturated fat intake, will intensify the requirement for dietary tocopherol as well. Products from the oxidation of fat-soluble substances may exert a number of harmful influences in the body ranging from the production of carcinogenic substances in the bowel to degenerative changes in cell membranes of the erythrocytes, the muscles, or the lungs. Dietary tocopherols may thus serve to neutralize the deleterious effects of environmental pollutants and other stresses that would otherwise produce destructive changes in these tissues. Some but not all of the vitamin functions of the tocopherols may be replaced by synthetic anti-oxidants such as butylated hydroxyanisole, and the mineral selenium also complements protective actions of vitamin E on membranes. It is known that selenium plays a role as the prosthetic group of glutathione peroxidase, an enzyme of erythrocytes and other tissue cells that protects the membranes from peroxidative membrane damage. To date however, no specific coenzymatic role of the tocopherol derivatives has been found.

Much of the original excitement about the applications of therapy with vitamin E in humans arose from the dramatic symptoms seen in experimental tocopherol deficiency in animals (sterility, muscle lesions) and the striking parallels with conditions seen in human patients (habitual abortion, muscular dystrophy, or cardiac myopathies). However, despite considerable expenditure of time and effort in clinical investigations there is no evidence that human infertility in male or female patients may be related to vitamin E deficiencies. Similarly, attempts at treatment of progressive muscular dystrophy with vitamin E supplements in the nutritional range or higher have also proven to be disappointing. In certain instances of cardiovascular disorders there have been reports of encouraging findings on administration of very large doses of α-tocopherol. The conditions required to show such benefits of megavitamin E therapy for patients with heart disease have proven to be quite elusive, and several clinical trials have failed to substantiate the curative or preventive effects for atherosclerosis or coronary artery disease. In one syndrome, intermittent claudication, there seems to be unanimity regarding the clinical value of vitamin E therapy. This condition, which is characterized by pain in the calf muscles while walking and diminished blood flow accompanied by low tocopherol levels in the affected muscles, has been shown by several clinical groups to be relieved by high doses of vitamin E (300– 400 mg daily).

Many people have been lured into self-medication with megavitamin doses of tocopherols because of the enticing claims made that it restores sexual powers, rejuvenates and protects from aging and degenerative diseases, and encourages supernormal muscular performance in athletic events; these are all fallacies surrounding the use of pharmacologic doses of vitamin E. As is true for the other fat-soluble vitamins, the excess tocopherols are not readily excreted by the body but may be stored in large amounts. Nonetheless, few serious consequences of hypervitaminosis E have been recorded, even in individuals who have taken the supplements for extended periods. Among the deleterious side-effects from megavitamin E intake include interference with normal blood clotting, elevations of serum lipids, and decreases of serum thyroid hormones. While none of these are life-threatening it is likely that the disadvantages of massive ingestion of vitamin E outweigh the unproven merits.

VITAMIN K—PHYLLOQUINONE

The last of the fat-soluble vitamins was recognized because of its ability to correct certain acquired hemorrhagic lesions, and hence was termed the *K*oagulations vitamin. The naturally occurring compounds are derivatives of 1,4-naphthoquinone with various long-chain alkyl substituents in the quinone ring, the most active and common form being vitamin K_1 with a substituted phytyl chain (C20) (Fig. 2– 18). Vitamin K is quite stable to heat, but may be decomposed by oxidation or exposure to light, particularly ultraviolet irradiation. Significant amounts of vitamin K are found mainly in green leafy vegetables such as spinach, cabbage, cauliflower or lettuce; most other vegetables, fruits, cereals, meats, milk, and other dairy products are relatively poor sources. There is also a considerable amount of vitamin K that is synthesized by the intestinal flora. For this reason daily dietary requirements are difficult to establish. In the normal adult a recommended intake of 70 to 140 μg has been proposed. As noted below the needs of newborn infants are somewhat greater, and a single injection at birth of 1 mg may be given to prevent neonatal hemorrhagic disease.

The cardinal symptom of vitamin K deficiency is the tendency to bleed extensively from sites of wounding. Postsurgical hemorrhage is a common manifestation with prolonged oozing from incisions. Profuse bleeding fre-

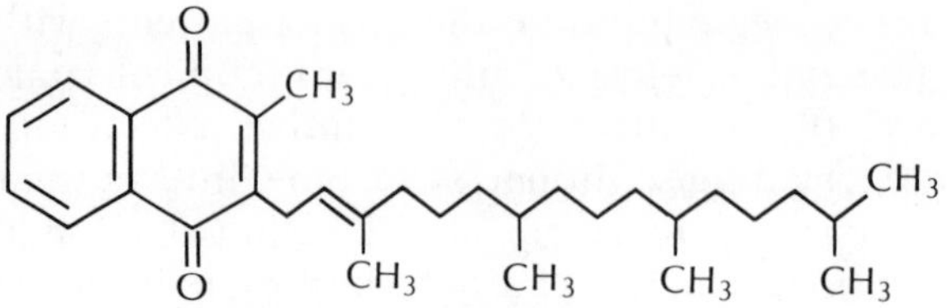

Vitamin K₁ - Phylloquinone

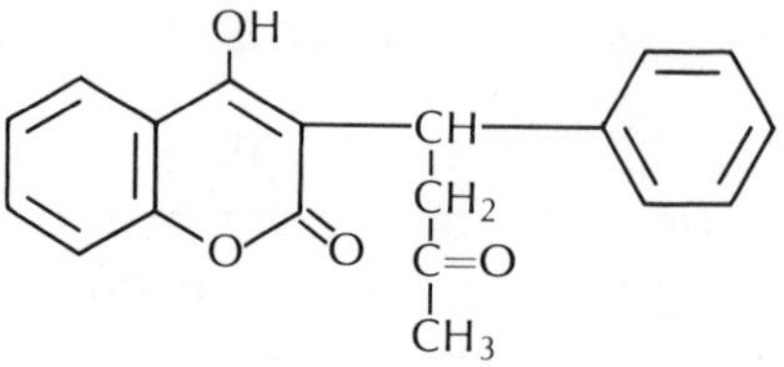

Warfarin - Anti-vitamin K

FIG. 2–18.   Structure of vitamin K and the antivitamin—Warfarin.

quently occurs from even minor wounds, with pronounced subcutaneous hemorrhages associated with areas of bruising. The concentrations of prothrombin and other blood clotting factors (VII, IX, and X) are markedly decreased to one-third or less of the normal values. Deficiency of vitamin K from inadequate dietary intake is virtually unknown. There are two sets of circumstances under which clinical expression of deficiency may occur: one is the failure to absorb the vitamin, and is seen in situations outlined earlier of fat malabsorption; the second is inadequacy of bacterial synthesis of the naphthoquinones in the intestinal tract. Obstructive jaundice with insufficient bile salt secretion for vitamin K absorption is an example of the first category. Vitamin K by injection, or oral intake of a water-soluble synthetic vitamin K derivative, such as menadiol sodium diphosphate (Synkayvite), can correct the hemorrhagic lesion in cases of malabsorption. Problems with the intestinal synthesis of vitamin K may arise from sterilisation of the gut, as in prolonged therapy with broad-spectrum antibiotics. This situation also occurs frequently in infants. At birth the gastrointestinal tract has not acquired its microfloral population, and since milk is a relatively poor dietary source the vitamin K stores of the body may be slow to reach their optimal level. In extreme cases when the prothrombin levels fail to rise the condition, known as hemorrhagic disease of the newborn, results. The latter should be treated by injection with fat-soluble vitamin K; the water-soluble forms may produce hemolytic anemia and jaundice in infants.

The mechanism of action of vitamin K has been discovered quite recently, and is described in some detail in Chapter I under Blood Clotting. In the plants and bacteria that synthesize them the vitamin K derivatives play an important role in redox reactions. Despite their superficial resemblance to benzoquinone (coenzyme Q of the mitochondrial electron trans-port chain) the substituted naphthoquinones do not appear to participate in mammalian respiration. The only metabolic role of vitamin K in hu-mans seems to lie in the post-translational modification of prothrombin and the other clotting factor proteins by carboxylation reactions in the liver. Warfarin (Fig. 2– 18), and similar antagonistic structural analogues of the naphthoquinones, act as anticoagulants by blocking this activation of prothrombin and the other vitamin K-dependent blood-clotting factors (VII, IX, and X). The inhibition, and the tendency to uncontrolled bleeding, as in the case of vitamin K deficiency, may be reversed by administration of vitamin K. Another cause of defective prothrombin synthesis that may produce hemorrhagic lesions is severe liver disease when the hepatic parenchymal cells lose their synthetic capabilities. If the pathology affects the biliary tree or the secretion of bile salts by the liver the lesions may be corrected by vitamin K injections as for obstructive jaundice. However, if the liver injury is more extensive and renders the hepatic cells incapable of prothrombin synthesis the administration of vitamin K is not able to re-store the normal coagulation process.

MINERALS

Inorganic elements and compounds play essential roles in the body, and must be considered to be of equal importance to the organic macro- and micronutrients in the diet. Several minerals are major constituents of tis-sue structures (calcium, phosphorus) or cellular and extracellular fluids (sodium, potassium, magnesium), and hence are required in bulk (100 mg daily or more). Others may participate as enzyme activators or prosthetic groups of proteins (iron, copper, zinc, fluorine, manganese), and are needed in minor amounts (a few mg per day). Still others are necessary for regula-tory or catalytic processes (iodine, chromium, selenium, molybdenum) but sometimes only in the merest traces (estimated in the range of micrograms per day). A final group of minerals (vanadium, silicon, tin, nickel) have been shown to be necessary for some animal species, but as yet have not been shown to be necessary for humans. Research is continuing on the importance of the minerals to human nutrition, particularly taking into consideration health maintenance, protection from diseases such as cancer and atherosclerosis, and pathologic effects from overdosage of nutrient minerals or exposure to toxic elements (mercury, lead, cadmium, be-ryllium).

SODIUM

Sodium ion (Na^+) is the major cation of the extracellular fluids in the body. Its main functions are to raise the osmotic pressure and thus to maintain the volume of blood and other fluids, to regulate the electrolyte and pH balance of the extracellular compartment, to control the electric potentials of excitable tissues such as nerve and muscle, and to facilitate the uptake of substances (sugars, amino acids) during active transport across cell membranes. Sodium is abundant in all manufactured foods, particularly cheese, processed meats, tinned vegetables, and tomato juice. A large portion of the average intake (5–10 g daily) comes from added table salt. A nutritional deficiency of sodium is highly improbable. The balance of sodium in the body is normally quite tightly controlled, with lowered blood sodium triggering the kidney to release angiotensin which causes the adrenal cortex to secrete aldosterone; the latter induces the renal tubules to reabsorb sodium from the glomerular filtrate and hence redress the imbalance. Decreases in plasma sodium thus may be related to defects of the kidney or adrenal cortex, or to prolonged sweating, vomiting, or diarrhea which can cause excessive losses of sodium-containing fluids from the body. Sodium presents a threat to health more commonly from over-consumption in the diet than from deficiency. The problem begins early in life with baby-food manufacturers who design their products to meet the salt-taste requirements of mothers. Children become adapted to highly salted foods and continue to eat more sodium than the body requires or than the kidneys can readily excrete to maintain the normal balance. The attendant increase in fluid retention and activation of the renin-angiotensin system may contribute to the development of high blood pressure. Recent studies show that the tendency to develop sodium-induced hypertension is greater in some individuals than others, with indications of a genetic defect of the Na^+–transport system which is required for normal renal excretion of sodium in those susceptible individuals. Patients with symptoms of hypertension are generally placed on sodium-restricted diets, and the general population is advised to reduce salt intake to 1.1 to 3.3 g per day.

POTASSIUM

Potassium ion (K^+) bears the same relation to intracellular fluid as Na^+ does to the fluids outside cells, and over 90% of the body's K^+ content is found inside cells. Nonetheless, the extracellular K^+ is important for its controlling influence upon neuromuscular irritability, and in particular its effects upon cardiac muscle. Intracellular potassium is taken up by an energy-requiring transport system, and is essential for a number of enzymatic processes including protein synthesis, glycogen synthesis, and glycolysis, as well as being essential for maintaining osmotic and acid–base balance as the main intracellular cation. Potassium is found in high concentrations in all unprocessed fruits (*i.e.*, oranges and bananas) and fresh vegetables (*i.e.*, tomatoes or potatoes) which are also low in sodium; milk products, eggs, and meat may be rich in potassium but have a variable accompaniment of sodium which may be undesirable. Potassium deficiency may occur when excess loss from the body by diarrhea or increased urinary excretion is not offset by sufficient intake to maintain the normal balance. Since the loss is from intracellular fluid the deficiency may not be reflected by any change of serum K^+. In severe deficiency, or when a sudden movement from extracellular to intracellular compartments occurs, the circulating potassium levels may drop to levels low enough to slow the heartbeat or interfere with vital muscles such as those involved in respiration. Minor deficits of potassium may be reversed by oral supplements, but extreme losses, as may occur in diabetic acidosis, must be made up more rapidly by intravenous injections to prevent a serious, even fatal outcome.

CALCIUM

Calcium is the major inorganic element comprising nearly 2% of the body mass (1300 g), 99% of which is in the skeleton. Functions of calcium are to provide the inorganic crystalline portion of hard tissues, such as bones and teeth, to promote conversion of prothrombin to thrombin during the coagulation of blood, to activate pancreatic lipase in fat absorption by the intestinal tract, to initiate the contraction process in muscle fibers, to regulate movements of ions and chemical transmitter substances across cell membranes, and to stimulate the activities of intracellular enzyme systems such as glycogen phosphorylase. Several excitation-responses of calcium ions are mediated by interaction with a receptor protein termed calmodulin. This small cytosol protein (molecular weight = 16,000) has four high affinity Ca^{2+}–binding sites. When Ca^{2+} is bound to calmodulin the latter undergoes a large conformational change which causes the activation of certain regulatory enzymes such as the protein kinases. In many instances these enzymes are also involved in cyclic AMP responses of tissues. Thus calcium ions and cyclic AMP play the role of coupled second messengers in the regulation of metabolism and cellular activity. The main dietary sources of calcium are dairy products such as milk, cheese, or yogurt; other minor dietary sources include vegetables, fruit, cereals, and fish. Average daily consumption of calcium by adults varies from 1000 mg in North American and most Western European populations, down to 500 mg or less in some Oriental and African populations where few milk products are consumed. Although the latter peoples seem to have adapted to lowered calcium intake without overt deficiency signs from an early age, dietary allowances are usually estimated at the higher levels which are required to maintain calcium balance in nonadapted subjects. There is some evidence that these higher levels of calcium intake may be beneficial, particularly in early life. Thus the supplementation of traditional diets with additional calcium during the growing period has been associated with a dramatic increase in stature of Japanese children over the past twenty years. An added margin is also advisable to compensate for the variability of losses in the feces owing to formation of insoluble calcium complexes (oxalates, phytates) in the gut, and the low efficiency of calcium absorption. Allowances for infants have been set at about 500 mg daily for the first year. For prepubertal children and adults 800 mg seems to be adequate, with extra supplements of about 400 mg recommended during rapid skeletal growth of adolescence, and during pregnancy and lactation. Extra calcium intake has also been suggested for the elderly who often show a progressive decline in calcium absorption.

Calcium homeostasis is a finely balanced process: Firstly vitamin D and its metabolites influence gastrointestinal absorption of calcium (as outlined earlier in this chapter); secondly parathyroid hormone vitamin D and calcitonin combine to control urinary excretion of calcium by the kidneys and mineral deposition or mobilization by bone as described in Chapter 3. These factors all contribute to regulate the plasma concentration at about 10 mg calcium per 100 ml within a narrow range. Approximately 50% is in the form of free ionized calcium (Ca^{2+}), 10% is bound to diffusible ligands (*i.e.*, citric acid), and the remaining 40% is nondiffusible, bound to proteins such as serum albumin. It is the ionized fraction which is active physiologically and which is subject to hormonal control. When ionized Ca^{2+} levels fall below 4.5 mg/100 ml there is an increased release of parathyroid hormone which stimulates the tubular reabsorption of calcium back into blood from the glomerular filtrate in the kidneys, activates the resorption of calcium from mineral deposits in the bone, and, by promoting the conversion of vitamin D to its active form, enhances calcium uptake in the gut. These processes, all of which tend to raise plasma Ca^{2+} back to normal, are accelerated only so long as the Ca^{2+} deficit exists. As the serum levels become normalized parathyroid hormone secretion declines and thus the movement of calcium into the blood is reduced. The hormone calcitonin operates in the opposite manner to parathyroid hormone by inhibiting Ca^{2+} reabsorption by renal tubules, and Ca^{2+} resorption by bone cells. The effect would be to resist any increase of serum calcium levels much above normal. However, it is not clear that calcitonin secretion responds to elevated Ca^{2+} directly; in fact, the primary control of calcitonin release may be exerted by gastrointestinal peptides (gastrin, pancreozymin). These peptides are secreted in response to food intake, and stimulate release of calcitonin. Since the calcitonin acts directly on bone to inhibit the resorption process it will allow the extra calcium from food absorption to be taken up by bone mineralization, and any excesses in the circulation would be excreted by the kidney.

An acute deficiency of extracellular Ca^{2+} will provoke a characteristic hyperexcitable state of the nerves and muscles known as tetany. The symptoms include numbness of extremities, emotional irritability, and tightness or spasms of the muscles leading, in extreme cases, to generalized convulsions. Besides the retardation of growth in young persons, the chronic deficiency of calcium will result in gradual depletion of mineral from mature bones in adults. With time this will produce a condition of osteomalacia similar to that seen in vitamin D deficiency, with a tendency to formation of pathologic fractures in the decalcified areas. The lesions are seen in persons suffering from chronic malabsorption of calcium, in women who have undergone multiple pregnancies without adequate calcium supplementation, and in the elderly, particularly postmenopausal females. In the latter, loss of mineral may commonly be accompanied by loss of the organic matrix component in bone (osteoporosis); such cases will require treatment with estrogens as well as dietary adjustments to ensure the maintenance of both organic and inorganic components of the skeleton.

PHOSPHORUS

The relative abundance of phosphorus in the human body is second only to calcium, representing 1% of the body mass (700 g), with about 80% residing in the bones and teeth in the form of insoluble calcium phosphate (hydroxyapatite). Phosphorus, in the form of phosphate anion, and its organic esters also play a vital part in cellular biosynthesis and energy metabolism. Inorganic phosphate salts constitute one of the important acid–base buffer systems in blood and inside cells. Organic phosphate esters, such as those of glucose and other sugar derivatives, act as key intermediates in metabolic reactions, and also constitute a useful trapping device for the sugars within cells. Other phosphate derivatives include the diesters, which form the backbone bridges between monomers of the nucleic acids and the polar outward-facing regions of phospholipids in membrane bimolecular leaflets. A variety of nucleotides and other specialized coenzymes require combination of vitamins with phosphate (pyridoxal phosphate, thiamine pyrophosphate, flavin mononucleotide, etc) to produce the active form. Chemical energy transfer and storage involve the reactions of the labile pyrophosphates (adenosine triphosphate) and phosphagens (creatine phosphate). Transmission of many hormonal effects requires the intracellular second messenger, cyclic AMP, which is an intramolecular phosphodiester. Phosphate in the diet is abundantly supplied in meats, cheese, eggs, and whole grains; in fact, its distribution in foods is so widespread that a dietary deficiency would be encountered only during prolonged starvation. The recommended allowance of phosphorus is the same as that for calcium, 800 mg as P per day for an adult, with the same extra allowances for growth, pregnancy, or lactation. The ideal Ca/P ratio near one is provided by ingestion of milk as the major dietary source of these minerals.

MAGNESIUM

Next to potassium, magnesium is the major intracellular cation. About 70% is found in skeletal tissues; the remainder is distributed generally throughout soft tissues with particularly large amounts in muscle and brain. Magnesium ions (Mg^{2+}) are essential activators of many enzyme systems especially those involving transfer of phosphate groups which include many kinases, phosphatases, and the mitochondrial process of oxidative phosphorylation. The enzyme, enolase, is highly dependent upon activation by Mg^{2+}, and thus the use of glucose by the glycolytic pathway is quite sensitive to magnesium deprivation. Obviously energy-generating reactions of all cells require magnesium; plants have a special requirement for the element which occupies the central position of the photosynthetic pigment, chlorophyll. Because of this key role, magnesium is ubiquitous among plant foods; good dietary sources include whole grains, vegetables, nuts, milk products, and meats. Recommended allowances rise from 50 mg in infants to about 300 mg daily in adults, with an extra 150 mg allowance during pregnancy and lactation. The average mixed diets provide amounts equal to or in excess of these figures.

Magnesium deficiency symptoms bear some resemblances to those seen in hypocalcemia, with muscular twitching, spasms, and the other manifestations of tetany. Other neuromuscular effects include numbness and muscle weakness, loss of memory, apathy, depression, confusion, hallucinations culminating in severe delirium, cardiac arrhythmia, tachycardia, ventricular fibrillation, and in critical cases convulsions, coma, and sudden death have been reported to occur. A deficiency from dietary causes alone is improbable, except in protein–calorie malnutrition or starvation. Acute, mild imbalances may occur in cases of heavy losses of magnesium in feces or sweat, or during fasting, but these loses are usually adjusted by mobilization of magnesium from bone. Malabsorption syndromes, chronic alcoholism, renal diseases, diabetic ketoacidosis, and pancreatitis are some of the disease states where symptoms of magnesium depletion have been seen. In hospitals, burn patients and those recovering from extensive surgery are susceptible to hypomagnesemia, particularly following prolonged maintenance on parenteral fluids, unless magnesium supplements are given prophylactically throughout the illness.

IRON

The average adult contains a total of 5 g of iron, mostly in the circulating red cells. High concentrations of iron are also found in liver and spleen, followed by heart and skeletal muscles, and then the kidneys and nervous tissues. The functions of iron center around oxygen utilization and the process of respiration. For these purposes the iron is bound in the complex with a porphyrin ring known as heme, and the latter is conjugated to a variety of polypeptides to form the diverse family of hemoproteins. Hemoglobin, the oxygen transport protein of blood, accounts for up to 70% of total body iron, and its synthesis and degradation dominate iron metabolism. The muscle oxygen-binding protein, myoglobin, has the same heme prosthetic group, and constitutes 3 to 5% of total body iron. The tissue hemoproteins, known as the cytochromes, are distributed widely through-out various organs and cell types where they play essential roles in mitochondrial electron transport and the aerobic generation of ATP by oxidative phosphorylation; they also play an important role in electron transfer reactions of the endoplasmic reticulum with associated hydroxy-lation or desaturation processes. A number of other enzymes may involve the heme group (catalase), nonheme iron–sulfur proteins (succinate dehy-drogenase), and inorganic iron (aconitase) as coenzymes. Iron storage pro-teins, ferritin and hemosiderin, normally comprise about 15% of body iron. Dietary sources of iron include liver, meat, fish, egg yolk, green vegetables, molasses, and whole wheat or enriched bread. Daily allowances should begin with 10 mg in infancy increasing to 18 mg in adolescence. Adult females should continue at this high level of intake to offset losses from menstruation. In child bearing large supplements beyond the iron content of habitual diets (*i.e.*, 30– 60 mg) are recommended to replenish stores lost by pregnancy. Adult males and postmenopausal females require 10 mg of iron daily.

Iron metabolism and functions are dependent upon interconversions between the ferric (Fe^{3+}) and ferrous (Fe^{2+}) valence states. Only the Fe^{2+} salts of inorganic iron may be absorbed in the gastrointestinal tract, and thus ascorbic acid, cysteine, and other reducing substances which can promote the ferric to ferrous conversion will facilitate iron absorption. Certain organic acids (citric, gluconic) and amino acids (histidine, lysine) form absorbable complexes with iron, while formation of phosphate com-plexes (phytate, orthophosphate) of iron retards the iron absorption. The iron from hemoproteins is absorbed as intact heme and is not influenced by ascorbic acid or anionic compounds. In general iron is absorbed more efficiently from foods of animal origin rather than plant origin; only 10% of food iron is ingested from ordinary mixed diets, but this proportion may double in iron-deficient individuals. Transfer of iron into the intestinal mucosal cells seems to be slowed down as iron stores in the individual are replenished, but the mechanism of this regulation is not clear. The Fe^{2+} form passes from the mucosa into the bloodstream where it is rapidly oxidized to the Fe^{3+} form by the copper-containing enzyme, ferroxidase, or ceruloplasmin as it has been termed in the past. In the Fe^{3+} form the iron combines with the transport protein, transferrin. From the serum the iron may be taken up by erythropoietic cells of the bone marrow for hemoglo-bin synthesis, by other tissue cells for hemoprotein enzymes or cyto-chromes, or by the liver for deposition in the storage proteins (ferritin or hemosiderin). Destruction of hemoglobin from nonviable red cells con-tributes a very large pool of endogenous iron (20– 25 mg daily), much greater than that absorbed from the diet (1– 2 mg daily). The endogenous iron is transported by the transferrin carrier from the reticuloendothelial cells which break down the senescent erythrocytes, and is rapidly recycled to restore tissue needs. Dietary iron is required only to offset the very minor quantities of iron that are lost from the pool by excretion or from hemorrhages.

Iron deficiency results in decreases of circulating hemoglobin and volume of packed red cells (hematocrit). The anemia is termed microcytic since the blood contains erythrocytes of reduced size, with pale complexions. The afflicted individuals also appear pale, "washed-out," and show signs of "tired blood." Other symptoms besides fatigue and listlessness include palpitations on exertion, gastric disturbances, lesions of the oral cavity and fingertips (spoon-shaped nails). In children suffering from iron deficiency, depletion of storage reserves occurs during rapid growing periods, and results chiefly from the low iron content of milk. It may cause growth retardation, decreased resistance to infection, and neurologic lesions. Iron deficiency symptoms are not uncommon in adults. The pronounced and repeated iron losses resulting from menstruation, pregnancy, and lactation render women of child-bearing age especially vulnerable to depletion of iron stores. The problem is exacerbated by poor diet or lack of hygiene in economically deprived females. Both the mothers and newborn offspring may require iron supplementation. It should be recognized that the anemia of iron deficiency is but one, albeit a most obvious sequela; in fact, deficient patients who receive supplements of iron may show subjective improvement before their hemoglobin levels are normalized, indicating that impairment of tissue cytochromes and iron-dependent enzymes may be of equal importance in development of manifestations of iron deficiency.

An excessive deposition of iron in tissues may produce toxic effects. The syndrome is termed hemochromatosis and appears to be of rare occurrence. In some populations the excess iron may be traced to drinking water or to preparation and storage of food in iron vessels. In other instances there may be accumulation of iron in the body from repeated blood transfusions over prolonged periods. There is also a genetic syndrome, an inborn error of iron metabolism, in which increased uptake and storage of iron occurs with a normal iron intake. The body has no mechanism to excrete the excess iron, and hence the accretion of even small amounts in excess of daily losses causes a cumulative buildup of the iron storage protein, ferritin, in various tissues, and spillover into the hemosiderin stores. The ensuing excess of hemosiderin deposition causes hyperpigmentation of the skin, cirrhosis of the liver, and damage to the pancreas resulting in diabetes mellitus. In such cases the best cure is the old remedy of the barber surgeons, phlebotomy (*i.e.*, blood-letting).

COPPER

Requirement for copper in human diets is linked to its functional role in several important cuproenzymes including cytochrome oxidase of the mitochondrial respiratory chain, dopamine β-hydroxylase of the catecholamine synthetic pathway, and ferroxidase (ceruloplasmin), which is also the major transport form of copper in blood plasma. Copper plays a key part in iron transport and metabolism into hemoproteins, and some cases of anemia in infants have been found to require combined supplements of copper and iron. Dietary sources of copper are fish, liver, nuts, green vegetables; milk products and most cereals are inadequate sources. Average intake values for adults have been estimated at 2 to 3 mg daily, and this seems adequate for maintenance of health. In adults symptoms of copper deficiency are rare, except in some malabsorption syndromes or during prolonged parenteral nutrition without copper supplements. In children the manifestations may appear during periods of rapid growth on exclusively milk-based formulas, particularly if copper stores are low as in the case of premature infants. An X-linked genetic defect in copper absorption, the kinky hair syndrome, characterized by defective keratinization of the hair, neurologic degeneration, and anomalies of bones and arteries, is fatal in early life unless treated immediately with parenteral administration of copper. Wilson's disease, or hepatolenticular degeneration is, another fatal inherited disease accompanied by decreased levels of copper in the blood. However, in this syndrome, an autosomal recessive trait, there is an excessive storage of copper in the liver, probably owing to defective synthesis of ceruloplasmin by the hepatic cells. In addition to cirrhosis of the liver, copper deposition in tissues produces damage to the nervous system resulting in neurologic and psychiatric disorders, and to the kidneys resulting in progressive loss of tubular reabsorption mechanisms for glucose, phosphate, and amino acids. The excess copper may be removed by ingestion of the chelating agent, penicillamine, which promotes urinary excretion of the copper complex, and the patients can be restored to normal if treated early enough.

ZINC

The importance of this metal in human nutrition is related to the large number of zinc-requiring enzymes that participate in a wide variety of metabolic processes. Included among these metalloenzymes are carbonic anhydrase of erythrocytes, serum alkaline phosphatase, several liver dehydrogenases (*e.g.*, alcohol dehydrogenases), and crucial enzymes of DNA synthesis including thymidine kinase and DNA polymerase. Zinc is involved in processes of tissue growth, development, and regeneration, and is necessary for production of the retinol-binding protein that is involved in transport of vitamin A in the blood stream. Major sources of zinc in the diet are animal foods (meat, eggs, fish), and intake should roughly parallel the amount of animal protein ingested; zinc is present in plant sources such as grains or beans, but is largely unavailable in these foods owing to the formation of a nonabsorbable complex with phytic acid. The recommended allowance in children is 3 to 5 mg daily, increasing to 15 mg in adults; an extra 5 mg is suggested during pregnancy and 10 mg during lactation. Zinc deficiency has been described among some middle eastern populations who subsist on unleavened bread with little or no animal protein. The symptoms are dwarfism, genital hypoplasia, loss of taste sensation, and impaired wound-healing. These pathologic changes and lowered levels of serum alkaline phosphatase may be corrected by administration of zinc.

OTHER TRACE ELEMENTS

Nutrition research is continuing in many areas of trace element metabolism in humans. In some cases it is difficult to pin-point the effects of individual elements outside of the context of generalized malnutrition. In others it is a question of providing appropriate ratios of minerals, and of walking the tight-rope between deficiency effects of undernutrition and toxic effects of overdosage. Frequently there is insufficient documentation to establish definitive allowance values, and it is necessary to cite current intake values in the average population with the rider that deficiency or toxic effects for the constituent do not seem to be rampant. Brief descriptions of some of the other trace elements follow:

Chromium

This metal acts in concert with insulin to maintain normal glucose utilization. Improvement of impaired glucose tolerance has been reported in some cases by administration of chromium, and it has been indicated that there is an abnormal utilization of this mineral in some diabetic subjects. Dietary intakes vary widely (5–100 μg per day), with lower levels resulting from ingestion of refined sugars and cereals. Molasses and yeast are good supplementary sources.

Selenium

This mineral complements the actions of vitamin E by preventing membrane lipid peroxidation in its role as prosthetic group for the enzyme glutathione peroxidase. Activity of glutathione peroxidase seems to be maintained on average intakes (100–200 μg daily), but deficiencies could arise in regions with low selenium content in the soil. Fish, whole grain, and meat provide most of the dietary selenium.

Iodide

This halogen is required by thyroid tissue for thyroxine synthesis, and dietary deficiency produces compensatory thyroid enlargement (goiter), and sluggish body metabolism from decreased thyroid hormone concentrations. Recommended intake is 150 to 200 μg daily in adults, and is readily met from ingestion of iodized salt plus iodine compounds in bread. Fish and seaweed products provide supplementary sources of iodide.

Fluoride

Traces of this element strengthen the enamel surface of teeth and render it resistant to dental caries. The daily requirement is 1 to 2 mg and is usually provided by supplementation of public drinking water to a level of 1 ppm. Local administration of additional fluoride to teeth at the point of and just after eruption provides additional cariostatic protection.

Molybdenum

The enzyme, xanthine oxidase, contains molybdenum as its prosthetic group. Purine catabolism and the exchange of iron with ferritin are catalysed by this metalloenzyme. No deficiency state has been ascribed to low molybdenum intake. The average diet contains 150 to 300 μg daily.

Manganese

Several enzymes including pyruvate carboxylase and superoxide dismutase (which may contain other metals such as Cu, Fe or Zn) have bound manganese, while others, such as the glycosyl transferases, are activated by free manganese. Average diets provide 3 to 5 mg daily which seems to be adequate for human requirements since no specific manganese-deficiency syndrome has been documented.

Nickel, Vanadium, and Silicon

These elements are required for several animals, but their role in human nutrition is not clear at present.

SUGGESTED READINGS

Altschule MD: Nutritional Factors in General Medicine: Effects of Stress and Distorted Diets, Springfield, Charles C Thomas, 1978

Anderson TW: New horizons for vitamin C. Nutr Today 12(1):6– 13, 1977

Anderson TW, Passmore R, Szent-Gyorgi A et al: To dose or megadose—a debate about vitamin C. Nutr Today 13(2):6– 33, 1978

Babior BM (ed): Cobalmin—Biochemistry and Pathophysiology, New York, John Wiley & Sons, 1975

Butterworth CE, Blackburn GL: Hospital malnutrition. Nutr Today 10(2):8– 18, 1975

Carroll KK: Lipids and carcinogenesis. J Environ Pathol Toxicol 3:253– 271, 1980

DeLuca HF: Vitamin D—Metabolism and Function, Berlin, Springer-Verlag, 1979

DeLuca HF: The vitamin D hormonal system: Implications for bone diseases. Hosp Pract 15(4):57– 63, 1980

Farrell PM: Deficiency states pharmacological effects, and nutrient requirements. In Machlin LJ (ed): Vitamin E/A Comprehensive Treatise, pp 520– 620. New York, Marcel Dekker, 1980

Fernstrom JD: Dietary precursors and brain neurotransmitter formation. Annu Rev Med 32:413– 425, 1981

Fernstrom JD, Wurtman RJ: Nutrition and the brain. Sci Am 230(2):84– 91, 1974

Goodman DS: Vitamin A metabolism. Fed Proc 39:2716– 2722, 1980

Hambridge KM, Nichols BL Jr (eds): Zinc and Copper in Clinical Medicine, New York, Spectrum Publications, 1978

Herbert V: The nutritional anemias. Hosp Pract 15(3):65– 89, 1980

Hodges RE: Nutrition in Medical Practice, Philadelphia, WB Saunders, 1980

Howard RB, Herbold NH: Nutrition in Clinical Care, New York, McGraw-Hill, 1978

Klee CB, Crouch TH, Richman PG: Calmodulin. Annu Rev Biochem 49:489– 515, 1980

McNeely MDD: Nutrition, vitamins and trace elements. In Gornall AG (ed): Applied Biochemistry of Clinical Disorders, pp 368– 377. Hagerstown, Harper & Row, 1980

Newberne PM, Zeiger E: Nutrition, carcinogenesis, and mutagenesis. In Flamm WG, Mehlmann MA (eds): Advances in Modern Toxicology, Vol 5, Mutagenesis, pp 53– 84. Washington, Hemisphere, 1978

Nordin BEC (ed): Calcium, Phosphate and Magnesium Metabolism, Edinburgh, Churchill Livingstone, 1976

O'Reilly RA: Vitamin K and the oral anticoagulant drugs. Annu Rev Med 27:245– 261, 1976

Passmore R: How vitamin C deficiency injures the body. Nutr Today 12(2):6– 11, 27– 31, 1977

Prasad AS: Nutritional zinc today. Nutr Today 16(2):4– 11, 1981

Schneider HA, Anderson CE, Coursin DB (eds): Nutritional Support of Medical Practice, Hagerstown, Harper & Row, 1977

Schnoes HK, DeLuca HF: Recent progress in vitamin D metabolism and the chemistry of vitamin D metabolites. Fed Proc 39:2723– 2729, 1980

Schrauzer GN: Vitamin C: Conservative human requirements and aspects of over-dosage. In Neuberger A, Jukes TH (eds): Biochemistry or Nutrition, Vol 1, pp 168– 188. Baltimore, University Park Press, 1979

Scott ML: Advances in our understanding of vitamin E. Fed Proc. 39:2736– 2739, 1980

Scrimshaw NS, Young VR: The requirements of human nutrition. Sci Am 235(3):50– 64, 1976

Shapcott D, Hubert J (eds): Chromium in Nutrition and Metabolism, Amsterdam, Elsevier, 1979

Sporn MB, Newton DL: Chemoprevention of cancer with retinoids. Fed Proc 38:2528– 2534, 1979

Story JA, Kritchevsky D: Dietary fiber: Its role in diverticular disease, colon cancer and coronary heart disease. In Newberger A, Jukes TH (eds): Biochemistry of Nutrition, Vol 1, pp 189– 206. Baltimore, University Park Press, 1979

Suttie JW: The metabolic role of vitamin K. Fed Proc 39:2730– 2735, 1980

Underwood EJ: New findings with trace elements: Including effects of processing on supplies and availability. In Neuberger A, Jukes TH (eds): Biochemistry of Nutrition, Vol 1, pp 209– 243. Baltimore, University Park Press, 1979

Underwood EJ: Trace Elements in Human and Animal Nutrition, 4th Ed, New York, Academic Press, 1977

Chapter Three
Hormone Action

The growth, differentiation, and physiologic functions of body cells are controlled by a family of chemical messengers called hormones. Because cells are influenced by a number of chemical agents (substrates, ions, vitamins, neurotransmitters, etc.) it is necessary to be somewhat restrictive in applying the term hormone. In the past this designation has been reserved for a class of regulatory chemicals which are synthesized and stored in specialized endocrine glands, so-named because they discharge their products into the circulation, and which are present and active at very low concentrations in the blood, with high selectivity for target organs that feel their effects. However, in recent years it has become evident that many important chemical regulators do not fit so neatly and specifically into the endocrine system. For example, a very important group of hormones (*e.g.*, factors that release the true endocrine secretions) have properties midway between classical hormones and neurosecretions, and so are classified as neurohormones. Another group that transcends the boundaries of classification are the metabolites of vitamin D (see Chapter 2), which exhibit properties intermediate between those of vitamins and hormones. Finally, there are the prostaglandins, a family of fatty acid derivatives that are generated within the cells they influence rather than extracellularly. Since most of their effects are mediated by cyclic nucleotide changes, the prostaglandins may be considered as modulators of the so-called second messengers within cells, rather than as true primary messengers coming from outside cells. Thus it seems wise not to maintain exclusive definitions of hormonal messengers based upon classical concepts of endocrinology. As future investigations unfold, the demarcations among regulatory substances will undoubtedly become blurred to an even greater extent.

The mechanisms used by hormones to exert their effects on cells are varied. Specific hormones can effect more than one cell type and produce quite different responses within different tissues. These pleiotropic man-

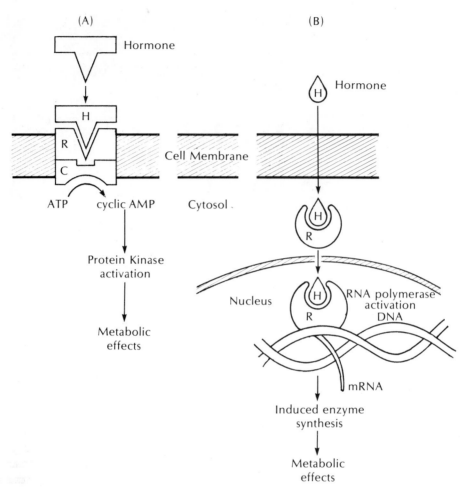

FIG. 3–1.   Hormone receptors: *(A)* nonpenetrating hormone (H) interacts with cell surface receptor (R), which activates catalytic protein (C) to generate a second messenger within the cytosol; *(B)* penetrating hormone (H) enters cell and reacts with cytosol receptor (R) which undergoes conformational changes, enters the nucleus, and activates a specific region of the genome.

ifestations of hormone action do not seem to be explainable by a single unitary hypothesis. For example, some hormone effects will occur through changes in cell membrane permeability to particular substrates, others will occur through changes in catalytic activity of existing cellular enzymes, while still others may involve new synthesis of enzymes or other cell constituents. Despite the wide divergence of hormonal effects, however, there is a unifying principle emerging to explain both the specificity and the diversity of cellular responses to hormones. This principle is the concept of hormone–receptor interaction (Fig. 3– 1). Cells in target tissues possess selective, high-affinity, protein receptors for the active hormones. The recognition process requires the combination of hormone molecule with its cellular receptor molecule in the specific manner that a substrate complexes with an enzyme. In the process, the three-dimensional config-uration or the subunit structure of the receptor is modified so that it will now direct some change within the cell.

There is also a general pattern concerning the location of hormone receptors depending on the ability of the hormone to penetrate cells. In general, the polar polypeptide hormones, such as adrenocorticotrophic hormone (ACTH), or the amine derivatives of amino acids, such as epi-nephrine, interact with receptors on external surfaces of cell membranes (Fig. 3– 1a). The membrane-bound hormone– receptor complex now exerts effects inside cells by eliciting the release of an intracellular second mes-senger, generally in the form of a cyclic nucleotide (cyclic AMP, cyclic GMP). Most of the subsequent effects produced by the second messenger, cyclic AMP, are explicable in terms of its activation of protein kinases, enzymes which modulate the actions of other proteins by phosphorylating them. On the other hand, the smaller, more lipophilic steroid hormones, such as cortisol, or the thyroid hormones, such as thyroxine, may pass through cell membranes and interact with cytoplasmic receptors inside cells (Fig. 3– 1b). The intracellular hormone– receptor complex may mi-grate to the nucleus where it will combine with particular segments of DNA in the chromosomes to alter the rate of transcription of certain genes. The main effect produced by this class of hormones is the induction of synthesis of specific enzymes and other proteins within target cells.

Biochemical effects of hormones in the body are extremely compli-cated owing to interactions between the substrates (carbohydrates, fats, amino acids), and the agents produced in various endocrine organs (polypeptides and steroids). These interrelations are currently being explored by studies of cultured target cells and by investigations of the isolated hormone-receptor complexes. The following account describes the present understanding of the biochemical actions and interactions of the major human hormonal agents.

PITUITARY HORMONES

An overriding influence on most of the endocrine systems in the body is exerted by a small gland situated below the brain, the pituitary. This tiny indispensable organ controls the growth, metabolism, and secretions of many other glands, and hence it has often been termed the "master gland" of the endocrine system. Moreover, hormones released from the pituitary may produce more general growth-stimulating (or trophic) ef-fects as well as regulating actions upon body cells. Thus this crucial con-trol center is an endocrine organ in its own right, but it may also be

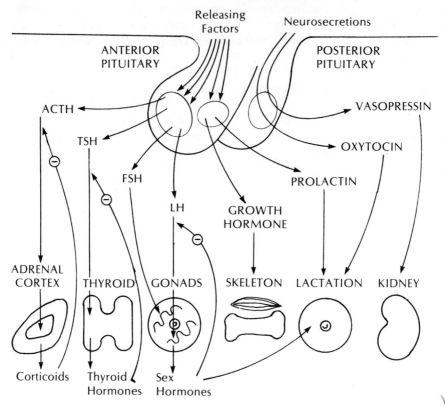

FIG. 3–2. Pituitary hormones, their target tissues, and negative feedback loops controlling their release.

considered as an intermediate way-station in the transmission of chemical signals between the nervous system, to which it has a direct physical connection, and other diverse endocrine glands, to which it communicates through the bloodstream. The pituitary has both direct and indirect actions in the growth and regulation of functions of the body as a whole.

There are three general categories of polypeptide hormones produced by the pituitary (Fig. 3–2). Those that stimulate other endocrine organs, such as the adrenal (ACTH), thyroid (TSH), and the gonads (FSH and LH), are each produced in response to a single releasing factor that is generated in the hypothalamus; secretions of such hormones are subject to classical feedback-inhibition control by the products from the target organs, and are described in detail under the appropriate endocrine glands. The second group includes the more generally stimulating growth hormone, or somatotrophin, which acts on liver, skeletal tissues, adipose cells, and the mammary-stimulating hormone, prolactin. These hormones are sometimes considered together under the umbrella term, somatomammo-trophins, and their regulation appears to have the character of an open loop system; the control of this type of system is ensured by the presence of a different push–pull principle, namely two hypothalamic factors with op-posing actions on the secretion. A third category of pituitary hormones are the small peptides of the vasopressin–oxytocin class, which are really storage components in the pituitary that are synthesized by neurosecre-tory cells of the hypothalamus.

GROWTH HORMONE AND SOMATOMEDINS

Growth hormone is produced by the characteristic granular acidophilic cells of the anterior lobe of the pituitary (adenohypophysis). Synthesis of the hormone by cultured pituitary cells shows a requirement for the thyroid hormone, triiodothyronine, which binds to a nuclear receptor and stimulates the production of growth hormone messenger RNA. Storage of growth hormone is greater than for other pituitary hormones and consti-tutes up to 1% of the weight of the gland, a fortunate circumstance since this is the only source of hormone for clinical work. The human growth hormone is a single-chain polypeptide of 191 residues with a molecular weight of 21,500; other mammals produce a growth hormone of similar size but there are such differences in amino acid composition, im-munologic properties, and biologic activity that only the hormone isolated from human pituitaries may be used in treatment of children with defi-cient growth resulting from hypopituitarism.

The growth-promoting actions of growth hormone are most apparent on skeletal structures before puberty, and are most effective in the nutri-tionally well-fed state. Cellular hyperplasia and the biosynthesis of colla-gen and mucopolysaccharides in connective tissue are promoted. Evi-dently the hormone has many target tissues since the growth of bone, visceral organs (such as the liver), and the musculature are all dependent on the hormone. Moreover, there is a generalized increase of protein syn-thesis in these tissues that is related to increased numbers of ribosomes and associated membrane synthesis leading to proliferation of rough en-doplasmic reticulum; enhanced synthesis of all species of RNA and induc-tion of the enzyme ornithine decarboxylase, which gives rise to the polyamine, spermine, a stimulator of protein synthesis, are also seen in tissues under the influence of growth hormone.

In addition to the hyperplastic effects, growth hormone exerts multi-ple metabolic actions (Fig. 3–3). Some of these actions support growth promotion. In the well-fed state when plasma levels of amino acids are elevated the pituitary secretion of growth hormone is stimulated. The in-creased growth hormone accelerates amino acid uptake into tissues and represses the induction by glucocorticoids of the amino acid catabolic reactions that are involved in hepatic gluconeogenesis. In this way growth

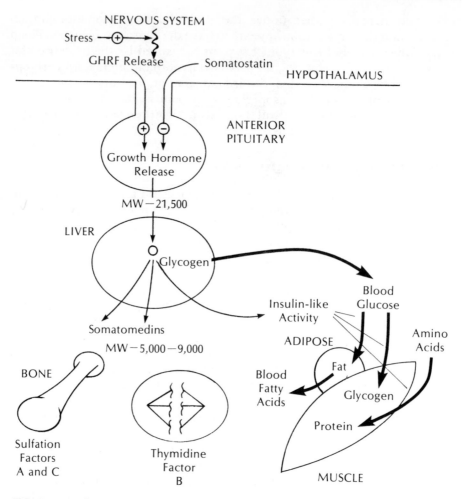

FIG. 3–3. Release of growth hormone and somatomedins, and actions in peripheral tissues. Heavy black arrows show processes which are stimulated.

hormone accentuates the channelling of amino acids into net protein synthesis and favors positive nitrogen balance in times of plenty. However, growth hormone may also serve an important protective regulatory role under conditions of inadequate nutrition. In this respect it promotes an anti-insulin response in the liver by stimulating hepatic glycogen breakdown and release of glucose into the bloodstream. At the same time growth hormone exerts an insulinlike action on extrahepatic tissues by stimulating glucose uptake and oxidation in adipose tissue and skeletal muscle; it also stimulates lipolysis and mobilization of fatty acids from adipose tissue in the presence of corticosteroids. The secretion of growth hormone by the pituitary increases when either blood glucose or fatty acid levels fall. Thus the combined effects upon carbohydrate and fat metabolism may ensure the energy needs during both tissue anabolism, as well as in times of fasting.

Despite the broad spectrum of affected tissues and processes it now appears unlikely that growth hormone acts directly at all these locales of metabolic action. Instead a number of smaller intermediating peptides, termed somatomedins, have been implicated as the circulating extracellular messengers for most of the somatotrophic and insulinlike actions of growth hormone (Fig. 3–3). Thus, although it is clear that the growth of cartilage in the body is dependent on elaboration of growth hormone by the pituitary, the exposure of the same tissue directly to the hormone *in vitro* is without effect. Isolated cartilage responds, however, to factors from the blood which are generated secondarily by secretion of growth hormone. These factors that stimulate reactions of proteoglycan synthesis, such as the incorporation of radioactive sulfate (sulfation factor) or the incorporation of thymidine into new DNA during proliferation of the cellular matrix (thymidine factor), have been found to be polypeptides with molecular weights in the 5000 to 9000 range. The liver, which has specific plasma membrane receptors for growth hormone, seems to be instrumental in the generation of the somatomedins, which then pass on in the bloodstream to affect other tissues. The mechanism of production of somatomedins is unclear but they do not seem to be produced by simple cleavage from a portion of the growth hormone molecule since they do not share antigenic determinants.

Control of the secretion of growth hormone in the hypothalamus is effected by secretion of two antagonistic neurohormones, one stimulatory and the other inhibitory, to the somatotroph cells of the pituitary. The stimulatory agent, termed growth hormone-releasing factor, (GHRF) has not been characterized. Its production seems to predominate under stressful conditions (hypoglycemia of fasting, heavy exercise, emotional stress), and is stimulated by the catecholamines, norepinephrine and dopamine, and by serotonin. The inhibitory agent has been isolated and identified as the tetradecapeptide, somatostatin, which also suppresses secretion of thyroid-stimulating hormone from the pituitary, and influences the release of insulin and glucagon from the pancreas as well. Somatostatin appears to counteract the effects of elevated cyclic AMP levels in somatotrophs induced by hypothalamic stimulatory factors; it blocks both the stimulated and the basal secretion of growth hormone. It is of interest that children with protein–calorie malnutrition have low levels of somatomedins but elevated growth hormone levels in blood, suggesting that somatomedins may exert a negative feedback on pituitary growth hormone secretion under normal circumstances.

The mechanisms of growth hormone actions are difficult to untangle because of the fact that some effects are exerted by the hormone directly and some by the somatomedins. In the latter category are the stimulation of glucose uptake by adipose and muscle cells and the increased use of glucose by these tissues for lipid or glycogen synthesis. These effects and the mitogenic activity of somatomedins have a considerable overlap with the anabolic and growth responses produced by insulin in tissues; nonetheless it seems that there are discrete somatomedin receptors on the surfaces of responsive cells that are separate from the insulin receptors. The physiologically important actions of the somatomedins are on growth; the effects on carbohydrate homeostasis *in vivo* are of a minor nature unless the somatomedin levels are exceedingly high. Conversely, insulin acts primarily to maintain metabolic homeostasis and serves as a growth-promoting hormone secondarily when its concentrations are greatly elevated. Concerning the direct actions of growth hormone, attempts are being made to find the active regions in the polypeptide sequence and to determine whether multiple actions of the hormone could be produced by different regions of the molecule. To this end various fragments of the hormone have been studied. Growth potentiation can be demonstrated with fragments lacking the last 57 residues at the carboxyl end; in fact removal of residues 134 through 150 produces a super growth hormone that is four to five times more active than the native hormone. Other fragments have been produced that can activate lipolysis, hyperglycemia, or other acute metabolic effects, but which may lack the ability to promote somatomedin responses of the long-term trophic variety. Possibly the multiple activities of growth hormone are packaged in the molecule as multiple active sites.

Consequences of the deficiency of growth hormone in childhood are symmetrical dwarfism and delayed maturation of secondary sex characteristics. Without treatment such children are generally half the normal stature for their age, although they may eventually attain adult heights between 4 and 5 feet. Both circulating growth hormone and somatomedin concentrations are markedly depressed as the result of pituitary insufficiency. Following treatment with growth hormone isolated from human pituitaries appetite, nitrogen and calcium retention, and linear growth are rapidly restored. The major drawback is limited availability of human growth hormone; presumably genetic engineering techniques will soon provide an increased supply of the hormone by cloning the human gene into an appropriate bacterial producer or mammalian tumor cell in culture.

The excessive production of growth hormone, for example by a somatotroph tumor of the pituitary or by overproduction of growth hormone-releasing factors from the hypothalamus, causes overgrowth of the long bones and gigantism of the young patient. Chronic hypersecretion of growth hormone in adults after closure of the epiphyses causes the syndrome of acromegaly. Since further linear growth is blocked, the continued stimulus to cartilage and bone produces increased diameter of bones, particularly in hands and feet, enlargement of the jaws and facial bones, and thickening of skin and subcutaneous tissues. The visceral organs (liver, kidney, lungs) are also grossly increased in size in acromegaly.

Somatomedin concentrations in serum are greatly elevated (2 to 4 times normal), and usually show some correlation with the clinical course of the disease rather than the degree of increase in growth hormone concentration. Complications include disturbances of joint function, entrapment of spinal nerve roots, insulin-resistant hyperglycemia, and congestive heart failure. Therapy is risky, involving microsurgery or radiation treatment to ablate the hyperactive pituitary and normalize growth hormone levels. Bromocriptine, an analogue of dopamine, has been shown to reduce growth hormone levels and is effective in reversing the pathologic changes in acromegaly. The search continues for an analogue of somatostatin with longer biologic half-life and more selective action on the pituitary as an alternative therapeutic tool to reduce excess growth hormone secretion.

PROLACTIN

This hormone, a polypeptide of 198 amino acid residues, bears a considerable structural similarity to growth hormone, and the two hormones share common amino acid sequences and have some overlapping biologic activities. They are, however, produced by two separate cell types in the pituitary and are subject to different control systems. The prolactin-producing lactotrophs are regulated chiefly by an inhibitory factor from the hypothalamus; thus when the stalk connecting the latter to the pituitary is severed the secretion of prolactin is enhanced, whereas the secretion of growth hormone and most other pituitary hormones is decreased. The inhibitory effect of the hypothalamus is reproduced in isolated pituitary cells by added catecholamines such as dopamine, and the dopamine agonist, bromocriptine, which reduces growth hormone secretion and inhibits the release of prolactin. There is also evidence for a prolactin-releasing factor from the hypothalamus, with a stimulatory effect by serotonin or its precursor, 5-hydroxytryptophan; however, serotonin has no direct effect upon isolated pituitary cells and must evoke the release of some as yet unidentified factor from the hypothalamus.

As in the case of growth hormone, prolactin can elicit multiple actions in diverse biologic systems involving growth and reproduction in both males and females. The system that has received most attention is the process of lactogenesis in the mammary gland which involves extensive collaboration by prolactin with insulin, steroids, thyroid hormones, and hormones of pregnancy, and the effects of the intracellular hormonelike agents called prostaglandins. Prolactin with insulin and estrogenic steroids promotes the lobuloalveolar development in the breasts; after birth elevated levels of prolactin together with a decrease of ovarian steroids initiate the production of milk in the mother. Prolactin is necessary for the biosynthesis of the milk protein, casein, and of the regulatory protein, α-lactalbumin which is a subunit of the galactosyl transferase that produces the milk sugar, lactose. It is also necessary for the synthesis of milk fats and for regulation of milk electrolytes. The stimulation of milk protein synthesis by prolactin is preceded by increased transcription of ribosomal RNAs and messenger RNAs. As in the case of growth hormone action, the activity of ornithine decarboxylase and the intracellular concentrations of the polyamine end-products are greatly increased in lactating mammary glands by prolactin. Activation of membrane phospholipase A_2, which releases the polyunsaturated fatty acid precursors of the prostaglandins, is another early event in mammary gland cells exposed to prolactin, and added prostaglandins can mimick effects of prolactin in stimulating RNA synthesis. Other factors, such as the polyamines, are required for the concerted expression of gene translation to generate the milk proteins. The production of casein requires glucocorticoids in addition to prolactin, while synthesis of the α-lactalbumin regulatory subunit of lactose synthetase needs the addition of triiodothyronine; the stimulatory effects of prolactin on both milk proteins are blocked by the progestational steroid, progesterone. Prolactin potentiates both the use of dietary fat for milk synthesis by activating the mammary extracellular lipoprotein lipase that is required for uptake of lipoprotein fats from the circulation, and the *de novo* production of milk fat by activating the lipogenic enzymes of the gland in concert with insulin.

OXYTOCIN

As in the case of its companion hormone, vasopressin, oxytocin is a small peptide of nine amino acid residues (nonapeptide) which is stored in the posterior lobe of the pituitary gland (Fig. 3–4). Because of its content of nerve terminals and direct connection by nerve fibers with the brain, this segment of the gland is commonly referred to as the neurohypophysis or *pars nervosa*, as opposed to the *pars distalis* of the anterior lobe (now commonly termed the adenohypophysis) which produces the other pituitary hormones previously described, and which connects with the nervous system by portal blood vessels from the hypothalamus. Oxytocin produces its major effect on milk ejection through contraction of the mammary myoepithelial cells. The release of oxytocin from the neurohypophysis is triggered by the suckling reflex, in which stimulation of nerve endings in the nipple causes bursts of electrical activity in neurons of the hypothalamus that innervate the posterior pituitary, and thus leads to periodic secretion of the hormone. Secondarily, oxytocin has a function also in parturition; it is released in an abrupt surge in the final stages of labor, and is thought to participate chiefly in termination of delivery by contracting the uterus thereby preventing postpartum hemorrhage. The continued release of oxytocin during suckling may exert effects on the uterus or ovary that contribute to postpartum infertility in nursing mothers.

Hormones of the posterior pituitary, both oxytocin and vasopressin, are synthesized in the cell bodies of neurosecretory cells within the

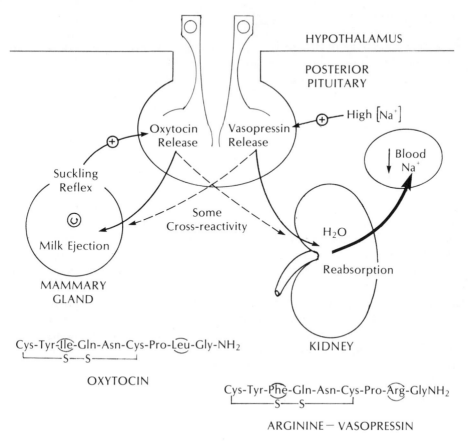

FIG. 3-4. Oxytocin and vasopressin structures and actions.

hypothalamus and are transported down the axons in tightly packed granules at the rapid rate of 3 to 5 mm/hr. The very high concentrations of hormones within the neurosecretory granules are maintained through formation of complexes with specific polypeptide carriers of molecular weights about 10,000 known as neurophysins. Each hormone is synthesized in a separate cell body simultaneously with its specific companion molecule of neurophysin, and the two form a one-to-one stable complex for packaging in the granules. Studies with labelled amino acid precursors are compatible with the production of a larger common protein precursor containing both the peptide hormone and its neurophysin. The complex remains associated during transport to and storage within the neural lobe of the pituitary. When the gland is stimulated both the labelled hormone and its neurophysin companion are released into the bloodstream, although they are not associated together in the complex form in the circulation.

VASOPRESSIN

This hormone is sometimes referred to as the antidiuretic hormone based on its ability to promote water conservation by the kidneys and thus reduce the volume of urine excreted. It differs chemically from oxytocin only by replacement of two amino acids, phenylalanine for isoleucine in position 3, and arginine for leucine in position 8 (Fig. 3–4). In view of the similarities it is not surprising that vasopressin exerts some, albeit weak, oxytocinlike responses, while oxytocin promotes a slight antidiuretic action. Vasopressin release is augmented on exposure of osmoreceptors in the hypothalamus to elevated plasma sodium ion concentrations. High environmental temperatures, decreased blood volume, decreased blood pressure, pain, other noxious stimuli, and heavy exertion may also trigger vasopressin secretion. On the other hand ethyl alcohol suppresses vasopressin release, which accounts for the polyuria, thirst, and dehydration of the "morning-after" syndrome. The antidiuretic effect of vasopressin is the result of its action in increasing the water permeability of cells in the collecting ducts and distal tubules of the kidneys. The peptide hormone combines with a tissue-specific receptor at the cell surface which stimulates adenylate cyclase in the membrane; the resulting increased levels of intracellular cyclic AMP presumably activate a protein kinase that catalyzes phosphorylation of membrane proteins which in turn enhance the permeability to water molecules. High concentrations of vasopressin elevate the blood pressure by inducing vasoconstriction, and stimulate the release of ACTH and TSH by the anterior lobe of the pituitary, but these effects may be pharmacologic rather than normal physiologic responses.

Deficiency of the antidiuretic hormone through lesions to the hypothalamic vasopressin-producing cells, or loss of the normal renal responses to the hormone (nephrogenic), result in the syndrome known as diabetes insipidus. In the absence of the normal concentrating process large volumes of dilute urine (20–30 liters per day) are excreted, and the patients suffer from continuous thirst. Unless water intake is increased to balance the urinary losses, the dehydration and hemoconcentration can result in cardiovascular collapse and death. The long-acting synthetic analogue with D-arginine at position 8 and lacking an amino group at position 1 is useful in reversing water losses of patients with vasopressin deficiency. The nephrogenic diabetes insipidus has shown beneficial responses to antidiuretic drugs such, as chlorpropamide, which acts to potentiate the sensitivity of the kidneys to endogenous vasopressin.

PANCREATIC HORMONES

Endocrine secretions of the pancreas represent one of the most important integrated systems for the control of metabolism in the body. Three distinct polypeptides have been characterized each arising from a different

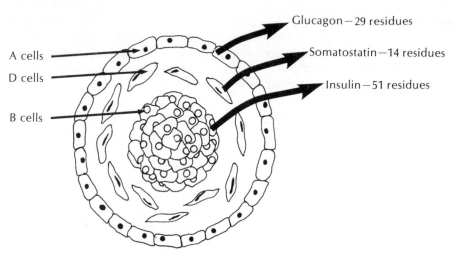

A cells

D cells

B cells

Glucagon — 29 residues

Somatostatin — 14 residues

Insulin — 51 residues

FIG. 3–5.   Pancreatic hormones and the cell types within islet tissue which produce them.

cell type within the pancreatic islets (Fig. 3– 5): the first comprised of 51 amino acids, is the anabolic hormone insulin, which is produced in the centrally assembled β or B cells; the second comprised of 29 amino acids, is the counteracting catabolic hormone, glucagon, which originates in the peripheral rim of α or A cells; the third comprised of a chain of only 14 amino acids is somatostatin, which has the capacity to restrict secretion of the other two hormones and is generated by the D cells which lie beneath the outer mantle of A cells. In each case the polypeptide hormones are released to the extracellular spaces by exocytosis, and make their way into the portal blood. Recent ultrastructural studies indicate the existence of numerous intercellular communications by gap junctions within the islet tissue so that functions of the different cell types may be coordinated within the gland. Whether intercellular communication takes place directly in this manner, or indirectly by the bloodstream, it is apparent that there is a high degree of orchestration of the hormones of the pancreatic islets. For convenience, we will consider the actions of individual hormones; however, we must keep in mind that it is not so much the isolated effects nor concentration of insulin and glucagon that determine the end result, but rather the interplay and balance between the pancreatic hormones and hormones of the adrenals, thyroid, and other endocrine organs.

INSULIN

Despite 60 years of intensive study of the chemistry and physiology of this major hormone there is still a good deal of mystery surrounding its actions in the body. In part this has arisen from the understandable emphasis on the hypoglycemic functions of the hormone in curing diabetes, and in part this is the result of the attempt to explain all insulin effects as the consequences of a single primary action on glucose metabolism. Moreover, although insulin has been considered the prototype of polypeptide hormones whose effects on cells are mediated by attachment to receptors on the cell surface, the subsequent responses with respect to intracellular cyclic nucleotides or other cytoplasmic "second messengers" are by no means clear cut. The effects of insulin seem to touch upon every aspect of cell function from membrane transport to differentiation. In addition to these scientific misconceptions about insulin there has developed an ill-founded popular complacency about its ability to eradicate the problem of diabetes. In fact diabetes, with its attendant complications, is still a serious contributor to sickness and mortality more than half a century after the epochal discovery of its "cure."

Insulin is synthesized on ribosomes of the rough endoplasmic reticulum in the precursor form called preproinsulin. This is a longer molecule which passes into the lumen of the endoplasmic reticulum, then into the Golgi system where the terminal end is cleaved off to yield proinsulin, which has a molecular weight of about 9000 (Fig. 3–6). Three –S–S– bridges are formed between cysteine residues in the molecule to produce a spiral-shaped single chain. In the next stage the prohormone is passed from the Golgi into storage granules, and simultaneously it is cleaved in two places by proteolytic attack to release a connecting segment (C peptide) from the middle of the molecule leaving the active insulin hormone. The latter consists of two polypeptides, the A chain of 21 residues from the carboxyl end of proinsulin, and the B chain of 30 residues from the amino end, still joined together by two of the –S–S– bridges and with a total molecular weight of 5734 (Fig. 3–6). In the storage granules 6 molecules of insulin are clustered around 2 zinc atoms to form a compact aggregate.

A major metabolic effect of insulin on cells is removal of sugars, amino acids, and fatty acids from the bloodstream for anabolic use. Accordingly, it seems logical that the surge of these nutrients in the splanchnic circulation after a meal is a major trigger for the release of insulin into the bloodstream. The main stimuli for insulin secretion by the pancreas are elevations of the blood glucose or amino acids, particularly arginine and lysine; other amino acids and fatty acids play a lesser role. Several of the gastrointestinal hormones (gastrin, secretin, cholecystokinin– pancreozymin) that are released in the process of digestion can also promote secretion of insulin, thus enabling the coordinated disposal of incoming nutrients. Stimulation of the parasympathetic nervous system (also associated with increased digestive activity) will, in turn, stimulate insulin release while the effectors of sympathetic activity, epinephrine and norepinephrine, are inhibitory. These controls allow for activation of mechanisms in the liver for the use of nutrients from the portal vein in anticipation of the high surges from absorption of a large meal. In addition

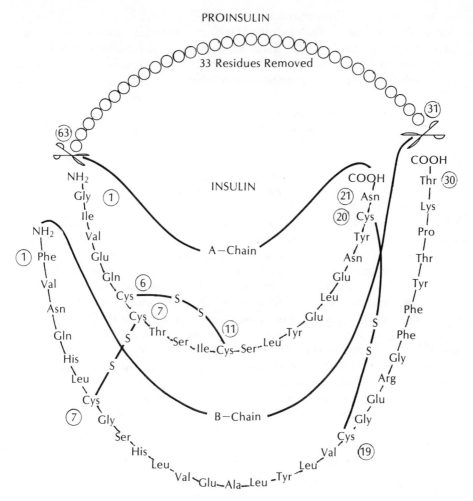

FIG. 3–6. Structure of insulin and its prohormone form—proinsulin.

to these interactions between the gut and the islet tissue, the so-called "enteroinsular axis," there is an integration of functions within the islets themselves. Thus glucagon stimulates the secretion of both insulin and somatostatin, while somatostatin decreases both insulin and glucagon secretion, and possibly the secretion of somatostatin itself. It is not clear to what extent these effects are mediated indirectly by the circulating hormones, or directly within the gland by intercellular communications of an internal or "paracrine" self-regulating system. Drugs of the sulfonylurea family, such as the oral antidiabetic agent tolbutamide, also have the property of effecting the release of insulin in increased quantities from islet tissue which may be unresponsive to normal physiologic stimuli.

In general, insulin acts on tissues to curb the breakdown of glycogen, fats, and proteins as evoked by catabolic hormones, notably glucagon, cortisol, and catecholamines; insulin also has a positive anabolic action in promoting biosynthesis of tissue macromolecules. Insulin also exerts effects upon the cell membranes to prevent glucose release, and enhances the cellular intake of glucose and amino acids. As noted in a later section (Chap. 4), cells that respond to insulin have specific receptors on the surface that bind the polypeptide. The nature of the internal signal arising from this hormone–receptor recognition is still problematic. There is evidence from studies with fluorescent-labelled insulin that some of the hormone may enter the cytoplasm of target cells, presumably following endocytosis of the hormone–receptor complex. Whether this internalization of intact hormone is involved in the transmission of insulin effects is not known. Instead, the process may represent a step in the removal of insulin, and activation of its intracellular proteolysis, which is quite rapid. Injected insulin is cleared from the circulation and degraded within minutes of its injection, most of the removal occurring within the liver and kidney. Internalization of insulin together with its receptor also constitutes a self-regulatory system in target cells whereby the numbers of surface receptors on the cells decrease as a consequence of elevated blood levels of the hormone. This process termed "down-regulation" renders the cells progressively less sensitive to further exposure to insulin. Responses of different organs to insulin are quite varied both in the quantitative and qualitative senses. For example, glucose uptake by muscle, fat, or white blood cells is regulated to varying degrees by insulin, but in liver, nerve, or red blood cells there appears to be no effect on glucose uptake. It is most convenient to consider the effects of insulin separately on the major tissue types whose metabolism is controlled by the hormone.

The liver is a focal point of insulin action (Fig. 3– 7). It is the direct receiver of nutrients, such as carbohydrates and amino acids, from gastrointestinal absorption, and of insulin and glucagon from splanchnic output of the pancreas. In fact, the concentration of insulin reaching the liver cells is several times greater than that in the systemic circulation bathing peripheral tissues since the liver degrades a substantial portion of insulin entering from the portal vein. Moreover, the liver, being the major seat of such metabolic processes as gluconeogenesis, glucose secretion, ketogenesis, ureogenesis, and fatty acid biogenesis, is the central integrative organ that balances the interconversions of major nutrients in relation to the physiological needs of the whole body. The following responses of liver cells to insulin are enhanced or accelerated: glucose phosphorylation, glycogen synthesis, glycolysis, reactions of the phosphogluconate pathway, citric acid cycle oxidations, and synthesis of fatty acids and triacylglycerols. The following responses of liver cells to insulin are blocked or inhibited: glucose release, alanine uptake, glycogenolysis, gluconeogenesis, lipolysis, fatty acid oxidation, and the formation of ketone

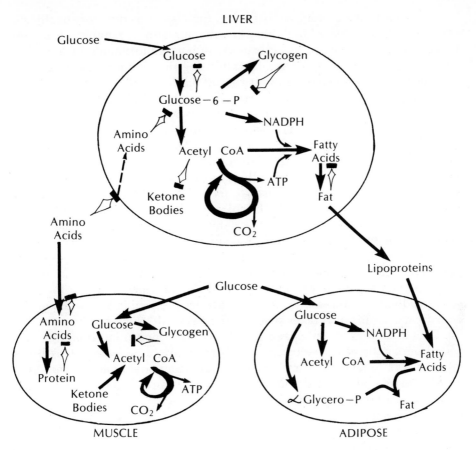

FIG. 3–7. Insulin effects in tissues. Processes indicated with heavy black arrows ➤ are promoted or accelerated by insulin; processes indicated with open arrows and a block ⤛ are inhibited by insulin.

bodies and urea. Thus, although insulin has no direct effect upon the transport mechanisms for glucose in the liver (hepatic cell membranes are freely permeable to glucose), the net effect is to draw glucose from the blood into the liver cytoplasm so that the oxidative use of glucose is intensified to generate ATP, NADH, and NADPH, and so that these energy sources are used to drive the conversion of glucose into glycogen and fats. Some of these insulin effects result from induced synthesis of enzymes (*e.g.*, glucokinase, glucose-6-phosphate dehydrogenase, enzymes of fatty acid synthesis), while some result from activation of existing enzymes (*e.g.*, glycogen synthetase). Insulin also represses the induction of enzymes of gluconeogenesis by cortisol (*e.g.*, phosphoenolpyruvate carboxykinase, fructose diphosphatase), and antagonizes the activation of some enzymes (*e.g.*, glycogen phosphorylase) by glucagon or catecholamines.

Adipose tissue is very sensitive to low concentrations of insulin because of the extensive, high-affinity receptors for the hormone on the adipocyte membrane. Insulin exerts an anabolic influence on fat storage in two main ways (Fig. 3–7): first, by inhibiting the mobilization of fat from the depots; second, by promoting the intake of circulating fats and carbohydrate to lay down more new fat. The primary effect, termed the antilipolytic action, results from antagonism of hormonal activation (by glucagon, catecholamines, ACTH, and cortisol) of the intracellular lipase in fat cells by very small amounts of insulin. This hormone-sensitive lipase cleaves triacylglycerols to form glycerol plus free fatty acids, the latter escaping from the cells and travelling in the blood complexed to albumin as a major energy-yielding substrate for liver, muscle, and other peripheral tissues. Insulin lowers the circulating free fatty acids by inhibiting this lipolytic process, and thereby shifts the energy metabolism of the tissues from fat to glucose use. Secondary anabolic effects are demonstrated with high carbohydrate diets that elevate blood glucose and insulin concentrations. Under these conditions glucose uptake and use in adipose cells are greatly stimulated, and the enzymes required for fatty-acid production from excess carbohydrate are induced in the fat depots. As noted earlier, this also takes place in the liver. Although the intracellular lipolysis is inhibited, an extracellular enzyme, lipoprotein lipase, is activated under the circumstances of high carbohydrate feeding. This lipase hydrolyses the triacylglycerols of the circulating chylomicrons that arise from intestinal absorption and hydrolyzes the triacylglycerols of the VLDL fraction of lipoproteins that arise from endogenous fat synthesis in the liver at the surface of the adipocytes to provide free fatty acids. The combined fatty acids from diet, from the liver, and from *de novo* synthesis in the depots, may now become esterified to form the storage triacylglycerols. Insulin also stimulates this latter stage of lipogenesis by accentuating glucose uptake and glycolysis, and by promoting the generation of energy and α-glycerophosphate. The latter is the precursor for the glycerol portion of triacylglycerols, and must be derived from the glycolytic intermediate, dihydroxyacetone phosphate, in adipose tissue.

Insulin also has anabolic effects in muscle cells by activating the uptake of glucose and amino acids following a meal (Fig. 3–7). Glycogen synthesis is activated in the same manner as it is activated in liver cells. The added influx of amino acids, particularly the branched-chain group (valine, leucine, isoleucine) that are poorly used by liver cells, stimulates the synthesis of muscle protein. At the same time insulin has a general anabolic effect upon ribonucleic acid synthesis, and a direct action upon the ribosomes to increase the efficiency of protein translation. Insulin also acts to oppose the effects of cortisol upon the breakdown of muscle proteins. In this way, it is not only anticatabolic with respect to muscle, but it is also antigluconeogenic with respect to liver since it curbs the release of alanine and other amino acid precursors from muscle for glucose production by hepatic cells.

Pinpointing the basic mechanism of insulin actions has proven to be difficult and contentious. For example, many effects, such as those upon glycogen metabolism and lipolysis, are antagonistic to those of hormones that act to elevate intracellular cyclic AMP. It has been shown in both liver and fat cells that insulin can lower the concentrations of cyclic AMP, possibly by stimulation of the phosphodiesterase enzyme which breaks down the cyclic nucleotide. The activation of protein kinase, which is necessary for expression of cyclic AMP effects in activating liver glycogenolysis and adipose lipolysis, would be blocked by insulin in this manner. However, the correlation between the lowering of cyclic AMP levels and insulin effects is not consistent with this being the sole mechanism of action of the hormone, particularly in muscle tissue where insulin seems to have no effect on cyclic AMP. An alternative explanation is that insulin–receptor complexes at cell surfaces may generate a different second messenger that antagonizes the cellular effects of cyclic AMP in a yin–yang fashion of two opposite principles. Such a postulate was proposed for the cyclic nucleotide, cyclic GMP, whose concentrations in liver and fat cells are augmented by exposure to insulin. A third suggestion is that insulin plays a role in modifying the intracellular composition of regulatory cations, such as Ca^{2+}. Some, but not all, of the effects elicited by insulin do require the presence of extracellular calcium ions, and there are also indications that insulin may alter Ca^{2+} flux and the cytoplasmic concentrations of this ion in several cell types. These studies are further complicated by the observation that binding of calcium to intracellular membranes and calcium uptake into cell compartments of the mitochondria or endoplasmic reticulum are also enhanced by insulin treatment. As for the findings with cyclic nucleotides, the effects of insulin on ion movements may be secondary rather than primary events.

It remains to be seen whether cyclic nucleotide or ionic interrelations can be rationalized into an integrated model for insulin action, or whether a still unidentified intracellular messenger is involved. In muscle tissue, insulin renders the protein kinase unresponsive to the normal stimulus by cyclic AMP. This could explain how insulin activates glycogen synthesis by keeping the synthetase enzyme in its active dephosphorylated form, and inhibits glycogenolysis by preventing conversion of inactive phosphorylase b to activated phosphorylase a without affecting the levels of cyclic nucleotide. Insulin also produces effects on adipose tissue that are not mediated by the adenyl cyclase system. The enzyme, pyruvate dehydrogenase, in the adipocyte mitochondria which is the major generator of acetyl CoA for fatty acid synthesis is inactivated by a cyclic AMP-independent protein kinase. As in the case of muscle glycogen synthetase, treatment of the fat cells with insulin converts the mitochondrial pyruvate dehydrogenase into the active dephospho-form to provide the carbon source for lipogenesis. In this instance the activation may be brought about not through effects on the kinase, but by stimulation of a specific phosphatase. It is interesting that the pyruvate dehydrogenase phosphate phosphatase is activated by low levels of Ca^{2+} inside the mitochondria. Current investigations indicate that the combination of insulin with its receptor at the adipocyte plasma membrane results in the intracellular release of a small polypeptide, presumably through activation of a specific protease. The polypeptide then may serve as the second messenger to stimulate the mitochondrial phosphatase which turns on pyruvate dehydrogenase, promoting the utilization of carbohydrates for fat synthesis.

The important effects of insulin on cellular transport processes may involve direct modification of the membrane components by processes different from those that influence the intracellular enzymes. The ability of insulin to stimulate glucose uptake in fat or muscle cells is prevented if certain of the free sulfhydryl groups on the cell surface are chemically blocked. The reaction of insulin with the membrane system appears to result in oxidation of the sulfhydryls to the disulfide form associated with the activation of glucose or amino acid transport into the cells. Other modifications of membrane components by insulin treatment have also been reported, including phosphorylation–dephosphorylation of membrane proteins and increased turnover of the membrane phospholipids, but as for so many insulin effects it is unclear whether such changes are responses to the primary signal or indirect secondary events in a chain of hormonal effects.

GLUCAGON

This pancreatic hormone consists of a single polypeptide chain with a molecular weight of 3647 (Fig. 3–8). It contains all of the amino acids in proteins except for proline, isoleucine, and cysteine; consequently, it is incapable of forming disulfide bridges such as those formed by insulin. Several larger polypeptides with glucagon-like immunologic properties (GLI) but without hormonal activity have been demonstrated in blood plasma, and constitute about half of the immunoreactivity. Most of the GLI seem to have an extrapancreatic origin, and can be demonstrated histologically in cells of the gastrointestinal tract. As for insulin, there is evidence that glucagon is synthesized as a larger precursor with molecular weight of about 9000. The active hormone is transported through the Golgi system and stored in granules of the A cells following cleavage from proglucagon.

Glucagon is secreted from the islet tissue under basal conditions at a rate much lower than that for insulin. The secretion of glucagon seems to be repressed by the normal levels of circulating glucose, may be decreased further when glucose concentrations rise following a meal, and is stimu-

H₂N -His -Ser -Gln -Gly -Thr -Phe -Thr -Ser -Asp -Tyr
       1 |
          Gln -Ala -Arg -Arg -Ser -Asp -Leu -Tyr -Lys -Ser
          | 11
          Asp -Phe -Val -Gln -Trp -Leu -Met-Asn -Thr -COOH
          21   29

FIG. 3–8.   Structure of glucagon.

lated under fasting conditions of hypoglycemia. The response is, therefore, opposite to that for insulin secretion, with the result that insulin/glucagon ratios in the portal circulation will be a greatly amplified signal of blood glucose levels. Glucagon secretion also responds inversely to insulin secretion with respect to the effects of the autonomic nervous system; stimulation of sympathetic nerves increases glucagon output while inhibiting insulin secretion. A similar signal for increased output of both pancreatic hormones is the concentration of circulating amino acids. Arginine markedly stimulates secretion of both hormones, leucine specifically elevates insulin secretion, and alanine activates secretion of glucagon but not insulin. The rationale for these controls is apparent in relation to the body's adaptation to food intake and activity. High carbohydrate meals, while evoking more insulin to promote glucose use by liver and peripheral tissues, will suppress the glucagon-mediated elevation of blood sugar. On the other hand, stressful stimuli resulting in catecholamine release will enhance glucagon secretion and thus raise the blood glucose as required for metabolism of peripheral tissues in order to meet the stress, while suppressing insulin secretion. Probably the most important single function of glucagon takes place in response to a high protein meal. In such a situation the entry of a large bolus of amino acids in the absence of ingested carbohydrate stimulates the β cells to release insulin and thus could lower the blood sugar to dangerous levels. However, the concomitant stimulation of α-cell glucagon secretion by amino acids ensures that liver glucose production will be adequate to prevent hypoglycemia following protein absorption, while allowing the increased cellular uptake of amino acids and their anabolic use as stimulated by insulin.

Glucagon exerts effects on the liver predominantly by interaction with specific receptors on the hepatic cell surface resulting in activation of the membrane-bound adenyl cyclase to produce intracellular cyclic AMP. The latter promotes conversion of phosphorylase from the inactive b form to the active a form by cascade activation of cyclic AMP-dependent protein kinases. Consequently, the major action of glucagon on liver is an immediate degradation of glycogen with release of glucose into the circulation. Additional effects of glucagon upon liver metabolism include an increased conversion of amino acids to glucose by the reactions of gluconeogenesis, and an increased oxidation of fatty acids to produce ketone bodies. These secondary effects are potentiated by the lipolytic activity of glucagon with increased mobilization of fatty acids for oxidative energy production to drive the reactions of gluconeogenesis. In this way glucagon not only promotes an instantaneous response to counteract acute hypoglycemia by glycogenolysis, but it also induces a long-term adaptation to prolonged fasting by accelerating gluconeogenesis. The former response depends on the insulin–glucagon balance, while the latter also depends on presence of the steroid hormones (*e.g.*, cortisol) that are essential for induction of gluconeogenic enzymes.

DIABETES MELLITUS

This term applies to a complex group of metabolic syndromes in which the use of carbohydrate, fat, and protein metabolites is disturbed resulting in the accumulation of glucose, fatty acids, ketone bodies, amino acids in the bloodstream in concentrations much above normal. The presence of hyperglycemia with spill-over of glucose in the urine is the characteristic diagnostic indication of this disease. In some instances blood glucose may not be increased in the fasting state, and the abnormal glucose use will then be detectable only following a test carbohydrate meal. In normal individuals blood glucose concentrations increase in the first 30 to 60 minutes after a large glucose load (50–100 g), but the stimulation of insulin secretion brings the level back to normal within 2 hours. The diabetic responds abnormally to this test by maintaining glucose concentrations at elevated values for several hours.

In general the condition of diabetes may be attributed to faulty insulin synthesis, insulin secretion by the β cells of the pancreas, or an impairment in the response of target organs to the hormone. A number of toxic chemicals (alloxan, excess iron storage) or virus agents may play a role in the damage to pancreatic islet tissue. A complex interrelation of genetic and dietary factors may also contribute to the etiology of anomalies in insulin production and function. Lack of peripheral response to insulin is a particularly insidious form of the disease and may remain latent for years. Moreover, long-term exposure to elevated circulating insulin causes further depletion of target cell receptors (down-regulation) thereby exacerbating peripheral insensitivity to the hormone.

Patients with diabetes can usually be divided into two classes: the "juvenile" form, which is a severe disease usually detected early in life and the "adult" or "mature-onset" form, which is a milder condition that may continue undetected for years. Juvenile diabetics typically suffer from marked insulin deficiency, and blood insulin levels do not respond readily to stimulation of islet tissue. Patients display hyperglycemia even in the fasting condition. They also tend to accumulate ketone bodies in the bloodstream, and often excrete these metabolites in urine, which is grossly increased in volume (polyuria). Juvenile diabetics are typically thin or wasted in appearance as though their bodies are catabolizing their own tissues. Control of carbohydrates and fluid intake with continuous insulin injections is required, and with time, even under the best of medical care, patients may develop serious pathological consequences of the vasculature causing blindness, atherosclerosis, and kidney involvement. Complications are less severe in the mature-onset cases, and the disease usually develops more slowly. In these adult patients the plasma often contains normal or supernormal levels of insulin in the basal state. The defect may lie in the target tissues, which are lacking in sensitivity to insulin, or in the islets, which fail to release more insulin in response to elevated blood glucose. Such patients will be easier to bring under control by dietary management or the administration of the oral insulin-secretory stimulants such as tolbutamide. There may be no need to administer insulin to such patients; in contrast to juvenile patients, adult-onset patients have fewer pathological complications, show lower danger of developing ketoacidosis, and usually appear well-nourished if not overweight. The correlation of obesity with adult diabetes is understandable in the sense that excess food intake over a period of years will place an excessive demand on a marginally functional endocrine pancreas and will lead to unmasking of the condition that was latent in early life. Chronic elevation of blood glucose leads to increased effort by the $\beta-$ cells to produce and release insulin. The peripheral tissues which may already possess low numbers of effective insulin receptors are rendered less responsive to the hormone by down-regulation, with consequent failure to utilize blood glucose. Thus a vicious cycle of accentuated hyperglycemia and further stimulation of insulin production ensues with resulting strain upon the secretory function of the pancreatic islets. Eventually, the obese adult diabetic will have a high probability of developing cardiovascular problems such as atherosclerosis.

Although traditionally considered to be a primary disease of carbohydrate metabolism, diabetes clearly involves every facet of the regulation of metabolic processes in the body. The characteristic hyperglycemia is chiefly attributable to an increased output of glucose by the liver relative to the needs of peripheral tissues. This is largely because of the unchecked actions of glucagon and epinephrine in activating glycogenolysis. With a high carbohydrate meal the failure of the islet tissue to release insulin in response to incoming glucose leads to failure of the liver to take up the excess. Incidents of infection, emotional stress, and other factors which activate the sympathetic nervous system will exacerbate the hyperglycemia. The unchecked gluconeogenic action of glucagon and cortisol results in enhanced uptake of alanine and other amino acids by the liver for conversion to glucose. The ensuing flow of amino acids from peripheral tissues to the liver leads to depletion of body proteins. This is most evident in the skeletal muscles where insulin deficiency interferes with normal ribosome function in protein synthesis; the uptake of amino acids into the muscles from the blood is also impeded. The latter accounts for

the enhanced level of branched chain amino acids in diabetic blood since
the liver metabolizes them inefficiently, and skeletal muscle is the normal
site of their disposal. Absence of insulin also prevents the normal uptake
and use of glucose by the muscles and adipose tissue. The decreased availability of glucose for generation of α-glycerophosphate also impairs the
deposition of triacylglycerol in fat cells. The anomaly in fat deposition is
apparent at several levels: first, in the liver where the ability to synthesize
fatty acids from glucose is impaired; second, in adipose cells, where the
unchecked stimulation of lipolysis by hormones that are normally antagonized by insulin leads to increased fatty-acid release into the
bloodstream; finally, in the liver where the increased mobilization of fatty
acids from the depots promotes a heavy reliance on β-oxidation as the
energy source for the accelerated gluconeogenesis. The increased circulating free fatty acids also inhibit glucose transport into peripheral tissues
contributing to the hyperglycemia. Not all of the above anomalies are seen
in every patient. The metabolic defects grow progressively worse with
duration of the disease. For example, the obese, adult diabetic at first
diagnosis may have normal fasting blood glucose, but is incapable of deal-
ing with carbohydrate meals, and shows abnormal tolerance in the postprandial period. Such a patient may also have normal or even increased
blood insulin levels following the meal. Accordingly, the adipose tissue
receiving increased glucose and insulin may deposit fat quite effectively
thus maintaining the obese state. Of course, in these cases the obesity is not
a consequence of the disease state *per se*, but rather may be viewed as an
indicator of the excess dietary intake that contributed to the development
of the disease state by placing increased demands on the malfunctioning
islet cells.

The condition of diabetic ketosis is a signal of failure to control metabolic imbalances, and is seen only in the severe juvenile form of the disease. Accumulation of excess ketone bodies (acetone, acetoacetate,
β-hydroxybutyrate) is the result of several interrelated factors. A major
consideration in the diabetic is the accentuated mobilization of fatty acids
from the fat depots, and the preponderant oxidation of fatty acids by hepatic cells in relation to the catabolism of glucose. Failure to provide sufficient regeneration of intermediates (*i.e.*, oxaloacetate) from carbohydrates
may thus allow for insufficient flux through the citric acid cycle to consume the acetyl-CoA derived from fat oxidation. In this way ketone bodies
may originate in excessive amounts from the build-up of end-products
from incomplete fatty acid catabolism. Apart from this indirect effect there
is also evidence to suggest that insulin can normally repress the reactions
of ketogenesis directly, so that insulin deficiency may play a double role in
promoting increased synthesis of ketone bodies by the liver. The muscle
tissue, normally the major consumer of ketone bodies, also shows evidence
of a defect in insulin deficiency. Thus the ketoacidosis may reach crisis

proportions in the uncontrolled diabetic as a result of combined over-production and under-utilization in the periphery.

The dangers of diabetes arising from the effect of either chronic excess glucose or lowered insulin include a thickening of basement membranes in blood vessels with attendant vascular damage, hyperosmolar coma caused by fluid being drawn from the extracellular spaces in acute hyper-glycemia, or ketoacidotic coma caused by excess accumulation of the ketone-body metabolites producing imbalances of acid–base regulation. An additional danger of diabetes arises from overtreatment with insulin resulting in a hypoglycemic coma. It is imperative to distinguish the cause of the comatose condition in a diabetic by determination of blood glucose levels before instituting the appropriate emergency treatment. Obviously insulin injection in the hypoglycemic patient would be fatal, while glucose administration would exacerbate the hyperosmolar or ketoacidotic condi-tion. Besides treatment with regulation of dietary carbohydrate intake and administration of insulin by injection or oral antidiabetic agents, the com-pensation for fluid and electrolyte imbalances is also essential in manage-ment of diabetes. Current investigations are examining the possibility of controlling the levels of glucagon in diabetics. While an excess of glucagon by itself is not a cause of diabetes, there is no doubt that with deficient or marginal insulin the presence of glucagon is a key factor in promoting hyperglycemia and in exacerbating the ketotic state. Since the normal restraints of glucagon production by insulin or somatostatin may be in-adequate in diabetes, approaches are being explored to expose the α cells of the islets to suitable analogues of these polypeptides so that the effects of excess glucagon can be eliminated. Much research remains to be done before medical science can cure the diabetic patient by replacement of a normal, fully automated, feedback-controlled endocrine pancreas.

ADRENAL– CORTICAL HORMONES

The outer portion, or cortex, of the adrenal glands produces a series of fat-soluble steroid hormones that are related structurally to and derived metabolically from cholesterol. The main secretion from the adrenal cor-tex in man is cortisol. Because of its primary role in carbohydrate metabo-lism this steroid is termed a glucocorticoid. Another class of steroid hor-mones, typified by aldosterone, is secreted in smaller amounts in response to imbalances of salt metabolism in the body, and hence they are termed the mineralocorticoids. A third class of steroid hormones, which is pro-duced in large quantities by the adrenal cortex, is the androgen, dehy-droepiandrosterone, that exerts weak male sex hormone effects. In addition to the physiological actions of the adrenal–cortical hormones when pres-ent in normal concentrations, cortisol and related synthetic steroids exert

H₂N -Ser -Tyr -Ser -Met- Glu -His -Phe -Arg -(Tyr) -Gly
       1   9 |

   Val -Pro -|Arg -Arg -Lys -Lys| -Gly -Val -Pro -Lys
      | 18   15   11

   Lys -Val -Tyr -Pro) -(Asp -Ala -Gly -Glu -Asp -Gln
   21   24   25 |

  HOOC -Phe -Glu -Leu -Pro -Phe - Ala -Glu -Ala -Ser
             39   31

FIG. 3–9.   Structure of adrenocorticotrophin (ACTH).

profound anti-inflammatory responses when administered in supranormal pharmacologic doses.

ACTH EFFECTS

The adrenal cortex is under the controlling influence of the anterior pituitary. The latter produces a linear polypeptide, adrenocorticotrophic hormone (ACTH), that maintains the size and weight of adrenal– cortical tissue, and also stimulates the synthesis and secretion of the corticosteroid hormones. Although the natural ACTH molecule contains 39 amino acid residues (Fig. 3– 9), it appears that only half the molecule is required for the biologic responses to occur. The use of modified and synthetic derivatives of ACTH has shown that the "active" site is localized in the first 10 residues from the NH₂ terminus, with the tryptophan residue at position 9 as an essential component for activation of the adrenal gland; residues 11 through 24 furnish the "binding" site, with a particularly important contribution to the affinity from the basic Lys– Lys– Arg– Arg sequence in positions 15 through 18. The COOH terminus containing amino acids 25 through 39 enhances the biologic potency by protecting the ACTH molecule from proteolytic breakdown. Even so, intact ACTH is removed from the circulation quite rapidly, and has a half-life of only several minutes.

  Secretion of ACTH occurs in response to stimuli originating in the nervous system and finding expression in the secretion of a humoral factor by the hypothalamus. This factor (or factors) has been termed the corticotrophin-releasing factor (CRF). A number of small peptides with molecular weights between 1000 to 1500, and with the ability to stimulate release of ACTH from the anterior portion of the pituitary have been demonstrated in the hypothalamus but a number of CRFs with molecular weights up to 30,000 are also present. Possibly the multitude of peptides represent precursors of the true hormonal factor whose structure is awaiting definition. The CRF is produced when the individual is subjected to conditions of stress, either emotional (*i.e.*, anxiety, depression) or physical (*i.e.*, surgical trauma, cold exposure) in origin (Fig. 3– 10). Together with

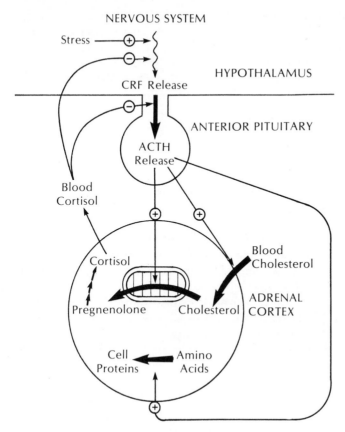

FIG. 3–10. Release of adrenocorticotrophin (ACTH) from the anterior pituitary. Factors which stimulate release are indicated by a positive sign; those which inhibit by a negative sign. Processes in adrenal cortex which ACTH stimulates are indicated by heavy black arrows.

the anterior pituitary to which it is connected by a portal blood system, the hypothalamus acts to integrate the production of the adrenal corticoids as needed for the person's adaptation to stressful situations.

The hypothalamic–pituitary–adrenocortical system functions normally as an efficient, self-regulated, servomechanism. The glucocorticoid hormones, when present at high levels in the circulation, act as negative feedback inhibitors of ACTH release, and turn off their own synthesis. The self-limiting action of the glucocorticoids occurs at several points: as a primary effect on the hypothalamus and higher centers in the nervous system to depress the discharge of CRF into the portal blood supply to the pituitary; as a direct effect upon the pituitary cells to render them less responsive to CRF stimulation and hence to block ACTH secretion; as an effect on the adrenal cortex itself to inhibit secretion of glucocorticoids. The feedback system has long-term effects in addition to short-term control of glucocorticoid levels. For example, loss of one adrenal gland or deterioration of the function of cortical tissue will lower circulating glucocorticoids, and hence remove the repression of ACTH stimulation. The result will be hypertrophy of the surviving adrenal glandular tissue with compensatory increases of glucocorticoid production. Conversely, the administration of large amounts of corticosteroids to minimize the inflammatory response of the body, or to suppress antibody reactions during tissue transplantation, will result in adrenal atrophy and a decreased ability of the patient to react to the stress of acute illness when steroid therapy is stopped. In order to maintain normal adrenal function it is necessary to provide intermittent doses of glucocorticoids, and to taper off treatments with steroids in order to minimize severe withdrawal symptoms.

The concentrations of ACTH and cortisol in human blood follow a daily rhythmic pattern that is dependent on activity habits. Normally the levels of both hormones are highest in early morning, conducive to mobilization of energy on awakening, and lowest during the night, conducive to sleep and synthetic processes of repair and restoration in the tissues. This diurnal fluctuation of ACTH secretion can be shifted in individuals who switch to nocturnal activity following some days of adjustment. No doubt some of the fatigue and malaise in long distance passengers ("jet-lag") is attributable to acute disruption of the normal relation between activity and circulating cortisol concentrations.

Actions of ACTH require a specific binding of the polypeptide to receptors on the exterior membranes of adrenocortical cells; there is no evidence for entry of ACTH into the cells, and potent steroidogenic activity can be evoked by ACTH while it is combined with large impermeable structures. However, shortly after its interaction with the membrane the ACTH molecule is inactivated, and at the same time it initiates a membrane response which results in the generation of a second messenger within the adrenal cell. A number of candidates for the role of second messenger have been considered, including cyclic nucleotides, divalent cations, and prostaglandins, but the situation is complicated by the multitude of biologic effects exerted by ACTH, and the difficulties in relating them to the primary response at the membrane.

Among the changes affected by exposure to ACTH are an acceleration of glycogen breakdown and increased catabolism of hexose phosphates in the pentose pathway. Since these metabolic effects could be related to activation of adenylate cyclase and cyclic AMP-stimulation of glycogenolysis, and since the use of glycogen by the pentose path could generate the NADPH required for steroid synthesis, it seemed that a consistent theory for ACTH action was at hand. However, later studies have shown that ACTH can stimulate steroidogenesis even when adrenal cells contain no glycogen or when NADPH is not rate-limiting. It has been verified that ACTH can promote rapid increases of cyclic AMP concentration in adrenal cells in the first few minutes prior to increased corticosteroid production, and that cyclic AMP and its active analogues, or agents such as theophylline that increase intracellular cyclic AMP by blocking phosphodiesterase, can mimic the action of ACTH in stimulating steroidogenesis. However, there does not appear to be a strong correlation between the effects of ACTH on adrenal hypertrophy and cyclic AMP levels. The growth-promoting influences of ACTH seem to require combined actions of cyclic AMP and cyclic GMP. Moreover, many of the effects of ACTH require simultaneous presence and uptake of calcium ions into adrenal cells. It may be that, as in the case of insulin action, the transport of Ca^{2+} within the intracellular compartments is an integral feature of ACTH effects; several enzymes of steroid hormone synthesis are activated by calcium ions. Finally, it has been shown that prostaglandins are formed in increased amounts by adrenal cells exposed to ACTH, and that prostaglandins can stimulate the secretion of corticosteroids without changing the levels of cyclic AMP in the cells.

Whatever the mechanism of ACTH-mediated effects may be, it is clear that ACTH promotes protein synthesis in the adrenal cells, and that these newly synthesized proteins are essential for the ACTH response; addition of inhibitors of translation, such as cycloheximide or puromycin, block the ACTH effect, while inhibitors of transcription, such as actinomycin D, have no effect. ACTH has also been reported to accelerate the uptake of cholesterol from blood plasma into adrenal cells, and from the adrenal cytoplasm into the mitochondria where initial steps in steroid synthesis occur. The rate-limiting desmolase enzyme in the mitochondria is also reported to be activated by ACTH-treatment of the cells. An interesting but unexplained action of ACTH is its ability to deplete the adrenal gland of its high content of ascorbic acid. Since no specific role for the vitamin has been demonstrated for adrenal function or steroidogenesis, and vitamin C deficiency leads to high–normal output of corticosteroids, the physiological significance of ascorbic acid depletion by ACTH is a mystery.

CORTICOSTEROID SYNTHESIS

The synthesis of corticosteroids occurs by a series of modifications of the cholesterol molecule. The latter may be derived either by *de novo* synthesis in the adrenal from acetyl-CoA or by uptake from the blood (Fig. 3– 10). Incoming cholesterol is the favored substrate for basal corticosteroid production in the unstimulated adrenal. However, when maximal rates of steroidogenesis are induced by ACTH, the endogenous fatty acyl esters of cholesterol stored in the gland as lipid droplets are hydrolysed to release the extra cholesterol required, as well as the polyunsaturated fatty acid precursors of prostaglandin formation. Some steps in the steroid pathway occur in the mitochondria, and some occur in the cytoplasm, with intermediates shuttling back and forth between these cellular compartments. The mineralocorticoid, aldosterone, is generated in the outer-most layer of cortical cells (*zona glomerulosa*) of the adrenal cortex while glucocorticoid and androgen production are confined to the inner layers (*zona fasciculata* and *zona reticularis*).

In the first stage cholesterol is taken up by the mitochondria and converted to the intermediate, pregnenolone, by cleavage of six carbons from the side-chain (Fig. 3–11). This complex reaction, catalyzed by the desmolase enzyme system, requires a cytochrome (P-450), NADPH, and molecular oxygen. It is the key reaction for regulation of steroidogenesis by ACTH in the adrenal, and by gonadotrophic hormones in the ovary and in the testis. Pregnenolone then migrates out of the mitochondrion to the endoplasmic reticulum where it is oxidized and isomerized to the important ketone intermediate, progesterone. In the production of cortisol three sequential hydroxylations of progesterone occur at positions 17 and 21 in the endoplasmic reticulum, and then, following another shuttle back to the mitochondria, at position 11. Each of these hydroxylase enzymes requires cytochrome P-450, NADPH, and O_2. It is noteworthy that the 11-β-hydroxylase enzyme is found only in adrenal cortex cells, the sole source of 11-OH glucocorticoids in the body. The 17-α-hydroxylase is found in inner zones of adrenal cortex and in the ovary and testis, but is absent from the *zona glomerulosa*. The latter layer of the cortex generates aldosterone by specific reactions of hydroxylation, and is followed by oxidation of the methyl group at position 18 to produce the aldehyde group of the mineralocorticoids. Androgens are formed prior to the introduction of a hydroxyl group at position 21, and, they are secreted in amounts comparable with the glucocorticoids. Congenital deficiency of the 21-hydroxylase in adrenal cortical tissues leads to disproportionate synthesis of the androgenic steroids with pronounced virilization at an early age. This autosomal, recessively inherited trait is termed the adrenogenital syndrome, and is the most commonly encountered disorder of adrenal function in pediatric practice.

GLUCOCORTICOIDS

The glucocorticoids exert a host of metabolic effects in the body designed to provide long-range adaptations to food deprivation; this is in contrast to the short-term regulatory effects of insulin–glucagon interactions. Virtually all cells possess receptors for glucocorticoids, and yet the responses are characteristically different for each tissue. In some instances the primary result is catabolic (lymphoid tissue, skin), in others anabolic (liver, mammary gland), while with respect to adipose tissue the glucocorticoids may cause mobilization of fatty acids from one site but deposition of fat in other sites. Other paradoxical effects of glucocorticoids are their well-documented role in the maintenance of muscular strength and integrity

FIG. 3–11.   Pathways and enzymes of corticosteroid biosynthesis.

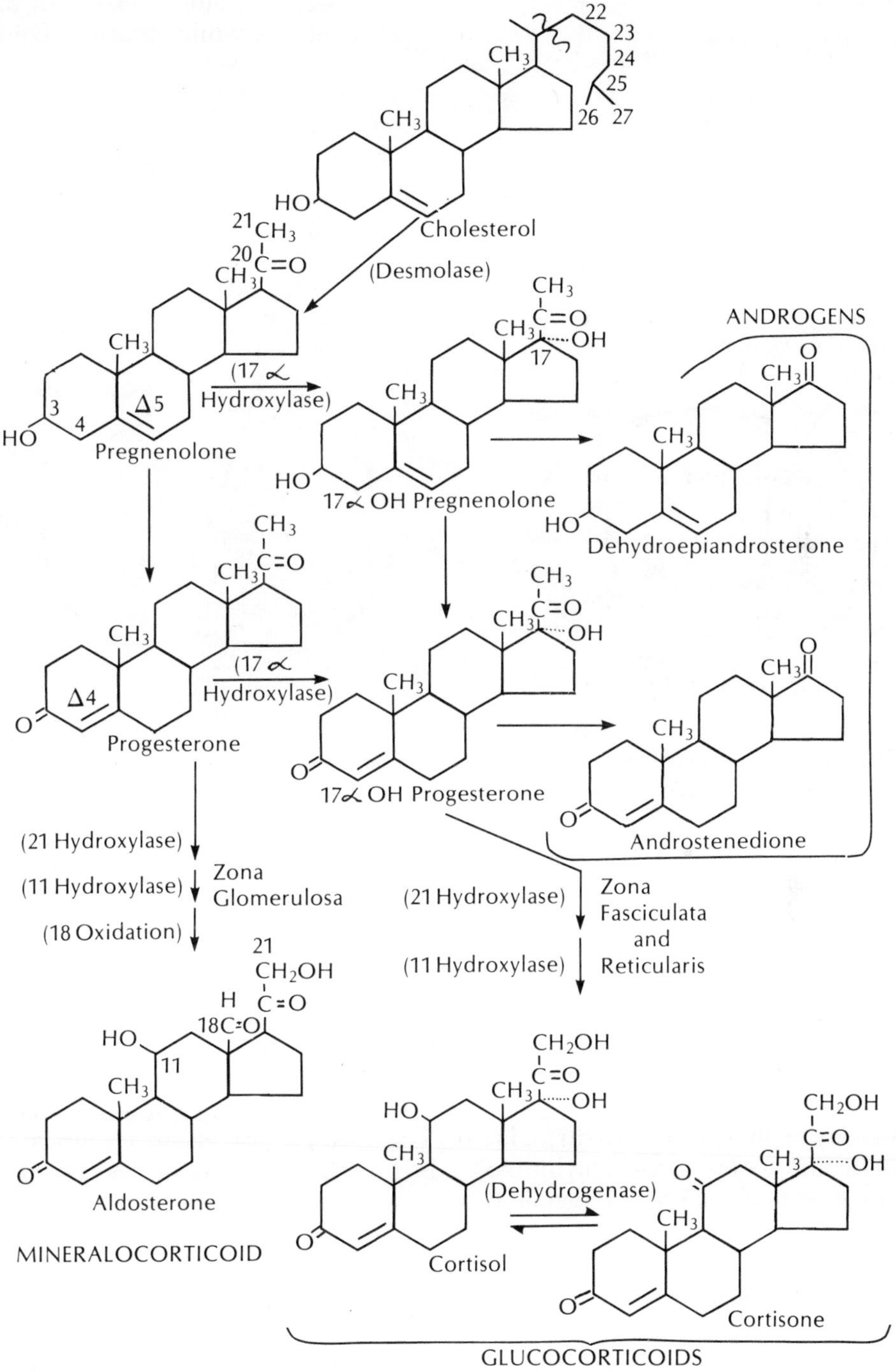

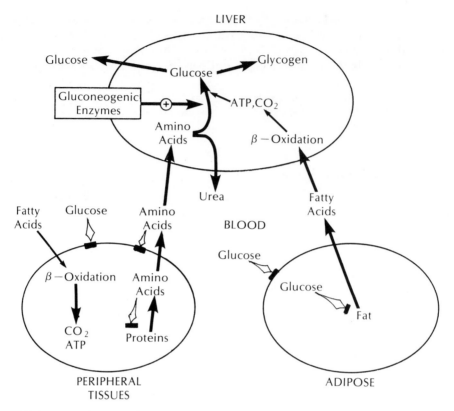

FIG. 3–12. Glucorticoid effects in tissues. Heavy black arrows show processes that are stimulated, open arrows show those that are inhibited.

while simultaneously promoting catabolic effects on the proteins of skeletal muscles. One may discuss the general effects of glucocorticoid actions on metabolism in terms of breaking down and releasing the endogenous energy stores from peripheral tissues to furnish the liver with adequate precursors for its glycogen stores and gluconeogenic reactions; this allows for the maintenance of blood glucose levels in the face of starvation or severe demands of stress.

The catabolic effects of glucocorticoids on peripheral tissues are not readily explained by any single factor. The most general effect is inhibition of the uptake of glucose and amino acids from the bloodstream (Fig. 3–12). In skeletal muscle the shortage of glucose results in a shift to fatty-acid oxidation to provide the energy for contraction, while the shortage of amino acids contributes to a slowing of protein synthesis. In addition there is a direct effect of glucocorticoids in slowing protein anabolism, and in some manner activating muscle-protein breakdown with a release of free amino acids into the bloodstream. Similar effects result in the breakdown of proteins, such as collagen, in the skin and bone, and proteins in lymphoid tissue (lympholytic response); the thymus gland is particularly sensitive, and up to 95% of its mass may be lost within 48 hours of a large dose of glucocorticoids. Adipose tissue generally reacts with a lipolytic response to glucocorticoids (Fig. 3– 12). Ability of the hormone-sensitive lipase to respond to lipolytic agents, such as epinephrine, is potentiated by prior exposure of fat cells to cortisol. The inhibition of glucose uptake by glucocorticoids also impairs the ability of fatty tissue to synthesize triacylglycerol. The result is a net outpouring of fatty acids into the circulation.

The liver is thus provided with increased amounts of the amino acids necessary to provide the carbon source, and of the fatty acids required to generate the ATP and reducing energy for gluconeogenesis. As part of the adaptive change in the liver the glucocorticoids induce the synthesis of a number of enzymes that are either directly or indirectly involved in the gluconeogenic process (Fig. 3– 12). In the latter category are enzymes required for catabolism of the aromatic amino acids, (e.g. tyrosine aminotransferase and tryptophan pyrrolase). Both enzymes are increased several-fold a few hours after the administration of cortisol. The increases can be shown to be the result of de novo synthesis of new enzyme molecules rather than the activation of existing enzymes. The induction is prevented by inhibitors of mRNA synthesis and of protein synthesis. Adaptive increases are also seen in the activities of enzymes of the urea cycle required to remove the nitrogen waste from excess amino acid breakdown. Several liver enzymes that are on the direct gluconeogenic pathway are also increased by glucocorticoids, (including glucose-6-phosphatase, fructose-1,6-diphosphatase, pyruvate carboxylase, and phosphoenol pyruvate carboxykinase). In addition, the conversion of carbohydrate to glycogen in liver is enhanced by the glucocorticoids. The effect appears to be mediated by stimulation of the phosphatase that converts glycogen synthetase into its active form. Thus the combination of increased substrate flux and activation and induction of the enzyme systems results in conversion of amino acids to glucose and then to hepatic glycogen. The latter may be broken down to provide free glucose when blood sugar levels fall, and the resulting elevation of glucagon in the circulation triggers glycogenolysis by the liver.

The model that has been proposed for the mechanism of glucocorticoid actions is similar to that proposed for estrogens and other steroid hormones: (1) the hormone enters the cell and combines with a specific cytoplasmic protein receptor molecule; (2) the receptor, when complexed with glucocorticoid, is activated to a form that will translocate into the nucleus; (3) the receptor–hormone complex interacts with selective regions of the genome to induce transcription of mRNAs corresponding to effector proteins. While the overall outline of this model is entirely consistent with glucocorticoid actions on target cells, there are still many gaps in our knowledge of the different receptors of various tissues and the nature of the discrete interactions of these molecules with particular stretches of the DNA molecule in the gene. Great advances in our understanding of the molecular basis of these effects have recently been achieved through the study of enzyme induction in hormone-sensitive liver cells growing in culture. Despite this fine resolution of the events surrounding control of the genetic machinery by steroids, there remain many perplexing mysteries at the physiologic level of the whole organism (*i.e.*, the molecular bases for protein catabolism in the muscles and lymphoid tissues).

The liver cytosol-receptor for glucocorticoids has been isolated and characterized as a protein with specific, high-affinity, saturable binding for those natural and synthetic steroids that possess biologic activity. The receptor interacts with the β side of the steroid molecule, requiring the 3-keto group, the 11-β-OH, and the α-ketol $C_{20,21}$ side-chain of cortisol. Substitution of the 11-OH by a ketone, as in cortisone, interferes with binding as it does with glucocorticoid potency in whole animals, while removal of the 11-oxy function eliminates both binding and biologic activity. Synthetic derivatives with enhanced potency also show good correlation between their biological activities and binding-affinities to the receptor. Introduction of a second double bond in the A ring, as in prednisol, and a fluorine atom at position 9, as in dexamethasone, specifically potentiates both the glucocorticoid activity and binding to the receptor protein (while eliminating the mineralocorticoid activity). The degree of saturation of binding of the receptor by a steroid molecule *in vivo* corresponds with its inducing effects upon tyrosine aminotransferase and tryptophan pyrrolase.

Binding of steroid by the receptor molecule results in a temperature-dependent change in the conformation of the protein. It is only in this steroid-induced form that the glucocorticoid–receptor complex is capable of binding to isolated DNA or to the liver nucleus. An increased rate of RNA synthesis can be detected coincident with interaction of nuclear DNA with the receptor–glucocorticoid complex. In several instances specific mRNA species have been isolated. For example, after treatment with cortisol *in vivo* the mRNA, which directs the synthesis of tryptophan pyrrolase, can be obtained in increased amounts from liver tissue. The time-course of the rise in enzyme activity, as well as the dose-response to increasing concentrations of cortisol, follows the same pattern as for the rise in the mRNA. Taken with the evidence that the rise in enzyme is prevented by inhibitors of RNA polymerase, the studies to date pinpoint the site of action of glucocorticoids at the level of transcription of certain mRNA species corresponding to the inducible proteins.

MINERALOCORTICOIDS

Mineralocorticoids act on the kidney to promote retention of sodium ions and elimination of potassium ions, and so these hormones are produced by the adrenal under a different set of stimuli from the glucocorticoids. The secretion of aldosterone is only increased momentarily by ACTH. In keeping with its role in electrolyte balance, aldosterone is secreted in response to changes in Na^+ and K^+ of the bloodstream. If K^+ concentrations in blood rise by more than 1 mEq above normal values there is a direct effect on the *zona glomerulosa* cells causing them to secrete additional aldosterone; the latter will stimulate the kidney to excrete more K^+ to correct the imbalance. The response to changes in Na^+ is opposite to that for K^+ (decreased Na^+ causes increased aldosterone production which in turn causes the kidney to reabsorb more Na^+ and thus excrete less Na^+ to restore the balance). The effect on the adrenal gland is indirect (a second hormone system, the renin–angiotensin system of the kidney, is involved as an intermediary control).

Renin is a proteolytic enzyme that is released into the blood by the kidney when Na^+ depletion causes the blood volume to fall; the decreased blood volume results in decreased blood pressure triggering renin-secretion by baroreceptor cells that surround the vessels entering the kidney glomerulus. The substrate for renin is a particular glycoprotein of the blood serum that is cleaved first into an inactive decapeptide, angiotensin I. The latter is converted to the active angiotensin II by a converting-enzyme that is present already in blood plasma. The mechanism whereby angiotensin II activates aldosterone production is thought to be by a specific receptor on the *zona glomerulosa* cell membrane which generates intracellular cyclic AMP; this process is analogous to ACTH actions on the *fasciculata–reticularis* zone cells. There is evidence that angiotensin II interaction increases the rate of conversion of cholesterol to pregnenolone and also the rate of conversion of corticosterone to aldosterone.

Aldosterone receptors have been demonstrated in the cytoplasm of a number of secretory epithelial cells including the kidney tubules, salivary glands, and intestinal mucosa. The ability of various steroids to bind to the receptor proteins or to compete for binding with aldosterone is correlated with their mineralocorticoid activities. The model for mechanism of action of aldosterone in kidney is analogous to that for cortisol in liver. Thus there is evidence for temperature-dependent transfer of the cytosol aldosterone complexes to the nucleus, followed by activation of transcription of a specific mRNA that codes for an effector protein (Fig. 3–13). This correlates with the 60 minute latent period between the injection of aldosterone *in vivo* and the observed end-effect on decreased Na^+ excretion in the kidney, a response which is blocked by actinomycin D. The nature of the effector protein that is induced by aldosterone in the kidney is not certain. It has been suggested that the aldosterone-induced protein could act as a permease to facilitate Na^+ movement through the membrane, or that it could activate the energy-requiring Na^+ pump. There is evidence that aldosterone induces mitochondrial citrate synthetase, and the activity of the tricarboxylic acid cycle is enhanced by aldosterone, thus generating more ATP for the movement of ions against the concentration gradients in kidney cells. Whatever the mechanism may be for Na^+ movements by aldo-

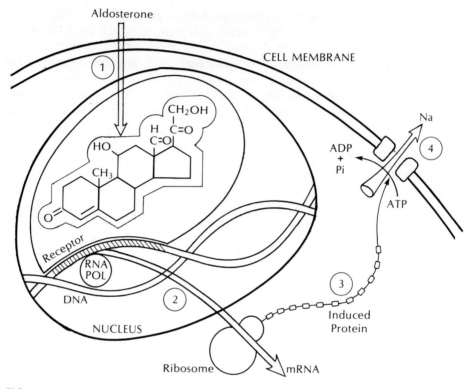

FIG. 3–13.   Actions of aldosterone.

sterone it is clear that the movements of K^+ are enhanced by a different process, which is genetically distinct from the sodium effect, and which is apparently not sensitive to inhibition by actinomycin D in the same way.

CUSHINGS SYNDROME

This eponym has come to be applied to the clinical entity hyperadrenocorticism that results from excessive glucocorticoid action. It may arise from overdoses of glucocorticoids administered as medication (in which case the ACTH levels of blood serum will be suppressed), or from adrenocortical hyperplasia (in which case the primary cause may be excess ACTH from a pituitary tumor). The effects of excess cortisol include a typical localized central obesity of the trunk, a moonface, and thin extremities; the excess adrenal androgens may result in precocious puberty in boys, and masculinizing changes of hair distribution in women. In the case of a pituitary tumor the production of ACTH may respond poorly to the normal feedback suppression by glucocorticoids. In many instances it appears that the normal diurnal rhythm of ACTH secretion is lost, and secretion is continued throughout the day at the rate normally reached during the peak activity period in the early morning. The severity and range of symptoms will vary depending on the duration and extent of hypercorticalism, but they may include muscle weakness, loss of connective tissue strength, resorption of bone with pain and curvature of the spine, and diabetes mellitus. The treatment is to reduce the excess glucocorticoids by discontinuing their administration, or by removal or irradiation of the tumors in the adrenal or pituitary.

ADDISON'S DISEASE

Hypofunction of the adrenal cortex leads to severe disorders of organic metabolism (deficiency of cortisol) and of electrolyte balance (deficiency of aldosterone). The principal symptoms include fasting hypoglycemia with muscle weakness, loss of Na^+ and retention of K^+ with hypotension, and characteristic bronze pigmentation in skin creases, mucous membranes and scar tissue. The pigmentation is a consequence of the unrepressed secretion of ACTH and its related fragment, MSH, the melanocyte-stimulating hormone that induces pigment deposition. At one time the major cause of adrenal insufficiency was tuberculous destruction of the glands; with advances in control of tuberculosis this cause has been virtually eliminated and, in most cases of Addison's disease, the cause is mainly from viral infections or auto-immune reactions of the adrenals. The chronic disease may be readily controlled by glucocorticoid plus mineralocorticoid supplementation (prednisol plus fluorocortisol taken orally). Larger amounts of steroids may be necessary when the patient is subjected to surgical operations or trauma, or when he demonstrates symptoms of acute disease (Addisonian crisis) accompanied by severe gastrointestinal upset and sudden hypotension leading to vascular collapse. Such patients require immediate infusions of several liters of isotonic saline in combination with several hundred mg of cortisol to correct the circulatory imbalances of fluid volume and sodium ion content.

GONADAL HORMONES

In both sexes the gonads (testes and ovaries) carry out dual roles: first, to produce, mature, and store the gametes (spermatozoa and ova), and second, to synthesize and secrete the sex hormones (androgens and estrogens). As noted below, cell types within the gonads are differentiated to carry out either the gametogenic or secretory functions. There are many similarities and parallels between these gonadal functions in the two sexes despite the apparent differences in the end results. The sex hormones produced are of the steroid family, generated by pathways analogous to those of the adrenal cortex. Male and female sex steroids are under comparable control by feedback regulated secretions from the anterior pituitary of polypeptide hormones termed the gonadotrophins. Steroid hormones producing male sexual characteristics are termed androgens, while those promoting the primary and secondary characteristics of feminization are termed estrogens. A third category of steroids, the progestins, play a key role in pregnancy and the menstrual cycle. While the androgens predominate in the male, the ovaries of the female and the adrenal cortex in both species also produce androgenic steroids; in turn, males generate some estrogenic steroids. Thus the differentiation and maintenance of normal sexual attributes and functions in the two sexes depends upon the appropriate interplays and balances of the sex steroids and their controlling gonadotrophins. In addition to their effects upon sexual characteristics, the male and female steroids generate important regulatory actions not only on protein, lipid, and carbohydrate metabolism in various organs of the body but also on the maintenance of mineral balance in the skeletal tissues.

TESTICULAR HORMONES

The male sex hormones, or androgens, are steroids that differ from the adrenal corticosteroids by removal of the two carbon atoms of the keto side-chain attached at position 17. They are thus C_{19} steroids as opposed to the C_{21} steroids of the adrenal cortex. They are also derived from cholesterol, but in contrast with the adrenal glands the testis is not able to use blood cholesterol effectively as the precursor. Instead, the principal substrates for testicular steroid synthesis appear to be the plasma fatty acids, which are first oxidized to yield acetyl-CoA as the starting material for endogenous cholesterol synthesis (Fig. 3–14). The most important endproduct is testosterone, which has local effects on spermatogenesis in the testis, and which also has distal influences on accessory male organs and characteristics such as hair distribution and timbre of the voice. The androgens also produce general anabolic actions on the musculoskeletal system, a consequence that has been exploited by competitors of both sexes in athletic events of the strength category. The male hormones, like female sex steroids, are under feedback control by the gonadotrophic secretions of the hypothalamus–pituitary system.

Gonadotrophin Effects

There are two gonadotrophins produced in the anterior pituitary that regulate both testicular and ovarian function: FSH (follicle-stimulating hormone) which promotes spermatogenesis by the cells lining the seminiferous tubules, and LH (luteinizing hormone) which stimulates androgen production by the interstitial cells located in spaces between the tubules. Both gonadotrophins are glycoproteins containing up to 30% carbohydrate with molecular weights of about 35,000. Each contains an almost identical nonspecific α subunit with 89 to 92 amino acid residues, plus a β

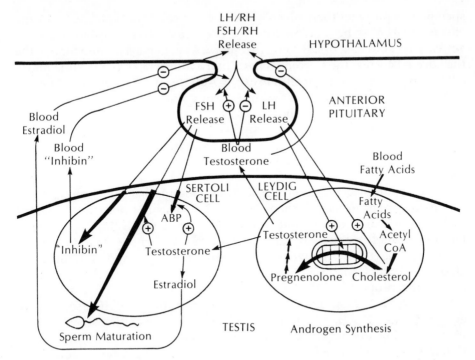

FIG. 3–14. Release of gonadotrophins and their actions in the testis. Factors which stimulate release are indicated by a positive sign, those which inhibit by a negative sign. Heavy black arrows show those processes in testis which are stimulated by gonadotrophins. ABP = androgen-binding protein.

subunit of 115 residues that is conformationally quite different in the two hormones because of the 80 residues at the carboxyl terminus.

The differentiation of the prepubertal testis, the production of mature sperm cells, and the metabolism of steroid hormones in the adult testis require the cooperation of two distinct cell types and regulation by the two gonadotrophins (Fig. 3– 14). The Sertoli cells of seminiferous tubules are the "nurse" cells which are responsible for guiding the sperm differentiation, and possess specific membrane receptors for FSH. The latter activates adenylate cyclase, cyclic AMP production, and protein kinase in turn. The phosphorylation of regulatory proteins may contribute to the accelerated protein synthesis and general hypertrophy of Sertoli cells by FSH. Many actions of FSH require the simultaneous presence of testosterone; one of the early actions of FSH is the induction of synthesis of a protein with high affinity and selective binding for androgenic steroids. Thus the production of this androgen-binding protein ensures a high concentration of testosterone in the vicinity of developing spermatocytes. The Leydig cells of the interstitial spaces are the primary steroid-synthesizing cells, and have specific receptors on their plasma membranes for LH. Again many of the effects of LH may be mediated by the adenylate cyclase system which is activated on binding of LH to its receptors. As previously noted for cells of the adrenal cortex, the steroidogenesis by Leydig cells of the testis is markedly activated when intracellular cyclic AMP levels are elevated. The site of action seems to be similar in both tissues, namely at the level of pregnenolone formation from cholesterol by the mitochondrial desmolase. Thus, acceleration of this rate-limiting step by LH results in stimulation of testosterone secretion by the Leydig cells, and in synergistic actions of the testosterone produced with FSH stimulation of the neighboring Sertoli cells.

The hypothalamus secretes a peptide-releasing hormone into the portal blood system to trigger the release of LH. This LH-releasing hormone (LHRH) has been isolated and characterized as a small linear peptide of ten amino acid residues. Because of its size and simple structure it has been possible to synthesize the releasing hormone and a number of analogues, which may be of value as potential birth control agents. The interesting property of the releasing hormone is its ability to stimulate release of both LH and FSH by the anterior pituitary, and it is sometimes designated as LH RH/FSH RH. This is somewhat surprising since LH and FSH secretion are regulated quite independently; for example, testosterone exerts a negative feedback on LH release as might be expected, but it does not prevent FSH secretion under physiological conditions. Instead, it appears that a substance, probably a peptide, generated by Sertoli cells and termed inhibin, may serve as a negative feedback regulator of FSH production. It is clear that LHRH secretion may be slowed down by a direct effect of androgens on the hypothalamus. However, the divergence in LH/FSH production is evidently expressed at the level of the pituitary. It has been shown with pituitary cells growing in culture that exposure to testosterone markedly inhibits the LH response to LHRH, while its effect upon FSH release is quite stimulatory. This allows for a separate, self-regulated LH– Leydig cell system to prevent overproduction of androgens, while providing a FSH– Sertoli cell system that is optimally active in the presence of high levels of testosterone.

Androgen Synthesis

The testis contains similar enzyme systems for steroid synthesis as those contained in the adrenal cortex, with exceptions of the 11-β-hydroxylase and 21-hydroxylase, which are absent. As in the adrenal, cholesterol is converted first to pregnenolone, and then to progesterone in Leydig cells of the testis. Both of these intermediates may be acted on by the enzyme, 17-α-hydroxylase, to provide two divergent pathways to the androgens, and Δ^5 pathway by 17α-hydroxy-pregnenolone, and the Δ^4 pathway by 17-α-hydroxy-progesterone (Fig. 3– 15). Although the latter is a normal

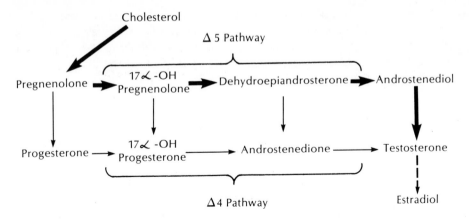

FIG. 3–15. Pathways of androgen biosynthesis.

secretory product of the testis, it appears to represent only a minor conversion, and the main route to testosterone in humans is the Δ^5 pathway. Testis contains an active 17,20 desmolase that catalyses oxidative cleavage of the two carbon ketol side-chains attached at carbon 17 to generate the 17-keto steroids. The major product is the weak androgen, dehydroepiandrosterone, from the Δ^5 path, with lesser amounts of the Δ^4 intermediate androstenedione. Reduction of the 17-keto group, plus the 3-β-hydroxy dehydrogenation and isomerization of Δ^5 to Δ^4 in the A ring leads to testosterone. Normally about 90% of the secretory products from the testis is in the form of testosterone.

In addition to androgen production, there is a small output of female sex steroids (estrogens) by normal testes. The estrogens result from aromatization of ring A in the steroid molecule by reactions discussed in the following section on female hormones. The main conversion from testosterone to estradiol occurs in Sertoli cells under control of FSH. As noted later, a similar two-cell hypothesis for estrogen synthesis in ovaries has been proposed. The significance of male estrogen production is not entirely clear, but it may play an important role in feedback effects in the hypothalamus and pituitary. Estrogens are responsible for positive induction of synthesis of the plasma-transport protein for sex steroids (sex hormone-binding globulin or SHBG) by the liver, an action opposed by androgens. Thus the male estrogens can facilitate normal transport of the androgenic hormones by SHBG in the blood. Extratesticular tissues can also carry out the aromatization reaction, and it appears that a significant proportion of the estrogens in males may arise by conversion from circulating testosterone, particularly in adipose tissue. Highly obese males may produce sufficient amounts of female sex steroids to induce notable enlargement of the breasts.

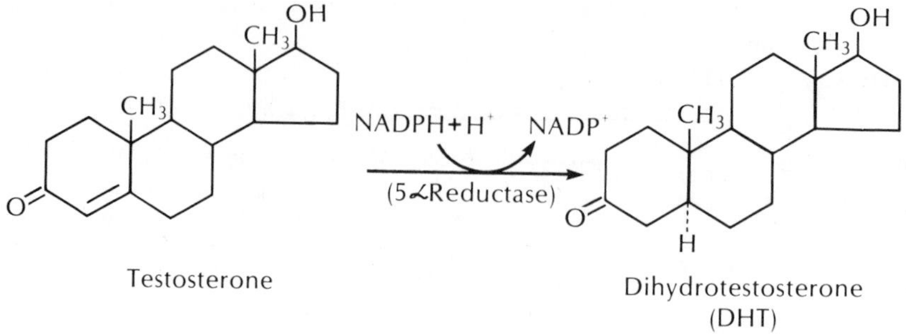

FIG. 3–16. Structure of testosterone and its conversion to dihydrotestosterone (DHT).

Testosterone and Dihydrotestosterone

Among the steroid hormones testosterone is unique in its actions since it must be metabolized by many of its target tissues prior to exerting its effects. Testosterone-sensitive organs (*i.e.,* the external male genitalia, the prostate, seminal vesicles, and pituitary) possess an enzyme, 5-α-reductase, that converts testosterone irreversibly to dihydrotestosterone (DHT) by saturation of ring A (Fig. 3– 16). In such tissue cells it appears that testosterone itself is inactive, but rather it serves as a prohormone. The 5-α-reductase is essential for the androgenic response. This is dramatically demonstrated by the recent discovery of an inborn error of male sexual differentiation resulting from inherited deficiency of the 5-α-reductase enzyme. Although the individuals are genotypically male with normal XY chromosome complement, they lack expression of the full male phenotype. The anomaly is owing to insensitivity of male accessory organs to testosterone and is termed testicular feminization. The external genitalia, prostate, and beard are undeveloped at puberty in the affected males. However, spermatogenesis, which in Sertoli cells is directly dependent upon testosterone, is normal, as is the development of male somatic psychosexual characteristics. Thus the anabolic and behavioral effects of male hormones may be separated from androgenic functions. It appears that testosterone acts directly as a hormone in neurons, bone, and skeletal muscle.

Those target tissues that contain the 5-α-reductase are highly sensitive to the circulating androgens, which act as potent stimuli for their further effects. The enzyme is induced by androgens; for example, in the male 5-α-reductase production is switched on by the surge in testosterone secretion at puberty; in the female estrogens and progesterone antagonize androgenic hormones at this stage. Further actions of androgens in these target tissues require complex formation with a specific receptor-protein, which, unlike the androgen-binding protein of Sertoli cells, is selective for DHT but has little affinity for testosterone. Studies with various steroid analogues indicate that the receptor-protein envelopes the DHT molecule in a narrow pocket, which requires the A ring to lie flat as in the disaturated structure of DHT. The production of the receptor for DHT, as for the 5-α-reductase activity, is quite dependent on prior levels of androgens reaching the tissues. A second type of inherited testicular feminization illustrates the consequences of the absence of DHT receptors in preventing normal androgenic responses. The patients are genotypic males, and they possess testes with normal Leydig cell function, and hence normal male testosterone production. In appearance, however, the affected persons are phenotypic women with normal breast development, but they have infantile female external genitalia and primary amenorrhea. Estrogen production is in the low normal range for females, and presumably originates from Sertoli cells of the testis. Because of deficiency of the gene for the necessary receptors for DHT in androgen target tissues the accessory sex organs fail to develop the normal male phenotype. In contrast with the patients with 5-α-reductase deficiency, those with testicular feminization arising from absence of DHT receptors are unresponsive to therapy with DHT.

Following the conversion of testosterone to DHT, and the binding of the latter to its receptor in the cytoplasm of androgen-sensitive tissues, the DHT–receptor complex undergoes a temperature-dependent conformational change. This change may involve disaggregation or proteolytic cleavage since there is a marked decrease in the sedimentation constant for the activated complex. The latter is taken up and retained by the nucleus where it becomes tightly bound to components of the chromatin. The interaction is complex and may involve both histone and nonhistone proteins.

Effects of the androgens at the nuclear level may be divided into two categories: early responses involving the protein-synthesizing machinery of the ribosomes and the synthesis of RNA; late responses requiring DNA replication. It is clear that many of these androgen effects may be correlated with the activation of dormant template functions in the DNA by interactions of androgens or their receptor proteins with particular acceptors along the chromatin. For obvious reasons most of the experimental studies on molecular mechanisms of androgen actions have been carried out in experimental animals thus far. The usual model system for such investigations is the ventral prostate gland in rats that have been previously subjected to castration. In castrated male rats the accessory sexual organs show regression and deceleration of macromolecular synthesis. Androgen-dependent functions may then be followed by measuring the synthesis of nucleic acids, proteins, and specific enzymes in the prostate at timed intervals after the administration of testosterone to castrated animals.

Among the early responses to androgens is an increased efficiency of the cytoplasmic translation system for protein synthesis (Fig. 3– 17). A few hours following castration the ability of prostate ribosomes to initiate

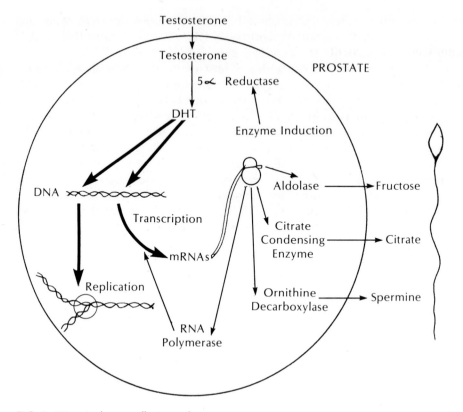

FIG. 3–17. Androgen effects on the prostate.

synthesis of polypeptides is reduced. The defect lies in the initiation factor IF$_2$, which loses its ability to bind the initiator transfer RNA. Within 10 to 30 minutes of intravenous injection of DHT in the castrated rats this defect is reversed. Other early responses include a stimulated synthesis of RNA species, ribosomal as well as messenger RNAs, and the syntheses of phospholipids and proteins of ribosome-containing "rough" endoplasmic reticulum of the prostate gland. Within 16 to 24 hours the androgen response is evident in the proliferation of ribosomal particles associated with poly A-rich RNA messengers (polyribosomes) and with functional membranes in the cytoplasm (rough endoplasmic reticulum). These active ribosomes demonstrate enhanced general protein synthesis, which is also seen in nonreproductive tissues such as the kidney (anabolic effect), and specific increases in the synthesis of enzymes such as 5-α-reductase (androgen response). A very wide range of enzymes in prostate and other accessory sex glands is induced by DHT (Fig. 3–17), including all the subcellular organelles—aldolase (cytosol), citrate-condensing enzyme (mitochondria),

β-glucuronidase (lysosomes), RNA polymerase (nuclei). Induction of enzymes for fructose and citrate production are important since these are significant components of seminal fluid. Ornithine decarboxylase, which plays a key role in synthesis of the polyamines, and spermine and spermidine, which are found in high concentrations in the prostatic secretion, are also regulated by androgens.

Detailed study of the induction of prostatic aldolase by androgens confirms that new mRNA synthesis is required, and that the rise in enzyme levels is specific for the tissue (liver aldolase is not affected) and for the steroid administered (glucocorticoids or estrogens have no effect). The poly A-rich messenger RNA fraction was isolated and shown to produce aldolase in a separated ribosomal protein-synthesizing system. However, although maximal amounts of functional mRNA were found about 12 hours following androgen administration, the maximal rate of aldolase synthesis was 6 to 12 hours later. Although the androgen–receptor interaction with nuclear DNA triggers the key transcription of aldolase-specific mRNA as a primary event, there is also an important androgen-mediated control of translation corresponding to the later promotion of functional ribosomes.

The preceding events may lead to greater metabolic rates or to increased growth (hypertrophy) of existing cells by restoration of androgens. The long-term response to androgens of tissues that have lost their normal complement of cells following castration can lead to the production of increased numbers of cells (hyperplasia). Obviously such a response requires a stimulus to the replication of DNA, as well as to transcription and translation. Cell renewal by activation of DNA replication and mitosis is evident only over prolonged periods of androgen administration.

The prostate and other secondary sexual glands in the male shrink to a minimum size about a week after castration when there is a severe reduction in cell numbers. Administration of testosterone produces a surge of mitosis after a lag of 48 hours, and the weight and cell populations return toward normal values thereafter provided that daily androgen supplements are given. During the lag period the prostate shows dramatic increases in DNA polymerase activity, as well as in the synthesis of tissue-specific, DNA-binding proteins. One of the latter events enables the DNA polymerase to form a stable complex with its DNA template by introducing areas of strand separation in the double helix (DNA-unwinding protein). Thus the lag period of the androgen-induced mitosis is explained by the prior formation of proteins required for the synthesis (DNA polymerase) and regulation (unwinding protein) of the daughter DNA molecules.

As for the early effects of androgens, DHT seems to be more potent than testosterone in stimulating proliferative growth. Moreover, in adult males over 40 years of age an increase in prostate size due to hyperplastic growth may lead to greater than fourfold weight increments accompanied by corresponding elevations of DHT concentrations in the enlarged glands and enhanced 5-α-reductase activity. As for the hypertrophic effects of androgens, the later hyperplastic stimulation appears to require conversion of testosterone to DHT, binding to the androgen receptor molecules, and translocation to nuclear binding sites prior to activation of DNA replication.

OVARIAN STEROIDS

Although it may seem paradoxical there are many similarities between the female and male reproductive systems with respect to production and regulation of their steroid sex hormones. Similar enzymatic pathways occur in both sexes; the main difference lies in the predominant end-products, C_{18}-estrogens in the female versus C_{19}-androgens in the male. Nonetheless, both types of steroids play significant physiological and behavioral roles in each sex, the balance between estrogen and androgen being the determining factor. Identical gonadotrophins (FSH and LH) act on the ovary and the testis, in controlling steroid biosynthesis and maturation of germinal cells. There is also a similarity in action of the two gonadotrophins on different cell types of ovarian follicles and a similarity in cooperative functions of the cells during steroid metabolism.

Despite these analogies there are as many obvious, profound differences in the chemistry and function of the hormones as there are in outward attributes in the two sexes. First is the extreme variation in hormone secretion in the female in relation to the rhythm of the menstrual cycle. Second is the production of larger amounts of progesterone as an end-product in the female rather than simply as an intermediate in the male. Finally, there is the extragonadal endocrine function, provided by the placenta during pregnancy, in producing gonadotrophins independently of the hypothalamus–pituitary system and in producing steroids independently of the ovaries.

The estrogens exert major metabolic effects on female reproductive organs, but also influence processes in other parts of the body. These include retention of sodium ions and water, increased production of growth hormone, antagonism of insulin effects, and increased production of serum lipids and blood-clotting factors. Such responses, if accentuated by prolonged over-exposure to estrogens, may uncover latent defects in carbohydrate and lipid metabolism, possibly developing into diabetic or vascular syndromes. Oral contraceptive pills with high estrogen content have been implicated in increased incidence of hypertension and thromboembolic disease. Estrogens play a role in normal calcium retention and bone maintenance as demonstrated by the reductions in bone mass of postmenopausal women and beneficial effects of administered estrogens for healing of bone fractures in elderly patients.

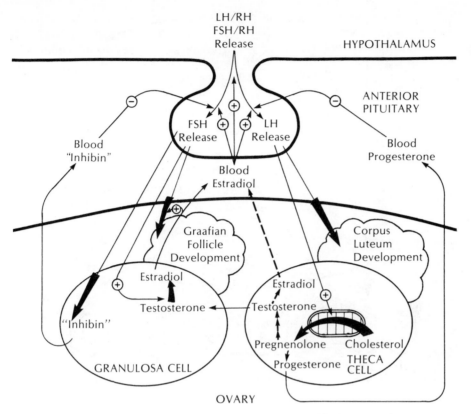

FIG. 3–18. Actions of gonadotrophins in the ovary. Heavy black arrows show stimulated processes.

Synthesis

As in the male gonads the synthesis of estrogens by the ovaries proceeds by the Δ^5 pathway of steroidogenesis, with testosterone acting as an intermediate. The reactions occur in the layers of cells surrounding the developing ova to form the structures called follicles (Fig. 3– 18). The "nurse" cells in the immediate proximity of the ovum are termed the granulosa cells and are a primary target for FSH action, while the theca cells, which are in contact with blood vessels as well as other interstitial cells, play a major role in steroidogenesis and respond to LH. This division of functions, as is true for the testis, is not completely specific; the responses to FSH require synergistic actions of estrogens from steroid-producing cells, while the synthesis of female sex steroids involves interactions between the cell types. It is clear that the granulosa cells can carry out part of the estrogen pathway, converting cholesterol to progesterone, or testosterone into estrogens, while theca cells can produce both androgens and estrogens. The

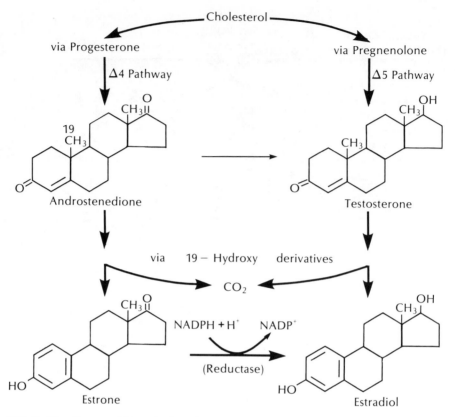

FIG. 3–19. Estrogen biosynthesis pathways.

gonadotrophin, LH, activates steroidogenesis in the theca cells in the same manner as it does for Leydig cells of the testis, namely by cyclic AMP and its stimulating effect upon the rate-limiting desmolase reaction. Maximal production of the major ovarian estrogen, 17-β-estradiol, is seen when both theca and granulosa cells are combined. One suggestion is that granulosa cells carry out the major part of the aromatization of androgens under the influence of FSH, while the theca cells generate the necessary testosterone precursor under stimulation by LH. As noted below, prior to ovulation the ovaries accumulate negligible amounts of progesterone. The progesterone may be carried from granulosa to theca cells to provide additional precursor for the production of testosterone, presumably by the quantitatively minor Δ^4 pathway.

The aromatase reaction is a complex, multistep process, and in the female, as in the male, may be catalysed by extragonadal tissues as well as by the ovaries (Fig. 3–19). The 19-carbon atom of testosterone, the CH_3 group between rings A and B, is first hydroxylated by a microsomal enzyme to produce 19-hydroxytestosterone, followed by NADP-dependent dehydrogenation. The next steps are unclear but seem to involve removal of the 19-carbon at the oxidation level of formic acid, with simultaneous removal of 2 hydrogens from the A ring by an O_2– NADPH-dependent reaction to produce estradiol. In isolated granulosa cells added testosterone is aromatized to estradiol only in the presence of FSH, which probably acts as it does in Sertoli cells of the testis because of the adenylate cyclase system. The minor androgen, androstenedione, may also arise from interstitial cells of the ovary or the adrenal cortex, and can be aromatized by the same mechanism to estrone. The latter is a weak estrogen which can be reduced to the potent estradiol; this is an important auxiliary estrogen source when ovarian estradiol production ceases, as it does in women after menopause.

Gonadotrophin Effects

The combined actions of FSH and LH on the developing follicle result in simultaneous maturation of the ovum and accelerating production of estrogens during the first two weeks of the menstrual cycle. Estrogen exerts positive feedback on FSH and LH secretion both by stimulating releasing-hormone (LHRH) production at the level of the hypothalamus, and by enhancing the response of the pituitary to the gonadotrophin-releasing hormone (Fig. 3– 18). There is also evidence for a specific "inhibin," analogous to that produced in Sertoli cells, that originates in the follicles and suppresses FSH release. The result of these interactions is a cumulative secretion of gonadotrophins and estrogens as the follicle matures with a differential surge of LH relative to FSH. The estrogens stimulate the tissue layer lining the uterus (endometrium) to generate macromolecules, activate the cells to divide, produce many-fold increases in the numbers of receptor molecules for progesterone, and strengthen the muscle components (myometrium) of the uterus. There are also marked changes of estrogens in the epithelial cells of the vagina, in production of secretions of glandular cells in the uterine cervix to favor sperm survival, and of both secretory and ciliated cells of the oviduct to favor migration of the ovum. At the end of this preparatory phase, timed to coincide with maturation of the ovum, the dramatic surge of LH triggers ovulation by facilitating rupture of the follicular wall.

The remaining cells of the follicle become transformed into the yellow body termed the corpus luteum. Under the influence of LH the corpus luteum cells resume the production of estrogens, but also generate large amounts of progesterone during the postovulatory two-week period of the cycle. Only the Δ^4 synthetic pathway is used at this stage. Subsequent events, designed to prepare the uterus for implantation of the fertilized ovum, are mediated by continuing actions of estrogens, and the ancillary, sometimes antagonistic, actions of progesterone.

Progesterone Effects and Pregnancy

Most actions of progesterone on the female reproductive system require the previous "priming" of the tissues by exposure to estrogen. In sum the effect of the progestins, as their name implies, is to promote the gestation of the fetus. Although it exerts a growth-promoting effect upon the uterus and breasts, progesterone exerts a net catabolic effect upon the body as a whole with increased urinary nitrogen excretion resulting from increased protein breakdown. In the uterus there is stimulation of RNA synthesis and activation of glandular secretion in the endometrium by the rise in progesterone content within 48 hours of ovulation. The endometrial surface architecture undergoes drastic changes to render it receptive to the fertilized ovum, and subepithelial capillaries proliferate to ensure adequate contact with the maternal bloodstream. While favoring the implantation of the early floating embryo, or blastocyst, the presence of progesterone produces an environment that is hostile to the survival and penetration of sperm. Use of natural or synthetic progestins as components of oral contraceptives may have a basis in this phenomenon; the continuous administration of progestins, particularly when combined with synthetic estrogens in the oral contraceptives, also inhibits ovulation, possibly by damping the surges of LH. With continuous exposure to progesterone the uterus eventually loses its receptivity to new blastocysts. Such mechanisms in the progesterone-dominated uterus prevent late development of over-ripe ova, or second implantation.

In the absence of fertilization, the corpus luteum recedes within 2 weeks after ovulation, the hormone levels of both estrogens and progestins decline markedly, and the failure to support the endometrium results in menstruation. However, the arrival of a fertile blastocyst in the uterus is heralded to the corpus luteum by an urgent signal to continue hormone production and thus sustain the nesting site. The signal is provided by a glycoprotein hormone, human chorionic gonadotrophin (HCG). This hormone contains α subunits that are identical to those of the pituitary gonadotrophins, and β subunits that possess many homologies of amino acid sequence with those of LH. Its biologic activity is identical with that of LH, although HCG and LH are sufficiently different to be distinguished by their reactivities to specific antibodies to their β subunits. This provides a basis for immunologic techniques to monitor HCG production during pregnancy.

The continuation of steroid hormone production beyond the normal two-week period in the corpus luteum requires the stimulation of HCG produced by the growing fetal tissue, or trophoblast, which eventually forms the supporting placenta structure in the uterus. However, the progesterone from the corpus luteum maintains the implanted blastocyst only for the first 6 to 8 weeks; thereafter, the ovary becomes a progressively less important producer of the steroids in pregnancy with the placenta and fetus assuming dominant roles in the synthesis of both progesterone and

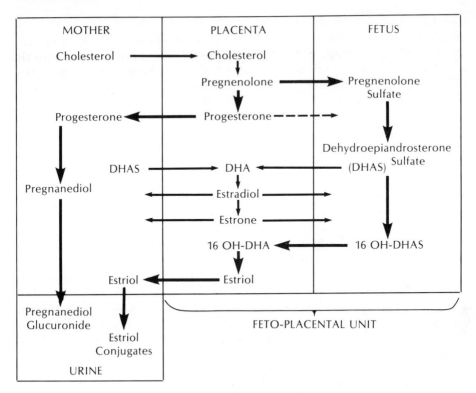

FIG. 3–20. Metabolism of steroids by the feto-placental unit in pregnancy.

estrogens. Unlike the ovary or the testis, the placenta is an incomplete endocrine organ and requires a collaboration with both the fetus and mother to carry out steroid hormone syntheses. The coordinated system is referred to as the fetoplacental unit.

The placenta has the capacity to convert preformed cholesterol to progesterone, but it does not synthesize the sterol *de novo* from small precursors, such as acetyl-CoA. Most of the incoming cholesterol seems to originate from the maternal circulation (Fig. 3–20). The total output of progesterone rises during pregnancy in concert with growth of the placenta. The high levels of progestin suppress ovulation, and interfere with sperm penetration and implantation to discourage double pregnancies. The role of elevated progesterone in late pregnancy is unclear; it stimulates growth of breast tissue and the muscular elements of the myometrial tissue in the uterus. Progesterone also exerts a suppressive or quieting effect on the myometrium by antagonizing contractions induced by estrogens or other muscle stimulants. In this way progesterone is thought to prevent premature ejection of the fetus. It has also been noted that progesterone/estrogen ratios decline markedly at term, suggesting a functional significance in the induction of normal labor. Much of the progesterone is transferred to the mother where its reduction to pregnanediol is carried out. The excretion of the latter metabolite is commonly measured as an indicator of progestogen production in pregnancy.

The synthesis and interconversions of the estrogens during pregnancy illustrate the complexity of the fetoplacental unit (Fig. 3–20). The placenta is incapable of producing either estrogens or androgens *de novo* since it is devoid of the essential 17-α-hydroxylase enzyme, but it converts significant amounts of maternal cholesterol into pregnenolone which may be taken up by the fetal circulation. Enzymes in the adrenals of the fetus carry out the 17-hydroxylation and subsequent reactions leading to the androgen dehydroepiandrosterone. The latter may be carried back into the placenta in its conjugated sulfate form for further metabolism to testosterone and the estrogens, estradiol and estrone. A similar amount of this androgen precursor in placenta is contributed from maternal sources. The fetus is protected from high levels of androgens by the efficient placental aromatization, a process that is stimulated by HCG.

The major estrogen produced as pregnancy progresses is estriol, the result of a unique 16-α-hydroxylation that occurs only in fetal liver (Fig. 3–20). The precursor is derived from dehydroepiandrosterone of fetal origin, and hence measurement of the levels of estriol in the maternal circulation provides a valuable indicator of the endocrinologic health of the fetus. A lowered estriol content of the mother's plasma or urine can serve as a signal of fetal distress, as in the extreme cases of anencephalic pregnancy or death of the fetus where adrenal or liver function on the fetal side of the placenta are incapable of furnishing the 16-hydroxylated precursors for estriol formation. The estrogens synergize with progesterone in stimulating growth of contractile components in the myometrium, and they also play a role in augmenting blood flow to the uterus. It is noteworthy that although the predominant fetoplacental steroid, estriol, has weak estrogenic effects on most tissues in comparison with estradiol, it is one of the most potent agents for increasing the uteroplacental blood flow, and this may be its predominant function.

Estrogen Actions

Actions of the estrogens at a molecular level appear to follow similar mechanisms to those outlined previously for androgenic responses. In brief, the reproductive target tissues contain specific cytoplasmic receptors for estrogens, and following modification of the steroid–receptor complexes they are translocated to the nucleus. The binding to selective regions of the chromatin then leads to activation of transcription of RNA molecules, triggering the events that lead first to cellular hypertrophy, and later to hyperplasia. The responses at target tissues are generally greatest with estradiol, which exhibits strong binding to the receptors and retention in the tissues. Estrone, which is weakly bound to estradiol receptors, has correspondingly weak estrogenic effects; it is considered to be an elimination product formed by dehydrogenation of estradiol, and estrogenic effects of administered estrone are chiefly results of reversibility of the 17-β-dehydrogenase. Estriol is bound effectively to estradiol receptors, but is not retained for any length of time in the bound form, a finding that also correlates with physiological effects of this estrogen which seems capable only of producing short-term responses in tissues. Studies with natural and synthetic estrogens have generated much interesting and valuable information about steroid mechanisms and applications to medicine.

As noted for the androgens, studies of estrogen mechanisms have depended chiefly on extrapolation from animal model systems. Responses are generally measured in relation to growth of the uterus of either immature or castrated female rats. As for other specific estrogen target tissues (vagina, hypothalamus, anterior pituitary), injected estradiol is rapidly taken up by the uterine cells and retained in a bound form. Estrogen-receptor proteins have been isolated from the cytoplasm of such cells, and seem to have comparable properties and actions in the different estrogen-sensitive tissues. In the uterus there are between 15,000 to 20,000 cytoplasmic receptors per cell. When bound to estradiol the receptor can undergo a temperature-dependent transformation into a different conformational state that is capable of entering the nucleus. Since the molecular weight of the nuclear estrogen–receptor complex (130,000–140,000) is approximately double that of the cytoplasmic form (70,000–80,000) it has been suggested that the former is a dimer of the latter. It seems paradoxical that activation to larger molecular size is obligatory for uptake, presumably through the nuclear membrane, and no satisfactory explanation has been presented. Within 15 to 30 minutes the hormone–receptor complexes are translocated to the nucleus where they become tightly bound to the chromatin and persist for several hours. Regardless of the total occupancy of cytoplasmic receptors by estrogen, there is a maximum of 2,000 to 3,000 sites within the nucleus, corresponding to the physiological levels of estrogens which are capable of saturating the hormone–receptor complexes.

Following interaction with the nucleus, the estrogens promote uterine growth stepwise at several levels. First, there is a rapid increase of cell volume which is not true growth but rather is caused by uptake of water into the cells. Next, there is hypertrophy of existing cells with associated synthesis of messenger and ribosomal RNAs. Finally, there is a period when DNA synthesis is accelerated resulting in cellular hyperplasia.

The difference between estradiol and estriol responses underlines the distinction between early and late effects. Thus estriol can promote only the accelerated water uptake but not the true increase of cell growth shown with estradiol. Although both estrogens are bound to receptors in the uterine nucleus estriol is removed within 4 to 6 hours, while estradiol persists as the nuclear receptor–hormone complex for 12 to 14 hours. It is only with the longer retention of estradiol in receptor sites on the DNA that sustained increases of both RNA polymerases I and II are observed; these increases correspond to accelerated synthesis of ribosomal and messenger RNA molecules respectively. The ensuing general increase of protein synthesis gives rise to true cellular hypertrophy in the period 24 to 48 hours postinjection. It is only when estriol is maintained at very high constant levels, as it is during pregnancy for example, that it would produce the true uterotrophic response that is normally evoked by pulses of estradiol. The stimulation of DNA synthesis and the mitosis of uterine cells, both of which occur 18 to 36 hours after estradiol administration, depend on the prior activation of RNA and protein synthesis, and are not produced by equivalent amounts of estriol. This estradiol effect is characterized by a shortening of the synthetic "S" phase of the cell cycle, and three- to fivefold increase of the α form of DNA polymerase which is involved in gene replication.

Progesterone Actions

Actions of progesterone at a molecular level have also been studied in the uterus of mammals. Specific high-affinity receptors for biologically active progestins have been described that are distinct from the estrogen receptors. Pretreatment with estradiol greatly augments (10– 20-fold) the number of progesterone receptors by a process that requires RNA and protein synthesis. Following exposure to progesterone the receptors are translocated to the nucleus where they stimulate the synthesis of messenger RNAs of selected proteins, such as uteroglobin, a polypeptide of 27,000 molecular weight that promotes growth and development of the blastocyst.

Progesterone receptors have been characterized more completely from investigations of the oviduct in birds. The chick oviduct is very sensitive to stimulation by estrogens and progestins, and responds by differentiation of the glandular tissue surrounding the ducts and by induced synthesis of several egg-white proteins. Treatment with estrogen primes the oviduct cells and specifically induces the production of large amounts of ovalbumin and its mRNA; studies of this process have yielded much basic information regarding the nature of messenger RNA as detailed elsewhere (Chap. 5). On treatment of estrogen-primed oviduct with progesterone a secondary response ensues involving other proteins of egg white, including the biotin-binding protein, avidin. The receptors for progesterone are abundant in the oviduct (35,000 per cell), and have been isolated in highly purified form. The receptor normally exists in the form of a dimer (molecular weight 200,000), and can be split into two dissimilar subunits (A—molecular weight 70,000– 80,000; B—molecular weight 115,000– 120,000). When combined with progesterone both subunits become translocated to the nucleus where the B subunits bind to acidic, nonhistone proteins of the chromatin while the A subunits bind directly to DNA. Full binding potential of the nucleus is attained only when both A and B por-

tions of the dimer are complexed with their respective loci on the chromatin, and the binding is followed by activation of transcription of the
mRNA's coding for specific egg-white proteins.

Studies of steroid hormone receptors have been important not only for
an understanding of basic endocrine mechanisms and gene regulation in
higher organisms, but also in the practice of clinical medicine. Thus
tumors of human breast, like the normal tissue, respond to steroid hormones, and it has been known since the last century that removal of the
ovaries can cause the regression of metastatic breast cancer. Unfortunately, only a third of patients respond in this manner, and until recently it
was impossible to predict which patients would not benefit from the drastic surgical procedure. It is now known that hormone-dependent tumors
have high levels of the receptor proteins for estrogens, while those which
are hormone-independent generally do not. A recent study of several hundred patients has demonstrated the value of measuring estrogen-receptor
levels in excised specimens of tumors. Of those patients with positive
tumor estrogen-receptor tests, 55 to 60% showed remission upon endocrine
therapy (surgical ablation of ovaries, and adrenals, treatment with anties-
trogen drugs), while very few (3–8%) of those with zero receptor values
responded. In future such patients may be spared fruitless treatment and
directed to alternative methods of chemotherapy.

THYROID HORMONES

Hormones of the thyroid gland exert powerful modulating effects on physical and mental development in the young, and on metabolism, particularly oxidative metabolism, and thermogenesis in the adult. The consequences of hypo- or hyperfunction of the thyroid, although less life-
threatening than for many endocrine organs, presents a very significant
medical problem since the total number of affected persons is greater than
for all other endocrine disorders combined except, for diabetes mellitus.
The thyroid, a large bi-lobed organ located in the neck, is under the trophic
and secretory regulation of the hypothalamo–pituitary system, and is subject to negative feedback control by its own end-products. The hormones to
be considered here are the iodine-containing derivatives of tyrosine—
thyroxine, or 3,5,3',5'-tetraiodothyronine (T_4), and 3,5,3'-triiodothyronine
(T_3); the other hormone produced by thyroid tissue, calcitonin, is involved
in calcium metabolism and will be discussed within that context.

Metabolism of the thyroid hormones is inextricably connected with
the use of iodine by the body, and ultimately depends on adequate dietary
intake of iodine. The thyroid serves as an effective trap and storehouse of
both inorganic and organic iodine. The latter is in the form of protein-
bound iodinated tyrosines and thyronines located in the extracellular lumens of the thyroid follicles as a viscous colloid. The colloid is composed of

a single protein, which is termed thyroglobulin, and constitutes a reservoir of thyroid hormones sufficient for at least a month's requirements. The thyroid is unique among the endocrine glands since it provides long-term storage of hormones and their precursors. Secretion of the thyroid hormones must be considered from the integrated viewpoints of iodine metabolism, thyroglobulin turnover, and the controlling influences of the specific trophic hormone from the pituitary, thyroid-stimulating hormone (TSH).

THYROGLOBULIN

Thyroglobulin is a large glycoprotein (molecular weight of 660,000) containing 10% carbohydrate and 3% of its amino acid residues as tyrosine (122 residues, of which a maximum of 40 may be iodinated). Synthesis of thyroglobulin takes place in a sequential manner in the follicular epithelial cells. First, the polypeptide chains are produced by polyribosomes attached to the rough endoplasmic reticulum, and then they pass through the membrane by way of the lumen of the endoplasmic reticulum to the Golgi apparatus where sugar units are added. The polysomes contain 30 to 60 ribosomes each and a messenger-RNA molecule coding for a polypeptide subunit of about 300,000 molecular weight; two of these subunits associate to produce the thyroglobulin molecule. There is no amino acid-activating system for the iodinated amino acids; the iodine is introduced into tyrosine residues of newly synthesized thyroglobulin after translation is completed, but possibly before all of the carbohydrate residues have been added. Of the sugars in thyroglobulin, mannose and N-acetyl-glucosamine are attached first while the polypeptide chain is still associated with the endoplasmic reticulum; the more distal sugars, fucose and sialic acid, are joined to the outer ends of the oligosaccharide chains while the glycoprotein migrates through the Golgi apparatus. Finished thyroglobulin molecules are secreted from the epithelial cells by exocytosis to form the viscous colloid deposit within the follicular lumens.

IODINE METABOLISM

The metabolism of iodine by the thyroid may be considered under the following sequential headings (Fig. 3–21): uptake of dietary iodine from the gastrointestinal tract into the blood; active transport of the iodide anions (I^-) from plasma into thyroid tissue; activation of I^- to a reactive species of iodine; use of a reactive species of iodine to iodinate tyrosyl residues in thyroglobulin; condensation of iodotyrosines to form T_3 and T_4 still attached to thyroglobulin; proteolysis of thyroglobulin with release of iodothyronine hormones into the blood; deiodination of the hormonally inactive iodotyrosines and retention of the released iodide within the

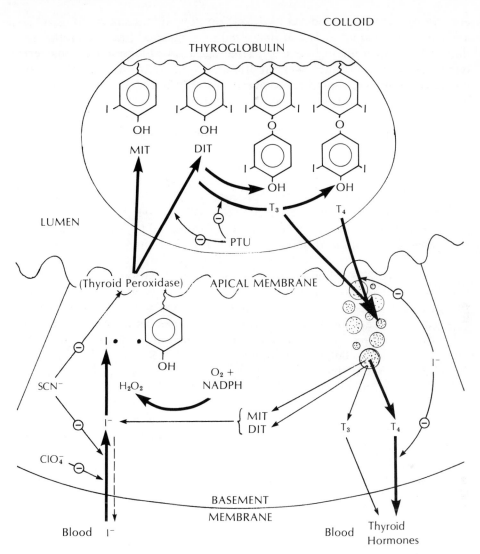

FIG. 3–21. Biosynthesis and secretion of thyroid hormones. T_3 = triiodothyronine; T_4 = thyroxine; MIT = monoiodotyrosine; DIT = diiodotyrosine; PTU = propylthiouracil. Heavy black arrows show the processes stimulated by thyrotrophin.

thyroid. As noted under each topic several of these steps are regulated by TSH, while some are autoregulated by levels of iodide ions in the blood.

Iodine is an essential nutrient in man. The daily diet for a healthy adult should include 0.1 to 0.15 mg or approximately 2 μg/kg body weight/day. Major natural food sources include seafood, eggs, milk, and water. Iodized salt, commercial bread with iodate, and other iodine-containing ingredients added as stabilizers are important sources of dietary iodine; most white bread marketed in the United States contains the daily adult requirement in a single slice! Excessive iodine intake may one day rival iodine deficiency as a nutritional problem. This is related to toxic effects on the thyroid of chronic overdoses of iodine, and is seen particularly among populations, such as the Japanese, who consume large amounts of marine fish and seaweed. The absorption of iodine from the gastrointestinal tract is complete, independent of the intake or the concentration in the blood.

The normal concentration of iodide in blood serum is 0.5 to 1.0 $\mu g/100$ ml. On reaching the thyroid gland, iodide is rapidly taken up by an energy-requiring active transport process so that the ratio of the concentration in thyroid relative to that in serum (T/S ratio) is $25-30:1$. In cases of iodine deficiency, or when the gland is stimulated by TSH, efficiency of the transport system is enhanced and the T/S ratio may rise to a value as high as $300:1$. These effects are complementary because the iodine-deficient thyroid is much more sensitive to stimulation by TSH, which promotes much greater generation of cyclic AMP in glands that are depleted of iodine. As the plasma iodide level rises with increased iodine ingestion, the trapping system eventually becomes saturated. As a result, the absolute uptake by the thyroid remains constant. At a very high intake level, the elevated content of iodine in the gland leads to the formation of an inhibitory agent that blocks further uptake by reducing the affinity of the iodide trap and allowing efflux of intrathyroidal iodide. Certain anions, such as perchlorate or pertechnetate, compete with the iodide anion for the transport system and stop the uptake of iodine by the thyroid. Thiocyanate anion also interferes with iodide uptake, but since it contains the SCN group it depresses iodination as well (Fig. 3–21). In a normal individual a third of the daily uptake (50 μg) is trapped by the gland; the other two-thirds are excreted in the urine.

IODOTHYRONINE SYNTHESIS AND SECRETION

Activation of the iodide molecule, its incorporation into aromatic rings of tyrosine residues, and condensations of the latter that produce iodothyronines, all seem to depend on a heme-containing enzyme, thyroid peroxidase. This enzyme is associated with the plasma membrane of the thyroid cell and is localized mainly on the apical surface where it reacts with the thyroglobulin molecule at the point where it is being secreted into the follicular colloid (Fig. 3–21). The peroxidase is very sensitive to inhibition by thionamide-($S=C-NH_2$) containing agents, such as thiourea, thiouracil, and derivatives, which block thyroid hormone synthesis, and like perchlorate, act as antithyroid drugs. Initial activation of I^- by the peroxidase involves the removal of an electron, and requires a supply of H_2O_2 as the oxidizing agent. The source of the latter is unknown, but appears most likely to be from reduction of O_2 by reduced pyridine nucleotides using the NADPH-cytochrome c reductase of the endoplasmic reticulum and vitamin K as intermediaries. Reaction of I^- with H_2O_2 raises it to the oxidation level of I_2, but it is unlikely that elemental iodine participates in the reaction. Instead, it appears that the iodination mechanism involves generation of an iodine-free radical (I·) bound to one site on the enzyme, and generation of a tyrosyl radical bound to a second proximal site. Iodotyrosine formation, then, would occur by combination of the two radicals while attached to the enzyme. The iodination proceeds sequentially by formation of monoiodotyrosine, then di-iodotyrosine. A number of other proteins, including serum albumin, may be iodinated by the thyroid peroxidase system, but thyroglobulin is the preferred substrate; free tyrosine is not subject to iodination in the thyroid. Coupling of iodotyrosines occurs within the framework of thyroglobulin, and requires the collaboration of the peroxidase enzyme. Possibly the latter is necessary to generate iodotyrosyl radicals, which would be able to join together to form the ether link between aromatic rings. The three-dimensional conformation of thyroglobulin is known to be very important for presentation of the iodotyrosines in the correct intramolecular juxtaposition for coupling. Thus, although denaturation of the structure of thyroglobulin does not interfere with the iodination of tyrosine residues, the native structure is required for efficient coupling to produce T_3 and T_4. The peroxidase inhibitor, propylthiouracil, exerts more potent effects on the coupling of iodotyrosines than on their synthesis (Fig. 3–21), possibly because of its ability to bind to thyroglobulin and thereby to interfere with the structural requirements of coupling. Analysis of thyroglobulin in normal humans reveals that each molecule may contain 5 to 6 residues each of mono- and di-iodotyrosines, and 2 to 3 of T_4, but that only one-third of the thyroglobulin molecules contains a T_3 residue.

The rates of the above processes and the ratios of their end-products are controlled by a number of physiologically important factors. First of all, TSH influences hormone synthesis at several points, as well as having a stimulating effect on I^- uptake by the gland. In the absence of TSH the thyroid shows reduction in the organification of iodine, and in production of T_4; since the latter is more profoundly affected than the synthesis of di-iodotyrosine it is concluded that TSH enhances the coupling of iodotyrosines by a mechanism that is independent of its stimulating effect upon iodination. Although neither of these effects of TSH is understood the increase in iodine fixation may be related to a greater rate of H_2O_2 formation in the thyroid cell. It is known that TSH markedly accelerates glucose oxidation in the thyroid by both the hexose monophosphate shunt and glycolysis, and there is a case to be made for the possibility that resulting increments of NADPH or NADH could bolster substrate requirements of the microsomal reductase for peroxide generation. Secondly, the concentration of I^- and various iodine species in the thyroid cell will modify the reactions. An excess of iodide in the gland will favor the production of I_2 from the free radical species and shift the reaction $- I_2 + I^- = I_3^-$ to the right, producing the I_3^- anion which is unreactive as an iodinating agent. The blockage of thyroid hormone synthesis by high serum I^- is temporary as a result of the inhibition of the iodide trapping mechanism and consequent decline of intrathyroidal I^- concentrations. A deficiency of iodide in thyroid cells, occasioned either by decreased dietary intake or by interference with the uptake system, will cause an increase in the ratio of T_3/T_4 produced by the gland. Importance of this T_3/T_4 ratio in the adaptation of thyroid function is apparent from the following considerations: with restrictions on the iodide available there will be more monoiodotyrosine formed relative to di-iodotyrosine; the larger proportion of monoiodotyrosine will accentuate its coupling with di-iodotyrosine to produce proportionately more T_3, and since T_3 contains less iodine than T_4 it economizes on the needs of the gland (most significantly, T_3 has at least three times the potency of T_4 in terms of its hormonal activity, thus effecting at least fourfold saving of iodine). As noted below a considerable portion of T_3 in the body is formed by deiodination of T_4 in peripheral tissues, and this conversion is also reported to be increased in hypofunctional states of the thyroid.

Secretion of thyroid hormones into the circulation involves the intracellular digestion of thyroglobulin (Fig. 3– 21). The process is stimulated by TSH, which triggers dramatic morphologic changes in the thyroid cells. Pseudopods form at the apical surface and engulf the colloid in small droplets by endocytosis. Lysosomes move out to meet and fuse with the droplets, and their proteases and glycosidases hydrolyse thyroglobulin down to its constituent parts. The free T_4 and T_3 pass through the basal surface of the cell into the blood capillaries where they combine with plasma proteins. The sugars, amino acids, and iodotyrosines remain in the cells for further metabolism. None of the mono- or di-iodotyrosine residues enter the circulation, but rather are acted on intracellularly by a thyroid deiodinase enzyme. The resulting iodide is mainly retained by the thyroid cells for recycling in iodination reactions, although a small component will leave the cells by the so-called thyroid leak. The secretion of T_3 and T_4 is blocked by ingestion of high doses of iodide, a fact that has been exploited in treatment of some cases of hyperactivity of the thyroid.

TSH EFFECTS

Control of thyroid function, as evident from the preceding discussion, is a complex interaction of external regulation by TSH and the internal process of autoregulation by iodide concentrations. Added to these considerations are the complications arising from the long duration of thyroid hormone effects in the body, the extensive storage of thyroid hormone in the gland, and the variable conversion rate of T_4 to more active T_3 in the target tissues. In all these respects the thyroid is quite unique as an endocrine organ, and possesses key homeostatic mechanisms to balance its actions.

Thyroid-stimulating hormone (TSH or thyrotrophin) exerts a number of trophic influences on the thyroid in addition to its role in controlling thyroid hormone synthesis and secretion. The chemical structure of TSH, a glycoprotein with a molecular weight of 28,000 to 30,000, bears close similarities to the gonadotrophins with two polypeptide subunits. In fact, the α subunit of TSH is the same as that of LH and FSH, while there are many similar or identical sequences in the β subunits. The mode of action of TSH on the thyroid cell is accomplished through binding to selective surface receptors and consequent activation of adenylate cyclase. The product, cyclic AMP, is the intracellular mediator of most, if not all, actions of TSH. An immunoglobulin present in serum from patients with Graves' disease, a type of hyperthyroidism, also promotes cyclic-AMP accumulation in thyroid cells, and mimics many effects of TSH. The ability of high concentrations of iodide to antagonize TSH effects can be related to inhibition of adenylate cyclase and of cyclic-AMP accumulation in thyroid tissue by iodide, or possibly an oxidized derivative.

The long-term effects of TSH on thyroid growth are superimposed upon its secretory stimulation (Fig. 3–22). Thus, although there is an immediate proteolysis of thyroglobulin after exposure to TSH, there is also an accelerated incorporation of amino acids into thyroglobulin and other thyroid proteins when the gland is stimulated. Some of these effects are mediated at the level of translation, either by increased rate of initiation of protein synthesis or by processing or unmasking of large mRNA molecules. The consequence is that TSH increases the formation of polyribosomes in thyroid cells without requiring new RNA synthesis. Later the synthesis of RNA molecules is enhanced by TSH stimulation, both as the result of increased RNA polymerase activity and greater availability of ribose and N base precursors arising from the acceleration of glucose metabolism. It is also noted that TSH increases the rate of biosynthesis of certain thyroid membrane phospholipids. At present it is difficult to determine which of these effects of TSH on macromolecular biosynthesis are related to hormone secretion or to the processes of hypertrophy and hyperplasia of the gland.

The secretion of TSH from the anterior pituitary is modulated both by the central nervous system, which exerts a positive influence, and by the levels of circulating thyroid hormones, which suppress the effects of higher centers in a negative feedback loop (Fig. 3–22). The stimulatory signal from the nervous system may be triggered by acute stress, particularly exposure of the body to extreme cold; cells of the hypothalamus respond by secreting the tripeptide, thyrotrophin-releasing hormone (TRH). The latter is carried to the anterior pituitary where it attaches to a specific receptor on the surface of the cells that produce TSH. By activation of

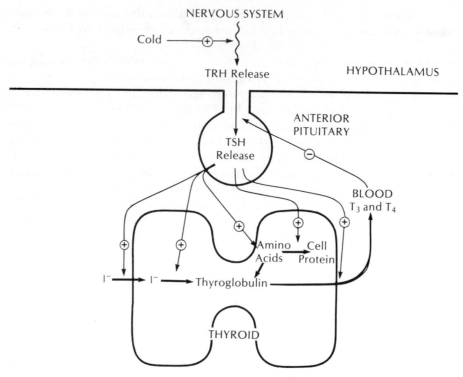

FIG. 3–22.   Release of thyrotrophin and its effects on the thyroid gland.

adenylate cyclase the TRH-receptor interaction promotes a cyclic AMP-dependent release of TSH, which leads in turn to rapid secretion of T_3 and T_4 by the thyroid. The latter hormones return through the circulation to the pituitary where they interfere with the responses to TRH and block TSH release. Thus the interaction of TSH and the thyroid hormones on the secretion of one another maintains the amounts of T_3 and T_4 that are available to peripheral tissues within narrow limits. The effect of T_3 is greater than that for T_4, as for most biologic responses. The feedback inhibition by T_3 or T_4 is not immediate but requires exposure of pituitary cells to thyroid hormone one hour prior to the administration of TRH. Since the block in TSH release is prevented if inhibitors of RNA or protein synthesis (*i.e.,* actinomycin D or puromycin) are present during that hour, it is clear that T_3 or T_4 must induce the synthesis of a polypeptide, which is the true antagonist of TRH action. In situations where there is a deficiency of circulating thyroid hormones, for example with the ingestion of antithyroid agents or diets low in iodine, the production of TSH by the pituitary will continue unchecked at quite elevated levels in an attempt to compensate for the lowered production of T_3 and T_4. Thus, in the long-run, the dysfunctional thyroid gland will be subject to repeated trophic stimulus, and it will enlarge greatly to produce a goiter, the visible swelling in the neck region.

METABOLISM OF THYROID HORMONES

Once they are released into the circulation, both T_4 and T_3 are largely bound in a reversible manner to serum proteins. Most of the binding occurs to a specific thyronine-binding globulin (TBG), an acidic glycoprotein with molecular weight of 63,000. TBG is a minor component among serum proteins (2 mg/100 ml) but it has a very high affinity for T_4 and secondarily for T_3; between 70 to 80% of both thyroid hormones is bound to TBG under normal physiologic circumstances. Part of the remaining T_4 and T_3 is complexed with a second thyronine-binding protein in the prealbumin region of serum electropherograms (TBPA). Although present in 20-fold greater concentrations than TBG in blood serum, its affinities for T_4 and T_3 are about 100-fold weaker so that the TBPA normally carries one-fifth as much of the thyroid hormones (10–15% of the total). A similar proportion is bound nonselectively to serum albumin, which has a 100-fold greater capacity than TBPA, based on its concentration, but a very low affinity reduced by the same factor. The consequence of these interactions is that there is a substantial extrathyroidal pool of the thyroid hormones present in the circulation to buffer fluctuations in secretion rates from the gland. Thus in the normal or euthyroid state in humans this extra thyroidal pool for T_4 amounts to 900 μg, which is ten times the rate of production per day. Because of its lower binding affinity the pool size for T_3 more nearly parallels its daily production rate of 40 μg. In replacement therapy for hypothyroid conditions daily maintenance doses of T_3 will be necessary, while a single weekly dose of T_4 will suffice owing to the great reservoir capacity for the latter.

On reaching peripheral tissues, the thyroid-binding proteins of the blood serum discharge their cargo of T_4 and T_3, which can diffuse freely into cells to exert their target effects, and which undergo metabolic conversions (Fig. 3–23). A percentage of these thyroid hormones is converted to glucuronides and other conjugated metabolites by liver, and is eliminated from the body in the bile, but the main metabolic fate of both T_4 and T_3 is deiodination. In the case of T_3, deiodination leads to formation of inactive metabolites; however the monodeiodination of T_4 in tissues will produce either the highly active 3,5,3'-T_3 if the outer phenolic substituent is attacked, or the inactive derivative 3,3',5'-T_3 [reverse T_3 (r T_3)] by removal of iodine from the inner ring. Normally about half of all T_4 produced each day is converted to rT_3 in the tissues, while a third is transformed into T_3. Since pre-formed T_3 present in thyroglobulin and released in secretions of the thyroid gland only represents about a twentieth of the T_4 output, it may be assumed that the greater part (80–90%) of circulating T_3 is generated outside the thyroid. Liver and kidney have the greatest conversion

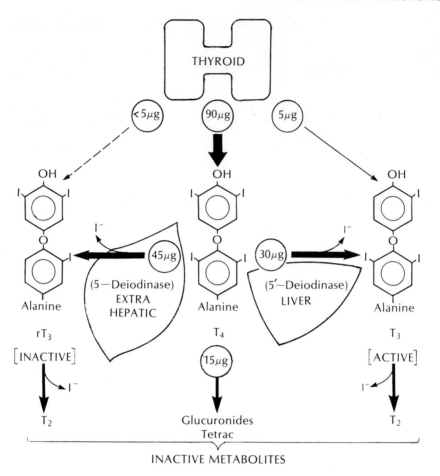

FIG. 3–23. Metabolism of thyroid hormones. T_3 = 3,5,3'–triiodothyronine; rT_3 = 3,3'5'–triiodothyronine; T_4 = thyroxine; T_2 = diiodothyronine.

capacity, but many other tissues, such as skeletal muscle which is significant because of its great mass, play a role in the deiodination reactions.

Two important considerations arise from these interconversions: the
first is the differential between T_3 and rT_3 production, and the possibility of
regulation of thyroid hormone action at the peripheral level by control of
$T_4 : T_3 : rT_3$ concentrations; the second is the question of whether T_4 should
be considered strictly as a prohormone serving as the precursor for the true
thyroid hormone T_3, or whether T_4 has separate, possibly even different,
hormonal effects as does testosterone in relation to its metabolite, dihydrotestosterone. The weight of present evidence seems to favor the latter
possibility, although it is clear that interference with conversion of T_4 to T_3

or diversion to rT_3 may constitute a significant adaptive response to depletion of the body's energy demands.

In individuals with liver diseases, such as chronic hepatic cirrhosis, the levels of circulating T_3 are depressed, while T_4 concentrations are normal, and rT_3 values are considerably higher than normal. Such findings suggest that the 5'-deiodinase enzyme system of liver plays a major role in peripheral conversion of T_4 to T_3, while the 5-deiodinase responsible for rT_3 production from T_4 is an independent enzyme that is localized in extrahepatic tissues. Reciprocal effects upon T_3– rT_3 levels in blood are noted in a number of other diseases and various physiological states. In the human fetus, for example, T_3 production is low while rT_3 production is greater than normal as indicated by comparison of levels in amniotic fluid or umbilical cord blood to those of adult blood. The values approach normality shortly after birth. Prolonged systemic illness such as febrile states, disseminated malignancy, or restriction of caloric intake also lead to reciprocal depression of T_3 and elevation of rT_3 in the circulation. In the case of fasting the effect is reversible on feeding, and is mimicked by the administration of glucocorticoids to well-fed individuals. The antithyroid drug, propylthiouracil, also blocks the peripheral conversion of T_4 to T_3 while it maintains elevated concentrations of rT_3. Since the T_4 levels in all of these situations are either normal, or slightly above normal, the normal feedback system to the pituitary is operational, and circulating TSH remains in the normal range. In other respects persons with low serum T_3 appear to be euthyroid, possibly because of compensatory increases in T_4. It has been suggested on the basis of such findings that the reduction of circulating T_3 levels may be a protective response of the body during the stress of systemic illness or caloric deprivation. In such states tissue catabolism could become excessive; suppression of production of the potent catabolic hormone T_3 could serve as a counteracting defense reaction. Presumably the anabolic and other thyroid hormone responses required for homeostasis would be provided by T_4.

The search for physiological regulators of the 5'-deiodinase that converts T_4 to T_3 is continuing. It is established that the enzyme has active – SH groups at the catalytic site and that it is inhibited by mercurials and other sulfhydryl reagents. Moreover, the low T_3 production in the tissues of fetal animals is associated with lowered levels of free – SH compounds, and is restored to values approaching those in adults by additions of reduced glutathione. In adult tissues the enzyme is inhibited in a competitive fashion by very low levels of rT_3 which suggests an important push–pull control linkage between the two deiodinase systems in the body. Another very potent blocking agent of the T_4 to T_3 conversion is the oral radiocontrast agent (ipodate), which is used in visualizing gallstones (cholecystography). At very low doses this substance inhibits the 5'-deiodinase in liver and causes drastic reductions of circulating T_3 when administered to patients. In the future ipodate may become a useful adjunct for the treatment of hyperthyroid conditions.

THYROID HORMONE ACTIONS

The thyroid hormones exert diverse effects which may vary with respect to the stages of development and differentiation of the specific target tissues. One of the more general characteristics of thyroid action is the acceleration of metabolic rate and generation of body heat. The consequence is striking since it is the respiration of major organs, such as muscle, liver, and kidney, that is enhanced, although the oxygen consumption of several organs, including lung, gonads, and brain, is unaffected. Nonetheless, the effect is detectable at the whole body level, and provided that the measurements of respiration are taken under standardized resting or basal conditions (the so-called basal metabolic rate), individuals with thyroid dysfunction may be diagnosed by this indicator. In the hypothyroid state the basal metabolic rate may fall to as low as one-half of the normal value, while in hyperthyroidism the basal metabolic rate may increase to almost twice the normal value.

The major effect of the thyroid status on tissue respiration and heat production has naturally focused attention upon the mitochondrion and the possible influences of the thyroid hormones on the electron transport system and oxidative phosphorylation. It has been found that T_4 is capable of acting as an uncoupler of mitochondrial oxidative phosphorylation, thus leading to elevated oxygen consumption and heat generation, but lowered ATP synthesis. Although this effect would be consistent with the inefficient production of work and associated thermogenesis that is typical of thyroid hormone excess, the uncoupling effect seems to require concentrations well above the physiological range, and subsequent work with mitochondria isolated from tissues of hypo- or hyperthyroid animals has detected little alteration of efficiency of energy coupling.

Administration of thyroid hormones promotes increases of a number of mitochondrial enzymes in sensitive tissues including cytochrome c oxidase of the respiratory chain, isocitrate dehydrogenase of the citric acid cycle, and α-glycerophosphate dehydrogenase of the shuttle system that is required for transfer of reducing equivalents from extramitochondrial NADH across the membrane to the respiratory chain (Fig. 3–24). The adaptive response of respiratory enzymes involves new synthesis of proteins, an increase of mitochondrial mass per unit weight of tissue, and increased amounts of the cytochromes, c and aa_3, per mitochondrion. Such studies have suggested an induced biogenesis of membranes of the mitochondrial cristae with their respiratory units accomplished by thyroid hormones.

A number of extramitochondrial enzymes and processes are also dependent on thyroid status. In several tissues the increase in oxygen con-

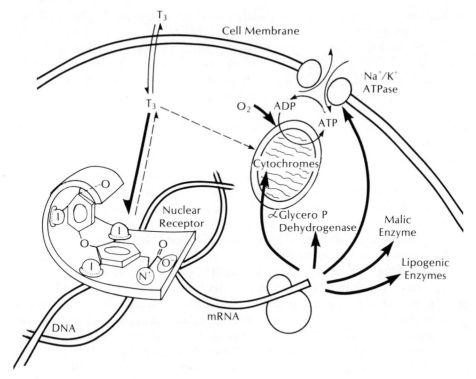

FIG. 3–24. Actions of thyroid hormones. Heavy black arrows show those enzyme proteins which are induced by interaction of T_3 with the nuclear receptor.

sumption produced by thyroid hormones is accompanied by increased active sodium transport and marked elevation of plasma membrane Na^+/K^+ ATPase. Moreover, ouabain, which selectively inhibits this enzyme and its function as a Na^+-pump, reduces the thyroid hormone-mediated increases in oxygen consumption. A significant part of the extra energy demands of cells exposed to thyroid hormones can be related to induction of the Na^+ $/K^+$ ATPase (Fig. 3–24). As a result, the intracellular ADP/ATP ratios will rise, and mitochondrial respiration will increase to restore the energy balance. Other cytoplasmic enzymes not associated directly with the respiratory response that show stimulatory effects with thyroid hormones include several anabolic processes associated with lipogenesis including NADPH-generating isocitrate dehydrogenase, glucose-6-phosphate dehydrogenase, and malic enzyme, as well as the citrate-cleavage enzyme.

The thyroid hormones exert their actions on target tissues by stimulating the synthesis of particular proteins as the consequence of their interaction with specific receptors in cellular nuclei (Fig. 3–24). Unlike the steroids, there is no necessity for binding of T_3 or T_4 to cytoplasmic receptors for movement from cytosol to nucleus. The nuclear receptors are nonhistone chromatin proteins with a molecular weight of about 50,000. The receptors have a very high affinity (10^{11} M for T_3), but low capacity (5000 binding sites/liver nucleus). When tested with a number of thyronine analogues, binding to the nuclear receptors was seen only with hormonally active derivatives. Thus, T_4 is bound only to one-fifth the extent of T_3 under *in vivo* conditions (which corresponds to relative potencies of the two hormones), while rT_3, which will only evoke a significant biologic response when present at 400 times the concentration of T_3, is bound to only 0.2% the extent of T_3. Moreover, the presence of nuclear receptors is found only in thyroid hormone-responsive cells. A strict correlation exists between biologic response and the occupation of nuclear binding sites by the thyroid hormones; the brain is the only exception to this rule. Despite the presence of substantial numbers of nuclear T_3-binding sites, nervous tissue exhibits no response to thyroid hormone in terms of oxygen consumption or Na^+/K^+ ATPase; presumably the true biochemical response of the brain to T_3 remains to be clearly defined.

Following the interaction of the thyroid hormones with nuclear chromatin there is a delay of several hours before the expression of biologic effects, as should be anticipated if intervening processes of transcription and translation are involved. Moreover, the responses are blocked by inhibitors of RNA polymerase, such as amanitin, or of protein synthesis, such as puromycin. Enhanced synthesis of poly A-containing nuclear RNA and certain species of particular mRNAs corresponding to inducible polypeptides in pituitary and hepatic cells has been observed as an early nuclear event prior to synthesis of the proteins. However, there is also a more generalized increase of mRNA and ribosomal RNA species following exposure of liver cells to T_3; this suggests a relatively general response of the genome to thyroid status in some tissues. In addition to these effects on nuclear chromatin, it has been indicated that thyroid hormones may also interact with receptors in the mitochondria. Whether such interactions may trigger direct effects on mitochondrial DNA, or exert analogous activation of the separate intramitochondrial protein-synthesizing system is a subject for further investigation.

HYPOTHYROIDISM

Deficiencies of thyroid function may arise from congenital defects in the gland, from inadequate intake of dietary iodine, from acquired toxic factors such as the goitrogens, or from diseases such as thyroiditis. The lack of thyroid hormones during critical stages of differentiation in infancy has severe effects, and if untreated will produce irreversible retardation of both physical and mental development. The resulting condition is referred to as cretinism, or Fagge's disease. In the adult, hypothyroidism is fairly common, especially among females, and has less serious consequences. The condition is typified by dry, flaking skin, and is referred to as adult myxedema or Gull's disease. The general metabolic rate is depressed, and serum cholesterol levels are characteristically elevated. The typical myxedematous patient exhibits a complacent, good-tempered disposition, lack of physical energy and activity, hypersensitivity to cold, thickening of the skin, and puffy edema of the face, especially around the eyes. In its severest state the lethargy may deepen to coma, hypothermia, and death unless prompt intravenous administration of thyroid hormones is started. In general, patients with hypothyroidism have lowered serum levels of T_4 and elevated TSH. The daily dose of replacement T_4 may be gradually increased until serum TSH falls to normal, the most sensitive indication of restoration of the euthyroid condition.

A number of inherited disorders of thyroid hormone biosynthesis have been discovered in congenital cases of hypothyroidism with goiter. In one group of patients there is a defect in the iodide accumulation mechanism with the results that iodine content of the thyroid is very low, and only a very small proportion of a dose of ^{131}I is taken up by the gland; such individuals may be treated simply by increasing iodide intake sufficiently to raise the plasma levels by 10- to 20-fold thus permitting entry of iodide by simple diffusion and providing for hormone synthesis. A second category of hypothyroidism is related to impaired iodide organification such that although the gland is capable of taking up ^{131}I initially, the labelled iodide is readily released from the thyroid (*e.g.*, following the administration of thiocyanate); these patients have a defect of the thyroid peroxidase system responsible for converting inorganic to organic iodide, and hence the inability to retain iodine in the gland. A third syndrome is seen in patients with a congenital defect in the coupling of iodotyrosine residues in thyroglobulin accompanied by impairment of iodothyronine formation; as in the second category treatment requires the replacement of thyroid hormones. Finally there is a lesion caused by deficiency of the deiodinase enzyme that normally recycles the thyroid iodide by breaking down free iodotyrosines, thus preventing their escape from the gland and their rapid elimination in the urine; such persons, therefore, suffer from inefficient retention and continued leakage of iodine stores, and, as in the case of patients with defective iodine-trapping mechanisms, these persons can be maintained with normal thyroid function simply by increasing their dietary iodide intake.

HYPERTHYROIDISM

The most common syndrome of thyroid hyperactivity is Graves' disease, a condition characterized by diffuse hyperplasia of the gland. The symptoms are the opposite to those seen in myxedema, namely acceleration of basal metabolism, lowering of serum cholesterol, and warm moist skin. These individuals may exhibit the characteristic eye changes called exophthalmos; this pop-eyed expression, accompanied with heart palpitations and a jumpy jack-in-the-box disposition, produces a striking resemblance to the typical patient suffering from thyrotoxicosis. In extreme cases the disorder may have a fatal outcome, generally as the result of a crisis triggered by acute infection that provokes uncontrolled hyperactivity of the gland known as thyroid storm. This severe exacerbation is accompanied by high fever, pronounced tachycardia, vomiting, and agitation leading to wild delirium followed by stupor.

The serum concentrations of T_4 and T_3 are usually elevated markedly (two- to fivefold) in Graves' disease, while the circulating TSH concentration is reduced to values that are barely measurable because of feedback inhibition of the pituitary by elevated thyroid hormone levels. Several abnormal thyroid stimulators (TSs) have been detected in the immunoglobulin fraction of thyrotoxic serum. The first of these to be described was termed the long-acting thyroid stimulator (LATS) since its effect on the gland was similar to that produced normally by TSH, but of longer duration. LATS seems to bind to the same membrane receptors, and produces activation of adenylate cyclase in the same manner as TSH. A different factor, HTS, specific to the human, has also been found in the immunoglobulins of Grave's patients, as has a substance which is implicated in exophthalmos production (EPS). The presence of these antibodylike agents and the proliferation of the thymus and lymphatic system has prompted much investigation and speculation concerning the role of immune responses in the pathogenesis of Grave's disease.

Treatment of thyrotoxicosis was initially restricted to surgical removal of the gland. Discovery of the selective avid uptake of iodide by thyroid cells and the availability of radioactive iodine isotopes has led to the technique of destruction of hyperactive glandular tissue by concentrative uptake following a drink of ^{131}I solution. The therapeutic dose is calculated on estimates of thyroid weight. Radioiodine therapy is less traumatic for the patient than surgical ablation, and leads to fewer complications (*i.e.*, impairment of calcium metabolism from parathyroid injury), but it is contraindicated in young patients or pregnant females because of the hazards of radiation-induced malignancy. In addition to ablation therapies, the administration of antithyroid agents is necessary. In the case of surgery a course of treatment with propylthiouracil for several weeks followed by high doses of potassium iodide just prior to the operation causes the gland to involute and improves the recovery. With radioiodine therapy it is not desirable to block glandular functions before administration of ^{131}I since this could reduce uptake of radioiodine into the gland, as in the case of potassium iodide, or render the tissue radioresistant, as in the case of – SH containing drugs such as propylthiouracil. Instead, the antithyroid agents are administered some days following the ^{131}I drink in order to provide a more rapid curtailment of thyroid function than that produced by ^{131}I alone.

PARATHYROID HORMONE — CALCITONIN

As noted in the preceding chapter, the balance of calcium in the body is subject to nutritional influences, the intake of vitamin D, and the mineral itself. Calcium metabolism is also regulated by two polypeptide hormones: the first to be described, parathyroid hormone, is produced by four discrete glands near the thyroid, and acts to elevate plasma Ca^{2+} levels; the second, termed calcitonin, is secreted by diffusely scattered cells known as C cells within the thyroid, and antagonizes effects of the parathyroid hormone.

Parathyroid hormone is synthesized as a preprohormone of 115 amino acid residues within the endoplasmic reticulum. This is cleaved first to the prohormone form inside the parathyroid cells, and then by removal of a hexapeptide to the active hormone for secretion. The active hormone is 84 amino acids in length, but it appears that only the first 34 are essential for activity. High levels of Ca^{2+} outside the parathyroid shut down the production and secretion of the parathyroid hormone, while a lowering of plasma Ca^{2+} acts as the signal to release more of the hormone into the circulation (Fig. 3–25). As with other polypeptide hormones it appears that parathyroid hormone combines with highly specific receptors at the cell membranes of target tissues where it elicits a stimulation of adenyl cyclase to produce increases in cellular cyclic AMP concentrations. The effects of parathyroid hormone stimulation described previously in connection with vitamin D metabolism include an activation of 1-α-hydroxylase in the kidney to produce the active 1,25-dihydroxycalciferols; the latter in turn promote calcium absorption by the intestine, increased reabsorption of calcium and decreased reabsorption of phosphate in the kidney tubules, and stimulation of calcium resorption in the bones. In addition to indirect effects through the active vitamin D metabolites, an elevation of parathyroid hormone exerts direct actions upon calcium balance, mainly at the level of enhanced bone demineralization and renal reabsorption of calcium. Some responses in the bone are long-term processes involving the proliferation and differentiation of the osteocytes and other cell types that participate in processes of mineral mobilization and bone remodeling. However, there are also more rapid effects of parathyroid hormone, acting in concert with vitamin D, that promote flux of calcium from the bone fluid compartment into the extracellular fluids. This compensatory contribution from bone calcium, together with the parathyroid hormone-mediated stimulation of reabsorption of Ca^{2+} in the kidney tubules, will provide minute to minute adjustments for the homeostasis of blood calcium. The combined effect is to increase the concentration of free calcium ions in the blood plasma, offsetting the hypocalcemia that triggered release of the hormone from the parathyroid glands.

In some individuals, the normal feedback inhibition of parathyroid hormone secretion by elevated Ca^{2+} is inadequate resulting in the syn-

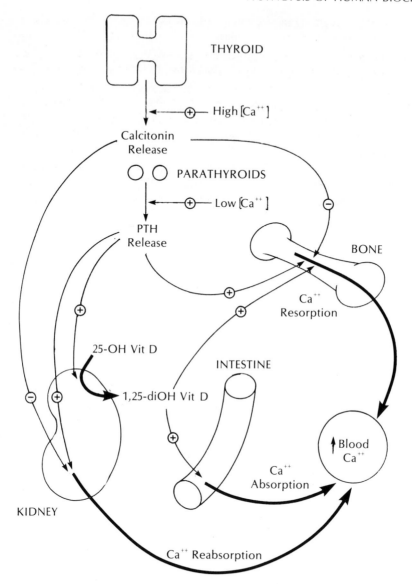

FIG. 3–25. Parathormone effects in tissues. Stimulating factors are shown by a positive sign, inhibitory factors by a negative sign.

drome known as hyperparathyroidism or von Recklinghausen's disease. Many of the symptoms (muscle weakness, fatigue, depression, peptic ulcers, pancreatitis, gouty arthritis, recurrent renal stones, and other signs of kidney dysfunction) may be consequences of the chronic elevations of serum Ca^{2+}. The other common features involving the skeleton, osteitis fibrosa, which is a form of localized, cystic bone resorption, or osteopenia, which is a more generalized or diffuse demineralization, result from prolonged stimulation of osteoclasts and osteoblasts by increased parathyroid hormone with dissolution of both the organic matrix and mineral of the bones. Skeletal pain, pathologic fractures, and undermineralization on radiologic examination are diagnostic signs of osteitis fibrosa cystica or osteopenia. The administration of 1 to 2 g of phosphorus daily in the form of inorganic phosphate salts can reduce the serum Ca^{2+} by promoting uptake into the bone. Since the cause of hyperparathyroidism is usually hyperplasia or tumor of the gland, surgical removal of abnormal parathyroids is the treatment of choice. In the postoperative period it may be necessary to provide supplemental dietary calcium and vitamin D to offset lowered serum calcium as in the treatment of cases of hypoparathyroidism.

Calcitonin is a 32 amino acid polypeptide that is secreted continuously at basal levels with normal physiological levels of Ca^{2+} in blood, and in increased amounts with elevations of blood Ca^{2+} concentration (Fig. 3–25). The hormone has two actions that are opposite to those of parathyroid hormone: first and foremost, it blocks the resorption of mineral from bone; second, it decreases the reabsorption of calcium by the kidney. Thus calcitonin can reverse the state of hypercalcemia, which triggered its own secretion by favoring retention of calcium in the bone and its elimination in the urine. The significance of this regulation in humans is obscure at present. There is no known state of hypercalcemia associated with calcitonin deficiency, and excessive secretion of calcitonin observed in some cases of medullary carcinoma of the thyroid C cells is not accompanied by hypocalcemia. Moreover, other secretogogues arising from the gastrointestinal tract, such as glucagon and gastrin, also stimulate calcitonin secretion. Thus during digestion of food these gastrointestinal hormones may provide an anticipatory surge of calcitonin to favor deposition of dietary calcium into bone as soon as it is absorbed. A pathologic condition, known as osteitis deformans or Paget's disease, is characterized by an excessive rate of bone resorption resulting in skeletal deformities, fractures, and pain. The cause of the disease is not known and does not seem to be due to simple deficiency of calcitonin, which is present at normal levels in serum; nonetheless, continued administration of calcitonin has produced significant remissions of bone pain and radiologic features of Paget's disease.

CATECHOLAMINES

The aromatic amines, epinephrine, norepinephrine, and dopamine act as major regulators of physiologic and metabolic processes in several sites of the body. They also play important roles as effectors of the sympathetic nervous system, and as neurotransmitters and modulators within the central nervous system. Imbalances and derangements of the catecholamines

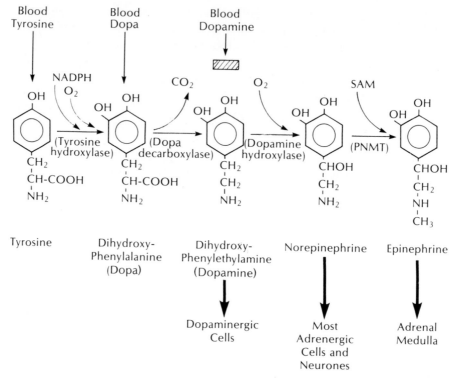

FIG. 3–26. Pathways and enzymes of catecholamine biosynthesis. SAM = S–adenosylmethionine; PNMT = phenylethanolamine N–methyl transferase.

can produce serious pathologic effects on cardiovascular and neuromuscular functions, and have been implicated in psychiatric disturbances and alterations of mood. As for the thyroid hormones they are derived from the aromatic amino acid, tyrosine; however, the catecholamines act more similarly to peptide hormones at their targets since they fail to penetrate membranes and produce their effects by binding to specific receptors on cell surfaces. Their actions are mediated, therefore, by the classical second messenger system involving ATP, GTP, and their cyclic nucleotide derivatives. In spite of a general understanding of this mechanism which has grown from 25 years of extensive study there are still many unanswered questions regarding the actions of the catecholamines.

METABOLISM AND SECRETION

Biosynthesis of catecholamines occurs within cell bodies and varicosities of adrenergic nerves, in sympathetic ganglia, and the adrenal medulla. The major precursor is tyrosine obtained in the diet or from hydroxylation of phenylalanine by the liver (Fig. 3–26). The formation and storage of

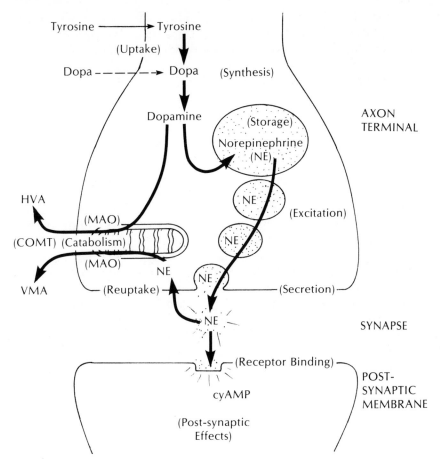

FIG. 3–27. Secretion of norepinephrine in neuronal terminals, DOPA = dihydroxyphenylalanine, HVA = homovanillic acid, VMA = vanillylmandelic acid, MAO = monoamine oxidase.

catecholamines by secretory cells is illustrated by the sequence which occurs in neurones (Fig. 3–27). An active transport mechanism in the cell membrane takes up tyrosine from the circulation. In the neuronal cell body the tyrosine is first converted to 3,4-dihydroxyphenylalanine (dopa) by a hydroxylase enzyme that is formally similar to the phenylalanine hydroxylase of liver; then it is converted to dopamine by a decarboxylase that removes CO_2 from the alanine side-chain. Tyrosine hydroxylase requires oxygen, NADPH, and a pteridine coenzyme, and is the rate-limiting enzyme of the pathway; it is subject to feedback inhibition by the end-product, norepinephrine, and it is also inhibited by the drug, α-methyltyrosine. The hydroxylase enzyme appears to be the major regulatory step in the pathway. Synthesis of the enzyme is induced following prolonged periods of release of the end-product catecholamines by repeated stimulation of adrenergic nerves. Dopa decarboxylase requires pyridoxal phosphate as a prosthetic group; the decarboxylase is present in excess and rapidly converts endogenous dopa, or that which is taken up into the cells from exogenous sources, into dopamine. The reactions to this point are catalysed by soluble enzymes localized in the neuronal cytoplasm; the newly synthesized cytoplasmic dopamine is packaged in the cell bodies into membrane-bounded storage vesicles. In some cell types, for example the dopaminergic cells of the basal ganglia and the hypothalamus, dopamine is the final end-product. A deficiency of dopamine in the basal ganglia has been demonstrated in the neurologic disorder known as parkinsonism. This condition, characterized by tremor, rigidity, and dyskinesia, shows dramatic improvement on administration of dopa, but not dopamine itself, indicating that the free amine cannot cross from the bloodstream into the brain. Presumably the defective neurons in the substantia nigra lack the tyrosine hydroxylase but contain sufficient dopa decarboxylase that they may take up exogenous dopa from the circulation to correct the dopamine deficiency. In other adrenergic neurons the dopamine is converted to norepinephrine, the major end-product, by the copper-containing enzyme, dopamine-β-hydroxylase, which is activated by ascorbic acid and is contained within the storage vesicles. The storage granules are transported to the terminal varicosities by axonal flow, an active process that requires energy. In the adrenal medulla a further conversion takes place whereby the norepinephrine is methylated on the amine group to form epinephrine. This reaction is catalysed by a specific enzyme, phenylethanolamine-N-methyltransferase, which requires S-adenosylmethionine as the activated methyl donor, and is feedback-inhibited by epinephrine. Induction of synthesis of the enzyme by the medullary cells requires the collaboration of glucocorticoids from the adrenal cortex. The adrenal medulla secretes small amounts of both dopamine and norepinephrine, but epinephrine is the main secretory product.

Catabolism of the catecholamines involves the sequential action of two enzymes, monoamine oxidase which catalyses oxidative removal of the amino group, and catechol-O-methyl transferase which adds a methyl group on the aromatic ring (Fig. 3– 28). However, the primary mechanism for the inactivation of catecholamines, once they have exerted their effects on the exterior surface receptors of target cells, is uptake into the intracellular compartment. In the nervous system this occurs mainly by reuptake back into the secretory neuronal cytoplasm (Fig. 3– 27). Here the catecholamines become exposed to the monoamine oxidase of the outer membranes of mitochondria and are oxidized to the corresponding aldehyde derivatives with release of ammonia (or methylamine in the case of epinephrine). In the central nervous system the aldehyde produced from norepinephrine is reduced to the alcohol, 3,4-dihydroxyphenylglycol; in

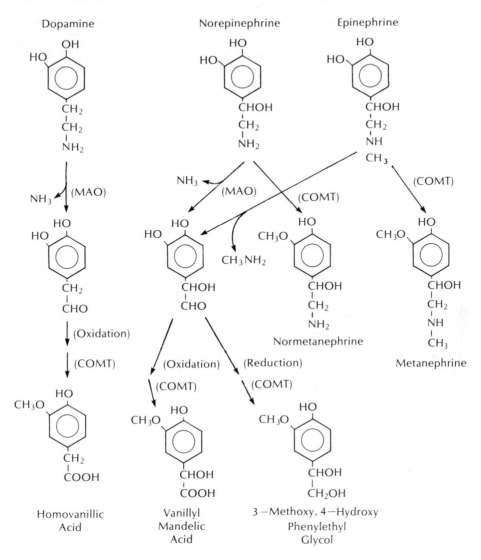

FIG. 3–28. Catecholamine catabolism. MAO = monoamine oxidase; COMT = catechol O-methyl transferase.

peripheral nerves the aldehyde is oxidized to the carboxylic acid derivative, 3,4-dihydroxymandelic acid. These deaminated forms may not be used by adrenergic cells and pass through the circulation to the liver, kidneys, and other tissues which contain the catechol-O-methyl transferase; the latter enzyme, using S-adenosylmethionine as the active methyl donor, converts the compounds to their 3-O-methyl derivatives, which are excreted in the urine. The principal urinary excretory product is the vanillylmandelic acid derived from peripheral neurons, while the minor 3-methoxyhydroxy phenylglycol excretion reflects, in part, the release from central neurons. The urinary end-product from the dopamine-producing cells is homovanillic acid. On the other hand, catecholamines that are released intact into the blood are inactivated mainly by methylation to produce metanephrine (from epinephrine) or normetanephrine (from norepinephrine) (Fig. 3– 28); the quantities excreted in these two urinary end-products, therefore, reflect the extent of turnover of the circulating catecholamines in the body.

Release of the catecholamines from their protected storage sites within the cytoplasmic vesicles occurs by a process of exocytosis, the vesicle membranes fusing with the cellular membrane and the contents of amines being discharged into the extracellular space. A small amount of catecholamine release occurs under basal conditions; stressful stimuli (anger, fear, trauma, cold exposure) greatly enhance the release by initiation of impulses in adrenergic nerves, or by stimulation of the adrenal medulla, to produce varying proportions of norepinephrine to epinephrine secreted. During the activation of the secretory mechanism the ion permeability of the plasma membrane increases with inward movement of both sodium and calcium cations. The increase of intracellular Ca^{2+} has been implicated in the migration of secretory vesicles along the microtubular framework by promoting the movement of contractile microfilaments, much in the same manner as it initiates contraction of actomyosin in skeletal muscle fibers. When the vesicles discharge their contents the membranes may be taken back into the secretory cell and refilled with norepinephrine or dopamine from the cytoplasm to undergo a new cycle of exocytosis. The recycled vesicles can be distinguished from those which have been newly synthesized by their much smaller size.

EFFECTS AND MODE OF ACTION

The effects of epinephrine are designed to prepare the body to react quickly to noxious stimuli, and to maintain the vigorous activity necessary to combat or escape harm (the so-called "fight or flight" response to danger). The primary metabolic effects are mobilization of energy-producing substrates from storage by activation of lipolysis in adipose tissue and of glycogenolysis in liver and muscle. The consequence is a rise in plasma fatty acid and glucose concentrations, and provision of extra sources of energy to sustain muscle contraction. Stimulation of the heart to increase the cardiac output, coupled with dilation of the bronchioles of the lungs and the vasculature supplying the muscles by epinephrine ensures adequate flow of oxygen and food during prolonged activity. At the same time nonessential visceral activities are shut down by constriction of the blood vessels to divert the flow away from the gut and kidneys, and the motility and secretory activities of the gastrointestinal tract are inhibited. While most nonmuscle tissues (skin, lungs, mucous membranes) respond to epinephrine with a reduced blood flow because of vasoconstriction, the situation is directly opposite for the liver. Increased hepatic blood flow by epinephrine ensures the rapid removal of circulating lactic acid produced by muscle activity, and the increased secretion of glucose from glycogenolysis and from gluconeogenic conversion of lactic acid.

Many effects of norepinephrine are similar to those of epinephrine, but there are notable differences. Both hormones raise the blood pressure, epinephrine chiefly through its stimulation of the heart, norepinephrine mainly by peripheral vasoconstriction. Both elevate metabolic rate and activate lipolysis, but epinephrine is more effective in counteracting hypoglycemia, while norepinephrine is more important in the response to cold exposure. In addition to its systemic effects it should be kept in mind that norepinephrine exerts localized effects on those tissues which it reaches directly by sympathetic innervation. Variable ratios of epinephrine/norepinephrine are secreted from the adrenal medulla into the circulation in different individuals in response to different stimuli (e.g., anger vs anxiety). Thus the physiological and metabolic adaptations to threatening situations vary greatly from person to person, and probably play a key role in the ability to cope with stress.

Different classes of catecholamine membrane receptors designated as α or β have been proposed to explain the differential effects of the hormones in various tissues. Synthetic analogues of the catecholamines have been discovered which activate either class of receptor (α- or β-agonists), while other drugs have been found to cause selective blockage of the activation of one or the other class of receptor (α- or β-blockers). In the smooth muscle cells, for example, the activation of α-adrenergic receptors promotes contraction and the activation of β-adrenergic receptors causes relaxation. Norepinephrine exerts strong α-adrenergic responses, as does the analogue, phenylephrine, while the drug phentolamine blocks these actions. The analogue, isoproterenol, produces the β-adrenergic effects which are blocked by the drug, propranolol. Epinephrine elicits both the α and β type of responses. Thus the typical α-adrenergic response is vasoconstriction produced by contraction of smooth muscle elements in blood vessel walls, and is evoked by catecholamines with the following order of potency: epinephrine > norepinephrine > phenylephrine; the β-adrenergic response is vasodilation by relaxation of vascular smooth muscle produced in the order isoproterenol > epinephrine $\geq$ norepinephrine. Some effects may be clearly placed in one or the other category, as for smooth muscle. For example, in the regulation of appetite in the hypothalamus α-adrenergic agents stimulate the "hunger receptors" to increase food intake even in the well-fed state, while β-adenergic agents produce the opposite effect to reduce appetite even during a fast. However, in other cases α

and β effects may not oppose one another, and, as noted below for catecholamine-induced release of glucose from the liver, may exert parallel, reinforcing influences on some cellular processes.

Mechanisms of the actions of β-adrenergic agents are clearly explicable in relation to the intracellular second messenger, cyclic AMP. In most systems the biologic efficacies of β-agonists show strong correlations with their affinities for β-receptors on the outer cell surface, and strong correlations with their abilities to activate adenyl cyclase on the inner cell surface. The actions of catecholamines in stimulating glycogen breakdown in skeletal muscle and liver and in activating hydrolysis of triacylglycerol in adipose tissue are accompanied by elevations of tissue cyclic AMP, and are potentiated by drugs, such as theophylline, which block hydrolysis of the cyclic AMP by phosphodiesterase. All of these effects of catecholamines may be mimicked by added dibutyryl cyclic AMP, a membrane-permeable derivative of the second messenger, and depend on the cyclic AMP-mediated dissociation of regulatory subunits from cellular protein kinases, which promotes an allosteric activation of their catalytic subunits. The activated protein kinases then phosphorylate the enzymes under their control either to activate (glycogen phosphorylase, hormone-sensitive lipase) or inactivate (glycogen synthetase) them.

However, mechanisms of α-adrenergic stimulation are not so clear-cut. In some instances where α and β stimulations produce opposite effects, as in the case of smooth muscle, it might be expected that α-adrenergic agents should cause a fall in cyclic AMP levels, but this has not been demonstrated unequivocally. An alternative model is based on the observation that cyclic GMP may be increased following α-adrenergic stimulation, and could exert effects antagonistic to those of cyclic AMP, the so-called yin–yang relationship of opposing forces. A third possibility views the changes in cyclic GMP as secondary to an elevation of cytosol calcium ions which may be the primary response to α-adrenergic stimuli. The situation is analogous to that for insulin actions, with many questions still to be answered. In the regulation of hepatic glucose release by catecholamines, factors other than cyclic AMP are involved. For example, the α-adrenergic agent, phenylephrine, activates liver phosphorylase without increasing cyclic AMP concentrations, and the α-blocker, phentolamine, can prevent the activation of phosphorylase by epinephrine without inhibiting the cyclic AMP accumulations. Identity of the intracellular mediator of the α-adrenergic stimulation of phosphorylase is unknown, but several lines of evidence point to calcium ions. The protein kinase that phosphorylates glycogen phosphorylase b to its active a form is stimulated by Ca^{2+} in the absence of cyclic AMP. Moreover, effects of α-adrenergic agents on glycogenolysis in liver cells are generally lower without the simultaneous addition of calcium; recent studies have suggested that the source of calcium for stimulation of phosphorylase kinase may be from mobilization of calcium that is sequestered within intracellular organelles rather than from influx of extracellular calcium. Thus, both α- and β-adrenergic mechanisms involving independent second messengers participate in the acceleration of glucose release by the activation of glycogen phosphorylase.

PROSTAGLANDINS

No single area within the field of endocrinology has witnessed such vibrant interest and frenetic progress in recent years as the study of the prostaglandins. These derivatives of the polyunsaturated fatty acids were first detected some 50 years ago as vasoactive principles in seminal fluid derived from the prostate gland, hence the name; however, only within the past decade has it become apparent that the main function of the so-called essential fatty acids (arachidonic, linolenic, and linoleic acids) in cells is to serve as precursors for the prostaglandins, and that the latter play pervasive roles in the control of physiologic and metabolic events including: (1) modulating the actions of other hormones such as the glucocorticoids, the gonadal steroids, gonadotrophins and ACTH; (2) affecting the secretions of a number of polypeptide hormones by the pituitary, of neurotransmitters by the brain, and of water by the kidneys; (3) reacting to stress and injury in the processes of hemostasis, inflammation, fever, and pain perception; (4) promoting contractile processes in the blood vessels, cardiac muscle, and the gastrointestinal tract. In many respects the prostaglandins act as true hormones, but they exert their major effects locally, often within the cells that produce them. Moreover, they are not stored within the tissues of origin and they are rapidly removed from the circulation, hence their actions depend on continued biosynthesis. Since they act intracellularly, frequently by modifying the cyclic nucleotide or calcium ion responses of other endocrine agents, the prostaglandins may be considered as second messenger modulators of hormone action rather than as primary effectors. Currently, attention has been drawn to the even more potent but unstable relatives of the prostaglandins, the thromboxanes and the prostacyclins.

Chemically the prostaglandins are C_{20} fatty acids with a cyclopentane ring near the middle of the chain formed by a bond between carbons 8 and 12: thus there is a 7-C chain terminating in the carboxyl group at position 1, and an 8-C chain with terminal methyl group at position 20. Variants in substituents on the ring are designated by capital letters A, B, C, and so forth, and numbers of double bonds in the chains are designated by number-subscripts 1, 2, or 3. Although all prostaglandins are fat-soluble, one series with 9-keto, 11-hydroxy substituents were originally found to be more soluble in ether (hence were designated E) than another more polar series with hydroxyls in both 9 and 11 positions (termed F from their differential partition into aqueous "fosfat" (Phosphate) buffers). These two major prostaglandin types (Fig. 3–29) frequently have opposing influences

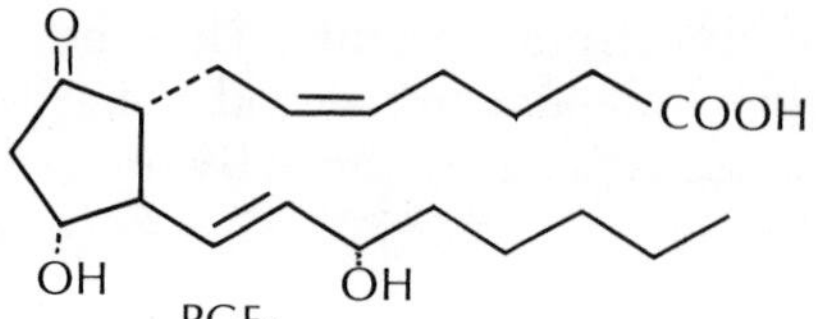

PGE_2

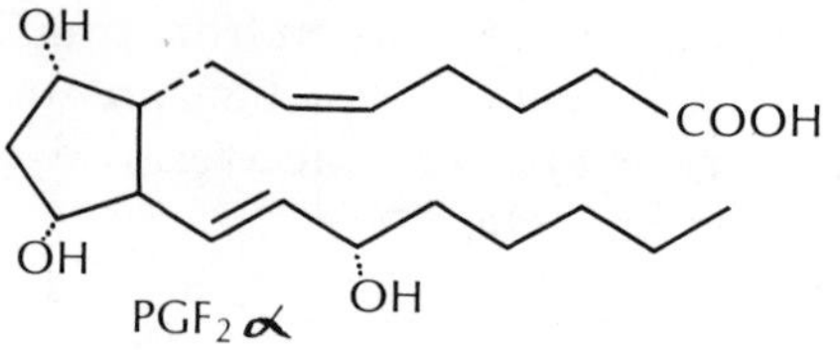

$PGF_{2\alpha}$

Prostaglandins

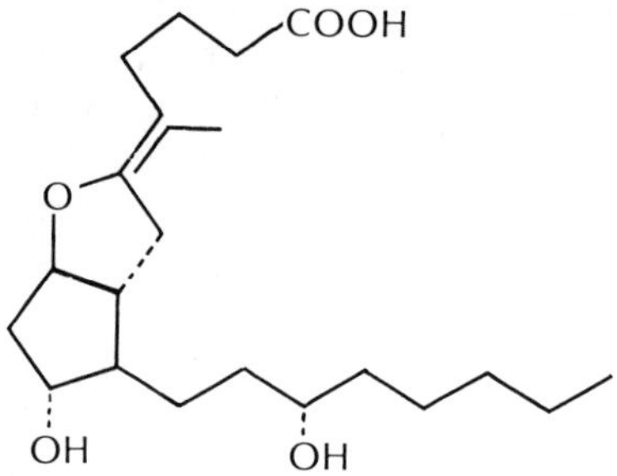

PGI_2
Prostacyclin

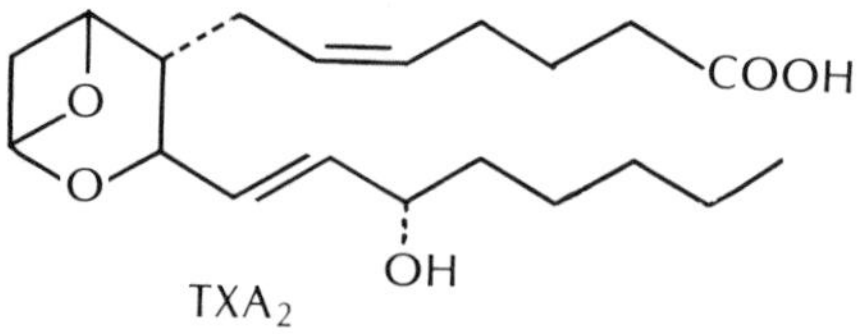

TXA_2
Thromboxane

FIG. 3–29. Structures of prostaglandins, prostacyclin, and thrombroxane.

on cell processes, therefore their ratios will be the determining factor in the effects observed. An additional group of prostaglandins designated A, B, C, D, G, and H, which are structurally and metabolically related to the E and F series, have subsequently been found under certain conditions. Recently, two more series with different ring structures, called prostacyclins (or prostaglandin I) and thromboxanes (Fig. 3–29), have been implicated in clot formation during thrombosis of the blood; these highly active derivatives may prove to be of immense importance in the pathology and the development of therapeutic measures in thrombotic diseases, and they may act as mediators in responses of other tissues as well.

METABOLISM

Biosynthesis of the majority of prostaglandins and related agents starts with the C_{20} tetraenoic essential fatty acids, arachidonic acid, and since the process results in the elimination of two double bonds, forms the end-products with two unsaturations (PGE_2, PGF_2, etc.). Lesser amounts derive from the C_{20} trienoic fatty acid which arises from linoleic acid (PGE_1, PGF_1, etc.) and traces from the C_{20} pentaenoic acid (PGE_3, PGF_3, etc.). The enzyme responsible for prostaglandin synthesis, cyclo-oxygenase, produces endoperoxide intermediates, PGG and PGH, which may go on to form the typical prostaglandins, PGE and PGF, or the related thromboxanes, TXA

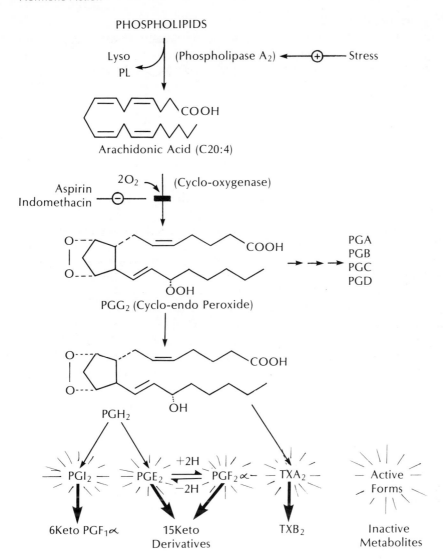

FIG. 3–30. Pathways of formation of the active prostaglandins and their inactive metabolites.

and **TXB**, and prostacyclin, **PGI**, by the subsequent action of the appropriate synthetase enzymes (Fig. 3–30). In addition there are a number of isomerases that can produce other derivatives (PGA, PGB, PGC, PGD) and a reductase, which is able to interconvert PGE and PGF in the presence of pyridine nucleotides. The cyclo-oxygenase enzyme requires oxygen as well as free polyunsaturated fatty acid as substrates, and heme as a cofactor; it is an important control point in prostaglandin biosynthesis.

Regulation of the cyclo-oxygenase is effected by concentrations of heme in the cell, and by a somewhat paradoxical response to oxygen levels. For example, it is clear that prostaglandin synthesis is enhanced by the lowered oxygen concentrations associated with tissue ischemia, yet oxygen is an essential substrate for the reaction. It appears that although the enzyme would be inoperative in the complete absence of oxygen it has a sufficiently low Km (in the micromolar range) that it is quite active with oxygen levels found in partial tissue anoxia. The physiologic rate-limiting factor seems to be the concentration of essential fatty acid substrate, chiefly unesterified arachidonic acid. Under normal unstressed conditions in cells virtually all the fatty acids, including polyunsaturated species, are tied up in ester linkages in complex lipids such as the triacylglycerols of the depots or phospholipids of plasma and intracellular membranes. Accordingly it requires the activation of lipid hydrolysing enzymes (lipases) in order to provide optimal substrate levels of arachidonic acid for the production of prostaglandins, thromboxanes, and prostacyclins. The key reaction seems to be that of a specific phospholipase which releases the polyunsaturated fatty acid selectively from position 2 of cellular phospholipids (Fig. 3–30). This enzyme, phospholipase A_2, is stimulated under stressful conditions of cellular injury which triggers prostaglandin formation; it appears to be activated by the action of an intracellular protease, and the activation is suppressed by the presence of the anti-inflammatory steroids such as cortisone. An important pharmacologic effect is also exerted on this control point by the nonsteroidal, anti-inflammatory agents such as aspirin or indomethacin. Aspirin (acetylsalicylic acid) acts as a potent irreversible inhibitor of the cyclo-oxygenase enzyme, involving removal of the enzyme from the cell. Following exposure to aspirin, the enzyme becomes acetylated, disappears, and only reappears after a period of 12 to 24 hours required for the synthesis of new enzyme molecules. Indomethacin, by contrast, is a reversible inhibitor of cyclo-oxygenase. A fascinating property of the isolated enzyme is its tendency to reach a climax of high activity, then self-destruct in the presence of oxygen. There may be a slow trickle of base-line prostaglandin formation in cells, but the cyclo-oxygenase is apparently designed for explosive generation of end-products in crisis situations.

Catabolism of the prostaglandins is an extremely effective and rapid terminator of their actions in cells, and eliminator of their presence in biologic fluids. Both the lungs and the liver possess highly active enzyme systems for conversion of active prostaglandins to biologically inactive metabolites, such that in one passage through the circulation 90% or more of the PGEs and PGFs may be eliminated. In the case of the thromboxanes the conversion from the biologically active A species to the inactive B species is so very rapid in tissues that the half-life of existence of TXA is estimated as only a few seconds, a formidable impediment to the analysis of reactions involving such transient agents. The main path for inactivation of the prostaglandins involves, first, an oxidation of the 15-OH group to a keto group, then, reduction of the 13, 14 double bond. In addition to the metabolites formed by this path, urine and other biologic fluids may contain derivatives that are 2 or 4 carbons shorter at the carboxyl end, indicating that β-oxidation can also play a role in removal of the prostaglandins from the body.

ACTIONS

Clarification of the actions of prostaglandins in tissues and in the blood has been hampered by technical difficulties and problems of interpretation arising from the highly unstable nature of many of the more potent derivatives. Recent work has shown that previous investigations concentrating on the prevalent PGE and PGF may have overlooked transient but important effects of the labile prostacyclin or thromboxane compounds, whose presence may be revealed only from their inactive, stable metabolites. In general the prostaglandins of the F series were shown to produce muscular contraction, while the PGEs caused relaxation of vascular smooth muscle and contraction of gut muscle. Thus PGF increased the blood pressure by vasoconstriction, while PGE, by inducing vasodilation, brought about a fall in blood pressure. Similarly, PGF constricted the blood vessels and bronchi of the airways in the lungs, while PGE had the antagonistic effect. PGF produced contractions of the uterus, and was used for the induction of abortion or labor at term in some instances. PGE enhanced blood flow through the kidneys by vasodilation, increased urinary volume, and the excretion of sodium ions. Prostaglandins F and E were, therefore, postulated to produce important and often countervailing influences upon the cardiopulmonary, excretory, and reproductive functions, possibly through interaction with other hormones (the PGFs typically accentuated adrenergic responses, and the PGEs opposed them) or the cyclic nucleotide second messengers (PGEs generally increased cyclic AMP levels in cells, and PGFs elevated cyclic GMP).

The first crack in this simple yin–yang theory of prostaglandin action came about from careful analysis of events in the aggregation of the blood platelets, an important accompaniment of coagulation during hemostasis. Activation of prothrombin to thrombin initiates the production of the blood clot by catalysing conversion of fibrinogen to fibrin, as detailed in Chapter 1; thrombin also acts on the minute thrombocytes or blood platelets in the vicinity of the clot rendering them sticky so that they will aggregate and reinforce the fibrin clot. The thrombin-mediated aggregation is associated with enhanced phospholipase release of arachidonate; moreover, added arachidonate can mimic the stickiness that is induced by thrombin, an effect which is counteracted by aspirin or other cyclooxygenase inhibitors. This anti-aggregating action of aspirin, or sulfinpyrazone (Anturane) on blood platelets provides the rationale for the extensive clinical trials which have been undertaken to assess their possible prophylactic use in patients with threatened cerebral or coronary thrombosis. An additional correlation was seen when it was observed that an increase in the cyclic AMP content of platelets was associated with inhibition of aggregation, and that added prostaglandins, which could induce platelet aggregation (*i.e.*, PGE_2) lowered the cyclic AMP levels, while others (*i.e.*, PGE_1), which stimulate platelet adenyl cyclase, act as anti-aggregators. However, later work established that PGE, PGF, or other "classical" prostaglandins were not produced in platelets at high enough levels to account for the aggregation phenomenon of the "arachidonate cascade." Instead of being channeled into the prostaglandin series of compounds by prostaglandin synthetase the endoperoxides PGG_2 and PGH_2 were diverted into production of a compound with a different ring system, the oxane structure; because of its potential role in blood thrombosis this new substance was named thromboxane. The active form, TXA, disappeared as rapidly as it was formed with a half-life of only 32 seconds giving rise to its inactive metabolite, TXB. In addition to being the most powerful aggregating agent and suppressor of cyclic AMP levels within the thrombocytes, thromboxane, on release from the platelets, could be seen to exert a strong constrictive action on the smooth muscle elements of neighboring blood vessels (Fig. 3–31).

It was the latter phenomenon and the ensuing quest for parallel vasoconstrictive agents within vessel walls that led to the discovery of the second new class of prostaglandin derivatives, the prostacyclins. The vascular tissue was found to lack enzymes for production of the oxane structure but instead converted major quantities of endoperoxides, such as PGG, to a polar end-product, PGI, with a 6, 9-epoxy ring. The isolated prostacyclin, PGI_2, was not only an active vasodilator, but it also could be shown to antagonize the actions of TXA_2 as the most effective stimulator of adenyl cyclase and inhibitor of platelet aggregation (Fig. 3–31). Thus in the cooperative interacting system of blood platelet–blood vessel wall the yin–yang antagonism resolves into two distinct cell types for the control of vascular tone and aggregation: platelets containing thromboxane synthetase to produce TXA which is analogous to the F series of prostaglandins, and which is vasoconstrictive and pro-aggregatory; cells of the vascular wall containing prostacyclin synthetase to produce PGI which is analogous to the E series of prostaglandins, and is vasodilatory and antiaggregatory. The clinical use of nonselective cyclo-oxygenase inhibitors in therapeutic regimens for vascular thrombosis must be re-evaluated in light of this balanced control system, with the goal of finding specific drugs that could block the thromboxane synthetase without interfering with the natural antithrombotic prostacyclin production of blood vessels.

To date so many systems and events in the body have been shown to be associated with prostaglandin actions that it seems safe to conclude that these agents may prove to be as important and ubiquitous as the cyclic

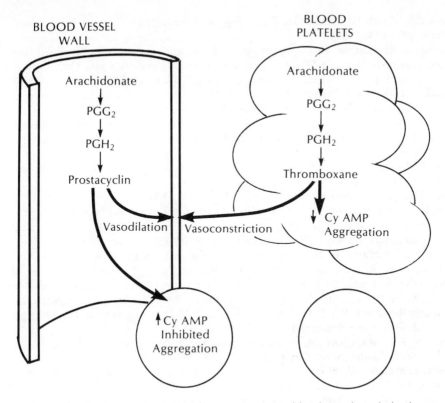

FIG. 3–31. Prostacyclin and thromboxane actions in blood vessels and platelets.

nucleotides. Refinements in the techniques for measurement of the trace concentrations of such active substances and for freezing their ephemeral existence in tissues will be required to assess their true physiologic effects. Until such developments occur the potential modulating role of prostaglandins in processes ranging from cardiovascular function to neurotransmitter release, from renal excretion to ovulation, adds new substance to the time-honored advice of physician to patient when all else fails— "take two aspirins and go to bed."

SUGGESTED READINGS

Anderson DC: Endocrine function of the testis. In O'Riordan JLH (ed): Recent Advances in Endocrinology and Metabolism, No. 1, pp 111–136. New York, Churchill Livingstone, 1978

Armstrong DT, Dorrington JH: Estrogen biosynthesis in the ovaries and testes. In Thomas JA, Singhal RL (eds): Advances in Sex Hormone Research, Vol 3, Regulatory Mechanisms Affecting Gonadal Hormone Action. pp 217–258. Baltimore, University Park Press, 1977

Austin LA, Heath H: Calcitonin: Physiology and pathophysiology. N Engl J Med 304:269–278, 1981

Axelrod J: Regulation of the synthesis, release and actions of catecholamine neurotransmitters. In Dumont J, Nunez J (eds): Hormones and Cell Regulation, Vol I, pp. 137–155. Amsterdam, North Holland, 1977

Bar RS, Harrison LC, Muggeo M et al: Regulation of insulin receptors in normal and abnormal physiology in humans. Adv Intern Med 24:23–52, 1979

Baxter JD, Rousseau GG: Glucocorticoid hormone action: An overview. In Baxter JD, Rousseau GG (eds): Glucocorticoid Hormone Action, pp 1–24. Berlin, Springer-Verlag, 1979

Beling C: Estrogens. In Fuchs F, Klopper A (eds): Endocrinology of Pregnancy, 2nd ed, pp 76–98. Hagerstown, Harper & Row, 1977

Bloom FE: Neuropeptides. Sci Am 245(4):148–168, 1981

Brodish A, Lymangrover JR: The hypothalamic–pituitary adrenocortical system. In McCann SM (ed): International Review of Physiology, Vol 16, Endocrine Physiology II, pp 93–149. Baltimore, University Park Press, 1977

Brownlee M, Cerami A: The biochemistry of the complications of diabetes mellitus. Annu Rev Biochem 50:385–432, 1981

Bruchovsky N, Lesser B: Control of proliferative growth in androgen responsive organs and neoplasms. In Singhal RL, Thomas JA (eds): Advances in Sex Hormone Research, Vol 3, Cellular Mechanisms Modulating Gonadal Action, pp 1–55. Baltimore, University Park Press, 1976

Cavalieri RR, Rapoport B: Impaired peripheral conversion of thyroxine to triiodothyronine. Annu Rev Med 28:57–65, 1977

Chan L, O'Malley BW: Mechanism of action of the sex steroid hormones. N Engl J Med 294:1322–1328, 1372–1381, 1430–1437, 1976

Chopra IJ, Solomon, DH, Chopra U et al: Pathways of metabolism of thyroid hormones. Rec Prog Horm Res 34:521–567, 1978

Clark JH, Peck EJ Jr, Hardin JW, Eriksson H: The biology and pharmacology of estrogen receptor binding: Relationship to uterine growth. In O'Malley BW, Birnbaumer L (eds): Receptors and Hormone Action, Vol II, pp 1–31. New York, Academic Press, 1978

Czech MP: Molecular basis of insulin action. Annu Rev Biochem 46:359–384, 1977

Czech MP: Insulin action. Am J Med 70:142–150, 1981

Daughaday WH, Herington AC, Phillips LS: The regulation of growth by endocrines. Annu Rev Physiol 37:211–244, 1975

Fink G: Feedback actions of target hormones on hypothalamus and pituitary with special reference to gonadal steroids. Annu Rev Physiol 41:571–585, 1979

Glass DB: Protein phosphorylation catalysed by cyclic AMP-dependent and cyclic GMP-dependent protein kinases. Annu Rev Pharmacol Toxicol 20:363–388, 1980

Goebelsmann U: Middle and late gestation. In Mishell DR Jr, Davajan V (eds): Reproductive Endocrinology, Infertility and Contraception, pp 121–134. Philadelphia, FA Davis, 1979

Gorman RR: Modulation of human platelet function by prostacyclin and thromboxane A_2. Fed Proc 38:83–88, 1979

Gornall AG, Luxton AW, Bhavnani BR: Endocrine disorders. In Gornall AG (ed): Applied Biochemistry of Clinical Disorders, pp 225–279. Hagerstown, Harper & Row, 1980

Gorski J, Gannon F: Current models of steroid hormone action: A critique. Annu Rev Med 38:425– 450, 1976

Habener JF: Regulation of parathyroid hormone secretion and giosynthesis. Annu Rev Physiol 43:211– 223, 1981

Habener JF, Potts JT Jr: Parathyroid physiology and primary hyper-parathyroidism. In Avioli LV, Krane SM (eds): Metabolic Bone Disease, Vol 2, pp 1– 147. New York, Academic Press, 1978

Harris RH, Ramwell PW, Gilmer PJ: Cellular mechanisms of prostaglandin action. Annu Rev Physiol 41:653– 668, 1979

Jackson RL: Insulin-dependent diabetes in children and young adults. Nutr Today 14(6):26– 32, 1979

Katzenellenbogen BS: Dynamics of steroid hormone receptor action. Annu Rev Physiol 42:17– 35, 1980

Labrie F, Borgeat P, Drouin J et al: Mechanism of action of hypothalamic hormones in the adenohypophysis. Annu Rev Physiol 41:555– 569, 1979

Lands WEM: The biosynthesis and metabolism of prostaglandins. Annu Rev Physiol 41:633– 652, 1979

Latham KR, MacLeod KM, Papavasiliou SS et al: Regulation of gene expression by thyroid hormones. In O'Malley BW, Birnbaumer L (eds): Receptors and Hormone Action, Vol III, pp 75– 100. New York, Academic Press, 1978

Leavitt WW, Chen TJ, Do YS et al: Biology of progesterone receptors. In O'Malley BW, Birnbaumer L (eds): Receptors and Hormone Action, Vol II, pp 157– 188. New York, Academic Press, 1978

Mainwaring WIP: The Mechanism of Action of Androgens, New York, Springer-Verlag, 1977

Martin CR: Textbook of Endocrine Physiology, Baltimore, Williams & Wilkins, 1976

McGiff JC: Prostaglandins, prostacyclin, and thromboxanes. Annu Rev Pharmacol Toxicol 21:479– 509, 1981

McGuire WL: Steroid receptors and breast cancer. Hosp Pract 15(4):83– 88, 1980

Munck A, Leung K: Glucocorticoid receptors and mechanisms of action. In Pasqualini JR (ed): Receptors and Mechanism of Action of Steroid Hormones, Part II, pp 311– 397. New York, Marcel Dekker, 1976

Nathanson JA, Greengard P: "Second messengers" in the brain. Sci Am 237(2):108– 119, 1977

Notkins AL: The causes of diabetes. Sci Am 241(5):62– 73, 1979

Olefsky JM, Kolterman OG: Mechanisms of insulin resistance in obesity and non-insulin-dependent (Type II) diabetes. Am J Med 70:151– 168, 1981

O'Malley BW, Schrader WT: The receptors of steroid hormones. Sci Am 234(2):32– 43, 1976

Phillips LS, Vassilopoulou-Sellin R: Somatomedins. N Engl J Med 302:371– 380, 438– 446, 1980

Pierce JG, Parsons TF: Glycoprotein hormones: structure and function. Annu Rev Biochem 50:465– 495, 1981

Reid IA, Morris BJ, Ganong WF: The renin– angiotensin system. Annu Rev Physiol 40:377– 410, 1978

Ross EM, Gilman AG: Biochemical properties of hormone-sensitive adenylate cyclase. Annu Rev Biochem 49:533– 564, 1980

Roth J: Insulin receptors in diabetes. Hosp Pract 15(5):98– 103, 1980

Samuelsson B, Goldyne M, Granström E et al: Prostaglandins and thromboxanes. Annu Rev Biochem 47:997–1030, 1978

Saxena BB, Rathnam P: Mechanism of action of gonodotrophins. In Singhal RL, Thomas JA (eds): Advances in Sex Hormone Research, Vol 3, Cellular Mechanisms Modulating Gonadal Action, pp 289–324. Baltimore, University Park Press, 1976

Schally AV, Coy DH, Meyers CA: Hypothalamic regulatory hormones. Annu Rev Biochem 47:89–128, 1978

Seif SM, Robinson AG: Localization and release of neurophysins. Annu Rev Physiol 40:345–376, 1978

Sherwin R, Felig P: Glucagon physiology in health and disease. In McCann SM (ed): International Review of Physiology, Vol 16, Endocrine Physiology II, pp 151–171. Baltimore, University Park Press, 1977

Sherwin R, Felig P: Pathophysiology of diabetes mellitus. Med Clin North Am 62:695–711, 1978

Simpson ER, MacDonald PC: Endocrine physiology of the placenta. Annu Rev Physiol 43, 163–188, 1981

Sönksen PH, West TET: Carbohydrate metabolism and diabetes mellitus. In O'Riordan JLH (ed): Recent Advances in Endocrinology and Metabolism, No. 1, pp 161–188. New York, Churchill Livingstone, 1978

Spelsberg TC, Toft DO: The mechanism of action of progesterone. In Pasqualini JR (ed): Receptors and Mechanism of Action of Steroid Hormones, Part I, pp 261–309. New York, Marcel Dekker, 1976

Sterling K: Diagnosis and Treatment of Thyroid Diseases, Cleveland, CRC Press, 1975

Sterling K, Lazarus JH: The thyroid and its control. Annu Rev Physiol 39:349–371, 1977

Tata JR: Growth-promoting actions of peptide hormones. In Ciba Foundation Symposium 41: Polypeptide Hormones, pp 297–307. New York, Elsevier, 1976

Unger RH, Dobbs RE, Orci L: Insulin, glucagon, and somatostatin secretion in the regulation of metabolism. Annu Rev Physiol 40:307–343, 1978

Vale, W, Rivier C, Brown M: Regulatory peptides of the hypothalamus. Annu Rev Physiol 39:473–527, 1977

Van Wyk JJ, Underwood LE: The somatomedins and their actions. In Litwack G (ed): Biochemical Actions of Hormones, Vol V, pp 101–148. New York, Academic Press, 1978

Werner SC, Ingbar SH: The Thyroid: A Fundamental and Clinical Text, 4th ed, Hagerstown, Harper & Row, 1978

Chapter Four
Cell Membranes

A universal feature of living things is their ability to segregate and protect themselves from the vagaries of their natural environment. Man, and other complex organisms, has evolved several layers of protection and insulation from external factors in the form of skin, hair, fatty tissues, and so forth. Even at the level of unicellular creatures, or the individual cells within our bodies, the "skin" of the plasma membrane serves to isolate each cell as a discrete unit; and within the cells of all eukaryotic organisms (as opposed to the prokaryotes such as bacteria), there are many intracellular membranes that serve the same protective function. These intracellular membranes envelop the nuclei, mitochondria, lysosomes, and other organelles, or secretory vesicles, to provide a complicated series of compartments inside the cells. Thus separation of functional entities within the human body is maintained by membranes at the levels of tissues, cells, or intracytoplasmic domains. This compartmentalization of human metabolic activities by cellular and intracellular membranes is of great significance in the control and integration of body functions.

Other than playing a role as protective barriers, membranes have special recognition properties that allow them to interact specifically with metabolites, ions, hormones, antibodies, or other cells. It is a mark of the individuality of cell types that each cell membrane exhibits a characteristic selectivity with respect to reactions with external molecules; this is the result of the presence of certain specific receptor groups on the outside surface of the cell membrane. The membrane receptors may transmit signals from the extracellular milieu, or they may facilitate the transport of molecules across the membrane-permeability barrier. Alternatively, the selective interaction with immunoproteins or other cells may result in aggregation phenomena.

Finally, membranes within the cells may be highly differentiated for particular metabolic functions. Examples include the electron transfer reactions of energy conservation in mitochondrial membranes or of drug detoxification in the endoplasmic reticulum; the reactions of lipid synthesis and organizing events in protein synthesis also in the endoplasmic reticulum membranes; the modification and packaging of proteins for export from the cell in membranes of the Golgi complex. Obviously, in each case each organellar membrane has adopted a characteristic pattern of enzymatic machinery that is intrinsic to the structure of that membrane system.

The major classes of membrane components—lipids, carbohydrates, proteins—play a significant part in facilitating each of the above functional roles. In dissecting the various features of cell membranes we will look at these components in turn, and show their chief contributions to the properties of various membrane systems.

STRUCTURAL FEATURES

Studies of membrane structure may be conducted by many techniques. The use of electron microscopy reveals the typical bilayer appearance in sections, or reveals a number of appendages to the bilayer by such methods as negative-staining. In combination with the use of larger particles attached to site-directed reagents, freeze–fracture techniques can provide information about membrane receptors and the topography of different surfaces of the membranes. The use of sophisticated physical analyses with fluorescent or other labelled probes can tell us a great deal about the fluidity of various sections of membranes. It is even possible to cross-link specific components of membranes *in situ* to determine their "nearest neighbors," and hence to find out which particular lipid–protein or protein–protein interactions are essential for membrane functions.

To date the sum total of these experimental studies has led to the conclusion that the framework of all cellular and organellar membranes in animals consists of a double layer or bilayer of lipids, which contains variable proportions of phospholipids and cholesterol. Most of the "barrier" properties of the membranes may be ascribed to this lipid bilayer. Embedded within the bilayer, and projecting to various extents on either side, are the different functional proteins of either transport or enzymatic processes. Superimposed on lipid and protein are covalently linked carbohydrates forming glycolipids and glycoproteins, which play the key roles in receptor and other recognition phenomena at membrane surfaces. Let us first consider the lipid skeleton of membranes, and later flesh out this description by showing how the other components are integrated with, and modified by the parts of the bilayer.

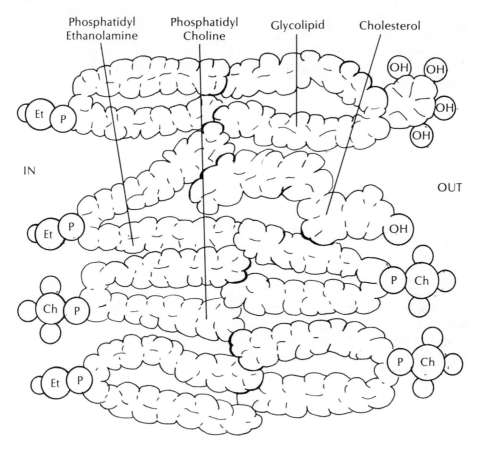

FIG. 4–1.   Membrane lipid bilayer structure.

LIPID BILAYER PROPERTIES

The lipids that make up the bilayer structure have a natural tendency to form membrane structures. The reason for this is the mutual affinity of the long hydrocarbon chains which associate together, pointing away from the watery medium of the cytoplasm. The phosphate-substituted ends of the phospholipids and the carbohydrate-substituted ends of the glycolipids, both being compatible with water, simultaneously line up to face the cytoplasmic or extracellular spaces quite spontaneously (Fig. 4–1). In fact, this self-assembly of membrane bilayers by phospholipids and glycolipids can be achieved very easily in the test tube without the agency of any cellular catalysts or structural proteins. The artificial lipid bilayers so formed have many of the permeability and resistance properties possessed by membranes of living cells, although they do lack, of course, any of the functional attributes conferred by membrane proteins. Recently it has even been possible to add proteins back to the artificial membranes to reconstitute active transport, electron transfer, and other natural membrane functions. Study of such model systems is contributing vastly to our bourgeoning knowledge of membranes.

Lipid bilayers and naturally occurring membranes share the property of being only slightly permeable to most water-soluble substances, charged molecules such as inorganic ions, or large molecules such as proteins. This is understandable when one considers that the close packing of the long alkyl chains produces a dense hydrophobic core within the membrane. Water and ionic substances may associate with the polar residues on the outer surfaces of the bilayer, but only lipid-soluble compounds can pass readily through the central fatty region. It is possible, therefore, to set up ion gradients on either side of artificial lipid bilayers and study their conductance and other electrical properties. Such models act with remarkable similarities to the membranes of nerves and other excitable cells. Thus the major contributor to the barrier function of natural membranes is undoubtedly the lipid portion.

A salient feature of lipid bilayers is the so-called degree of fluidity within the hydrocarbon core. Although attracted to one another to form quite densely stacked arrays, the fatty alkyl chains can be shown to possess considerable freedom of movement. Since they are firmly anchored by ester linkages only near the external surfaces, the long projecting hydrocarbons have considerable flexibility, particularly closer into the core of the bilayer. The use of various hydrophobic probes, which can be inserted into the lipid core to investigate the extent of order within the lipid bilayer, has demonstrated that the fatty alkyl groups undergo considerable motion, much as the densely stacked arrays of plants shimmy in a barley field.

The extent of ordered packing, and hence the degree of fluidity of movement, depends on the lengths and shapes of the fatty alkyl chains. The situation parallels that for the change from solid to liquid states of the constituent fatty acids: the shorter the chain length and the more unsaturation in the chain, the lower the melting temperature. Similarly in the bilayer structure, the long, highly saturated alkyl chains confer rigidity by associating in ordered arrays, while shorter chains, and particularly those with kinks produced by double bonds, allow for greater freedom of movement and hence more fluidity. At physiological temperatures the composition of the fatty acids of cell membranes is carefully regulated to maintain considerable fluidity.

Cholesterol by itself is incapable of forming bilayer structures; nonetheless, when phospholipids and glycolipids are included the steroid ring readily inserts itself into the membrane. The sterol molecule packs easily into crevices between fatty alkyl chains; the A ring bearing the polar hydroxyl group associates with the polar head-groups of phospho- and glycolipids on the external surfaces of the bilayer, while the opposite end of the molecule, the D ring bearing the nonpolar hydrocarbon chain, associates with hydrophobic groups in the core of the bilayer. Biologic membranes vary considerably with respect to the amounts of cholesterol they contain. Frequently there appears to be a correlation between the contents of sphingomyelin and cholesterol in different membranes. For example, both lipids are most abundant in plasma membranes (*i.e.*, red cell ghosts, or plasma membrane-derived structures, such as the myelin sheath investing the nerve cells), but they are almost nonexistent in some membranes (*i.e.*, the inner cristae of the mitochondrion).

Membranes that contain a high molar ratio of cholesterol differ qualitatively from pure phospholipid bilayers. The latter show discrete melting behavior resulting from transformation of the rigidly ordered alkyl chains packed in a crystalline array at low temperatures to the more flexible, fluid conformations of the liquid state as the temperature is raised. This phase change can be studied calorimetrically and exhibits a characteristic transition temperature corresponding to melting points of the constituent fatty acids. However, when the stiff sterol molecule is interposed between alkyl chains at the usual normal proportions of 1 mole per mole of phospholipids in the bilayer the two different phases can no longer be observed. Instead, an intermediate gel is formed that is more liquid than the crystal phase when it is below the normal transition temperature, but more viscous than the liquid phase when it is at higher temperatures.

The explanation for these effects is related to effects of the sterol molecule on neighboring fatty acids. On the one hand insertion of cholesterol into the bilayer perturbs the orderly stacking of alkyl chains in the solid crystalline state with the result that the membrane remains fluid far below the normal melting temperature of the constituent lipids. At higher temperatures, in the physiological range of the body, the close association of fatty acids with inflexible cholesterol rings restricts their mobility and therefore increases membrane viscosity. One important consequence is that the amount of cholesterol in natural membranes greatly influences those properties which are dependent on fluidity. For example, increased cholesterol content, like an increase of saturation relative to unsaturation of alkyl chains, will decrease the permeability of membranes at normal body temperature.

LIPOSOMES IN BIOLOGY AND MEDICINE

Much of our current background knowledge about membrane behavior has come from studies of model systems. Previous mention was made of the natural tendency (thermodynamic stability in scientific terms) of lipids to form bilayers (Fig. 4–2). Several sophisticated models of membranes can be constructed using this principle: application of a droplet of phospholipid solution to a tiny orifice in a divider between two water

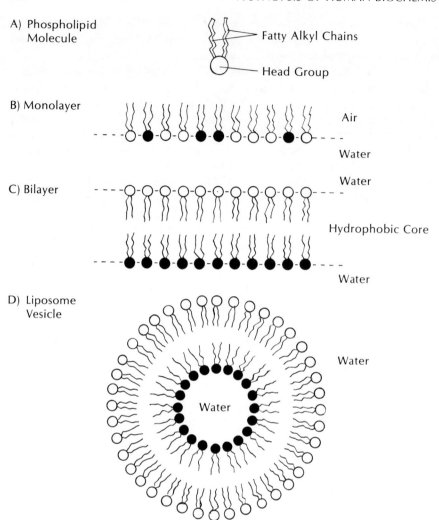

A) Phospholipid
   Molecule

Fatty Alkyl Chains

Head Group

B) Monolayer

Air

Water

C) Bilayer

Water

Hydrophobic Core

Water

D) Liposome
   Vesicle

Water

Water

FIG. 4–2. Intermolecular structures from amphipathic lipids: (*A*) phospholipid molecule; (*B*) monolayer formed by interactions of molecules at the air–water interface; (*C*) bimolecular leaflet with water interacting at outer polar head-groups; (*D*) single-lamellar vesicle or liposome of spherical lipid bilayer enclosing an aqueous phase. The last two structures can show lipid asymmetry.

solutions causes formation of a thin bilayer membrane separating the two aqueous compartments; alternatively the orifice-containing divider may be pushed down through two phospholipid monolayers spread on top of the two water solutions to form a bilayer with different lipids on either side. Such systems provide useful models for the measurement of rates of penetration of substances through the bilayers, and the measurement of electrical properties of membranes as a function of lipid composition. Another model system that has proven tremendously useful is the liposome. As in so many areas of science this simple system, which was originally developed to provide esoteric delight for the curious biologist, has led to a technological "spin-off" that is of great potential significance to the practicing clinician.

The liposome system is formed by shaking phospholipids vigorously with an aqueous solution to form a dispersion. Electron microscopic study reveals that the phospholipid dispersions consist of onion-like concentric bilayers with some of the aqueous phase entrapped. Further treatment of these multilamellar bodies (which closely resemble the structure of myelin) by high-frequency sound waves (ultrasonics) converts them into single membrane vesicles in which the bilayer surrounds a single aqueous space. By preparing liposomes or vesicles in a medium containing tagged water-soluble components (*i.e.*, radioactive ions $^{24}Na^+$, $^{40}K^+$), then removing the particles and washing them free of the medium, it is possible to obtain sealed membrane "cells" containing the entrapped substances. Subsequent measurement of the rate of leakage of radioactivity out of liposomes has proved important information about properties of membrane ion translocation in the presence of anaesthetics, ionophores, nerve-blocking agents, and other drugs. In other useful liposome model systems it has been possible to entrap enzyme proteins and measure their release by crystals of sodium urate in simulation of lysosomal activation in gouty patients. In reconstruction experiments documented in a later section much has been learned about active transport and about energy coupling through the integration of catalytic proteins in the bilayer of liposomes or lipid vesicles.

Study of these and other membrane models have contributed greatly to our knowledge of the dynamic nature of lipid components of the bilayer as well as their role in processes of natural membranes. In the first place physical studies of liposomes reveal that membrane lipids have a considerable freedom to slide or drift in the plane of the layer. This so-called lateral movement of whole lipid molecules is different from the flexible swaying of hydrocarbon chains, and occurs at extremely rapid rates so that a labelled lipid molecule introduced at one end of a cell could move to the opposite end on the order of one second. By contrast there appears to be only a very slow transverse movement of lipid molecules across from one side of the bilayer to the surface facing the opposite compartment. With the use of tagged molecules this slow "flip-flop" of lipids has been estimated to require several days to equilibrate the two sides of liposome bilayers, but more rapid "flip-flop" rates have been indicated from recent studies with natural membranes, suggesting that insertion of proteins may alter the behavior of lipids in the bilayer.

Other studies have revealed the tendency of mixtures of lipids to assume asymmetric distributions in the bilayer. Phosphatidylcholine favors the outer surfaces, phosphatidylethanolamine favors the inner surfaces in vesicles which are produced by sonicating equal portions of the two lipids. Similarly, in natural membranes, such as those of red blood cells, the choline-containing phospholipids, phosphatidylcholine and sphingomyelin, are shown by specific lipase degradations to face selectively outwards, while phosphatidyethanolamine and phosphatidylserine can react with enzymes and chemical reagents only when the cells are opened or turned inside out. An explanation for the asymmetry in lipid vesicles is related to their small radii and hence the sharp curvature of the bilayer. The bulkier choline residues have more space to pack together in the outer layer, which has a greater surface area than the inner layer. Cellular asymmetry must be maintained in part by additional interactions with proteins in the two layers since the curvature factor would otherwise be of lesser importance.

Finally, it has been established that the lipids from one liposome can readily interchange with those from another liposome, provided that a transfer protein is present. The protein carries a certain class of lipid in equimolar exchange between the two outer surfaces, and is referred to as lipid-exchange protein. So far several specific proteins have been characterized for exchange of different phospholipids or cholesterol, and they can transfer the lipids between natural membranes and soluble lipoproteins as well as liposomes. The importance of such exchanges lies in the fact that they occur fairly quickly so that within minutes the lipid components of diverse membranes of a cell would, for example, have similar fatty acid contents. (Differences in the fluidity and permeability properties of these various membranes must be controlled chiefly by their sterol content.) Although the exchange proteins have the capability for net transfer of lipids, for example to lipid-depleted membranes, it is not known as yet whether they play this role *in vivo* during the biosynthesis and differentiation of cellular membranes.

Practical applications of liposomes in medicine at present are centered around their use as encapsulating systems for intravenous administration of certain drugs, antigens or macromolecules. It is known that cell membranes will interact with liposomes by processes of endocytosis or fusion (as discussed later in more detail). In the process the occluded contents of the lipid vesicles are assimilated intact by the cells. This packaging of agents for intracellular delivery would have significant clinical applications (*i.e.*, the restitution of deficient enzymes or other proteins since direct administration of the macromolecules in the circulation will not lead to effective uptake into cells). In the future, it is a possibility that introduction of DNA to repair genetic damage might be accomplished by the same vehicle. The important feature of liposomes as delivery systems, in addition to their efficiency and protection of molecules from circulating antibody reactions and so forth, is the fact that surface properties of the vesicles can be tailored so that they can interact selectively with particular cell types. These modifications could involve, for example, altering the projecting glycolipid components or other antigenic determinants. The possibility of targeting liposomes loaded with toxic antitumour agents so they will be engulfed selectively by cancer cells will provide potent new weapons for the armamentarium of the cancer chemotherapist. Promising results have been obtained with the selective delivery of liposome-encapsulated methotrexate and other anti-cancer agents. One useful trick is to prepare the liposome-drug capsules at an elevated temperature using high melting point phospholipids. On cooling, the liposomes will remain solid at the body temperature, and following injection into the blood, there will be little tendency for the capsules to leak the toxic agent to normal tissues. Precise warming of the location of cancerous lesions, for example by infrared irradiation, will cause the liposomes to melt in the vascular bed of the tumor and discharge their lethal contents selectively to the cancer cells. An additional application arises from the great efficiency with which macrophages and other phagocytic cells of the reticuloendothelial system engulf foreign particles such as liposomes from the circulation. Thus, highly poisonous antimonial drugs have been selectively directed to parasitic organisms that infect the reticuloendothelial cells, as in the treatment of the common tropical disease, leishmaniasis. Moreover, the stimulation of macrophages to assist in the production of antibodies, as described in Chapter 1, or to attack invasive cancer cells, is markedly potentiated if antigens, lymphokines, or other macrophage-activating factors are administered in liposomes. The tumoricidal properties of macrophages have been exploited in this way by the systemic injection of liposome-encapsulated activating factor in order to eradicate cancer metastases by activation of the macrophages *in situ*.

PERIPHERAL AND INTEGRAL PROTEINS

Proteins may be associated with the lipid bilayer of membranes in a number of ways and conformations. Earliest theories of membrane structure assumed that the proteins would be mainly hydrophilic in nature, and thus it was hypothesized that only polar groups of the lipids would exert attractive forces on polypeptide chains or β-pleated sheets of protein lying in contact with outer surfaces of the bilayer. X-ray diffraction and other physical studies of the multilamellar structure of myelin suggested that there was little room for proteins in the membrane core. By extrapolation to other systems the general concept of a "unit-membrane" arose, in which the membranes of cells were depicted as lipid bilayers sandwiched between layers of protein. However, this picture is now agreed to be erroneous since there are many instances known where the membrane proteins penetrate the center of the bilayer to associate with alkyl chains of the lipids by hydrophobic interactions (Fig. 4–3). Such proteins are quite firmly embedded in the membrane and may extend completely through

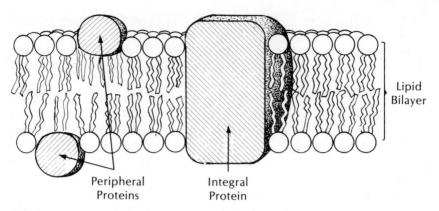

FIG. 4–3. Biologic membrane structure showing the interactions of peripheral and integral proteins with the lipid bilayer.

the bilayer; these are generally referred to as *integral* (or intrinsic) membrane proteins, and they may be isolated only by physical disruption of the lipid bilayer, or by chemical extraction with bile salts or synthetic detergents that disturb the lipid-protein interactions. Other proteins that form looser attachments to the outer polar ends of the lipids may usually be readily removed from the membranes, for example by extraction with salt solutions; these proteins are referred to as *peripheral* (or extrinsic) proteins.

Both categories of proteins may project on either side of the lipid bilayer, but they show an absolute asymmetry of topography as opposed to the lipids that exhibit only relative asymmetry. For example, in the red blood cell about 80% of the membrane phosphatidylcholine is in the outer layer of the bilayer; any particular protein, however, will always exist in the same orientation, so that 100% of its molecules face the same way. It is true that some molecules of protein that span the membrane will have one portion facing the cytoplasm and another facing the extracellular space, but it is almost invariably the same end (the NH_2-bearing terminus) that faces outward. As noted below this has implications for the synthesis of integral membrane proteins. The carbohydrate portions of membrane glycoproteins also project exclusively from the exterior surface, and the same may be true for the glycolipids of the bilayer.

The integral proteins of membranes are linked to the lipid portion by similar forces to those which hold the soluble lipoproteins of blood plasma together. The nonpolar amino acid side-chains (*e.g.,* leucine, valine, phenylalanine, etc.) can exist happily in the low water content of the core of the bilayer. In several cases where the amino acid sequence of integral proteins is known it appears that there is a sequence long enough to span the bilayer, which is rich in such lipid-loving amino acids. On balance, therefore, integral proteins may not be much more hydrophobic than others but they possess hydrophobic regions compatible with their lipid associations.

The behavior of integral membrane proteins is strongly influenced by the properties of the lipid bilayer. The fluidity of the membrane, determined by the amount of cholesterol or degree of unsaturation, will alter the catalytic activity or transport functions of integral proteins. There is every indication that the proteins have considerable mobility in the plane of the bilayer, and current visualizations of membrane structure at a molecular level depict the membrane proteins floating at varying depths of immersion in a lipid sea. However, there is no evidence for true "flip-flop" of proteins as there is for lipids which have the ability to move from one side of the bilayer to the other. Freedom of movement of the proteins through the membrane is probably quite severely restricted.

In some cases there is clear evidence for immobilization of certain proteins by attachments either to other integral proteins (as for the associations between complexes of the electron transport chain of mitochondria), or through peripheral proteins to fibers in the cytoplasm (as for those linkages which maintain the shape of the red blood cell). In other instances it is seen that the random "iceberg" concept of drifting membrane proteins must be modified to accommodate the observed dynamic contractions of membrane-associated proteins involved in changing the shape of migratory epithelial cells, and in forming aggregations of protein particles (as documented for surface activation of such cells as blood lymphocytes). There are also indications that certain integral proteins that span the bilayer exert an immobilizing effect on those phospholipids in their immediate vicinity. These lipids, which form a ring or annulus about the imbedded protein molecule, may number about 100 molecules and fail to show those properties which are characteristic of the fluid bilayer lipids. Thus, just as the behavior of the lipids can modify functions of membrane proteins, so can the protein components influence the state of membane lipids.

FUNCTIONS

The following section will summarize some of the basic features that are of general significance to cell membrane functions. These will include examples of interactions with external stimuli, means of transferring substances across the permeability barrier, and organized catalytic systems within membranes. More specialized aspects of membrane activities are considered under the later section on particular intracellular membrane systems.

MEMBRANE RECEPTORS

While some transfers of information between cells may occur by the passage of molecules into the cytoplasm (as for the steroid hormones), there are many known instances where signaling agents are polypeptides (*e.g.*, insulin, glucagon) or amino acid derivatives (*e.g.*, acetylcholine, epinephrine) that are incapable of crossing the hydrophobic permeability barrier of cell membranes. Such hormones, transmitters, and a number of drugs which are of great importance to clinical medicine seem to produce their very selective effects in the body by reacting with highly specific proteins facing outward on certain membrane surfaces. Such proteins which are designed for ultra-high affinity binding of effector molecules are referred to generally as *membrane receptors*. They are not randomly distributed throughout the body, but they are strictly limited to the membranes of the target tissues and hence determine the selectivity and sensitivity of hormone and drug responses at the cellular level. In this discussion let us consider an example of the membrane receptor system for each category of effector: a hormone (insulin); a transmitter substance (acetylcholine); and a series of drugs that act at the membrane surface (the opiates).

Insulin receptors have been studied in considerable detail in fat cells and liver cells. Radioactively tagged insulin is bound to the extent of several thousand molecules for each intact fat cell; when the cells are turned inside out by physical means the inverted membranes fail to bind to the radioactive hormone. Such experiments confirm the presence of insulin-binding to membrane receptors on the external surface, which is associated with the intracellular response to the hormone.

Isolation of the insulin-receptor protein has been achieved by an affinity-binding technique involving insulin molecules attached to insoluble polymer beads. The receptors, which are integral membrane proteins, are released by disruption of cell membranes with a detergent, and the detergent-solubilized proteins are poured through the insulin beads. All other proteins will pass freely through the insulin beads but the insulin-receptor protein will become attached to the beads. Mild chemical treatment suffices to release the receptor, which is a rodlike glycoprotein with a molecular weight of 300,000. The bound sugar residues are of importance to receptor function as shown by the fact that specific hydrolytic agents, which remove them, (glycosidases) or sugar-binding agents (lectins) interfere with the ability of the protein to bind to insulin.

Purification of the acetylcholine receptors, which are also integral membrane proteins from the nerve–muscle junction, has been achieved using similar technology and certain exotic animal species. It is well-known that the toxic component of certain venomous snakes, bungarotoxin, exerts its paralytic effects on the muscles of the snake's prey by binding tightly to the acetylcholine receptor on the muscle membrane, thus rendering the muscle refractory to the signal of acetylcholine release from its nerve supply. It is also known that the electric ray fish possesses a highly differentiated set of nerve–muscle connections in the "battery" of its electric tissue, which contain densely packed membranes bearing acetylcholine receptors. The receptors can be released from electric organs by detergent solubilization, and become bound selectively to bungarotoxin affixed to an inert supporting bead, as is the case for insulin–receptor isolation. Study of the purified acetylcholine receptor has revealed that it is also a glycoprotein consisting of 5 to 6 subunits, each of molecular weight between 40,000 to 50,000, which are packed as a hexagonal array of rosettes on the outer surface of the membrane.

Recent investigations of the effects of morphine and related substances on nerve membranes have revealed important information about binding sites for the opiate drugs, and have established the somewhat surprising discovery of opiatelike compounds that are produced within the body. Pharmacologists had assumed for some time that a receptor system for morphine and related analgesic drugs (known collectively as the opiates) had existed in pain centers of the brain. However, it has only been in recent years that technology using tagged opiates similar to the technology used for study of insulin receptors showed the selective high-affinity binding sites in certain synaptic membranes within the brain. As in the case of the acetylcholine receptor, antagonist molecules exist analogous to bungarotoxin (*i.e.*, N-allyl morphine) which can also bind to the morphine receptor and block the action of the opiate-agonist drugs. The degree of binding of the active painkilling opiates is greatly reduced by the presence of sodium ions; binding by the antagonists is not reduced. There is a good correlation between the sodium effects in the *in vitro* binding test and the extent to which opiates become addictive to the patient *in vivo*. Thus this simple application of the receptor studies to drug design may allow the production of powerful painkillers that do not lead to drug addiction.

The existence of receptors suggests the natural occurrence of opiatelike substances within the brain itself; after all, it seems unlikely that membranes of the nervous system were constructed in anticipation of the chance encounter between man and the poppy. Recent work has established the presence of several peptide substances in the brain that mimic binding properties of morphine drugs, and competitively displace the opiates from their membrane receptors. Such endogenous morphine-like substances have been termed the "endorphins." The simplest endorphin structure that is effective is the pentapeptide known as encephalin, and it produces potent analgesic effects when injected into the blood or cerebrospinal fluid bathing the brain. In the future, tailormade derivatives of the encephalin–peptides may be designed as nonaddictive replacements for the opiate drugs.

It is perhaps too early to speculate on the mechanisms whereby binding to membrane receptors may produce the effects of extracellular agents on the cell. Receptor proteins themselves do not seem to perform catalytic functions; rather they seem to modulate the activity of membrane-bound enzymes or transport proteins, somewhat analogous to the way in which regulatory subunits exert controlling influences on the catalytic subunits of allosteric enzymes (Fig. 4–4). Binding of hormone or other agent to the exposed region of the receptor, probably involving carbohydrate residues near the NH_2 terminus, presumably induces a profound conformational change of that portion of receptor that is buried in hydrophobic reaches of

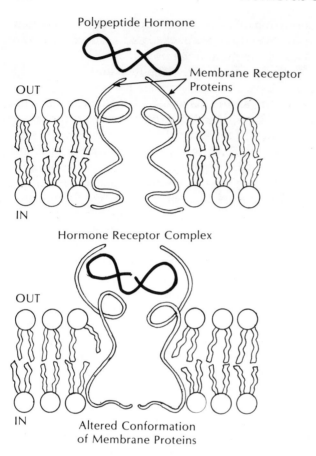

FIG. 4–4. Membrane receptor model. Interaction with an extracellular signal molecule (hormone) may be transmitted by a conformational change in the receptor protein to convey an intracellular signal across the membrane.

the bilayer. In its altered state the receptor–effector complex could affect the activity of neighboring integral proteins in the membrane. The insulin-receptor complex thus may induce negative modification of the action of adenylate cyclase in some target cells to lower the intracellular levels of cyclic AMP, or activation of guanylate cyclase in others to raise the levels of cyclic GMP. The acetylcholine-receptor complex is postulated to interact with adjacent ion channel proteins that span the membrane, and in the activated state open up to allow the uptake of sodium ions and outflow of potassium ions that cause depolarization of the cell. Opiate-receptor effects also seem to involve the reaction of sodium ions with nerve cell membranes, and have also been associated with changes in intracellular cyclic nucleotides similar to those noted for insulin. It seems likely that studies in the near future will show a universal linkage between receptor-mediated effects and intracellular alterations of ions (Na^+, K^+, Ca^{2+}, Mg^{2+}) and nucleotides (cyclic AMP, cyclic GMP).

MEMBRANE PERMEABILITY

The barrier to diffusion of water-soluble components across cell membranes is attributed to the hydrophobic properties of the core of the lipid bilayer. In general, only those compounds which are sufficiently hydrophobic to intermingle happily with the fatty alkyl chains can penetrate rapidly through the core. This explains, for example, why the hydrophilic hormones that are derived from amino acids (*e.g.*, epinephrine) require reactions with receptors at the cell surface, while steroid hormones can enter the cytoplasm to react with intracellular receptors. Charged substances in particular, such as simple inorganic ions (K^+, Ca^{2+}, $H_2PO_4^-$, etc.), organic ions (citric acid cycle intermediates, sugar phosphates, etc.), nucleotides (ATP, cyclic AMP, etc.), and a host of other important metabolites are reluctant to enter the lipid bilayer, and hence tend to be avidly retained within intracellular compartments. Studies with liposomes composed of phospholipids of varying degrees of saturation have shown that the simple diffusion of substances across the bilayer will depend on the fluidity of the membrane. Increased flexibility, associated with the kinks introduced by double bonds, presumably provides more spaces for penetration between alkyl chains. Conversely, the introduction of sterols to fluid bilayers by facilitating tighter packing of the fatty acid chains reduces permeability. This probably accounts for the observation that the outer plasma membrane of cells (such as the erythrocyte membrane), which functions as the permeability barrier to external substances, contains a higher proportion of cholesterol than intracellular membranes (such as mitochondrial membranes).

An important tool in the study of membrane permeability and in the design of models for membrane transport is the class of antibiotics known as ionophores. The term refers to the antibiotics' ability to carry ions across lipid bilayers or natural membranes that are normally ion-impermeable. Two types of ionophores have been characterized (Fig. 4–5). The first, the carrier molecule, is epitomized by valinomycin, a cyclic peptide which forms a specific complex with potassium ion on a 1 : 1 basis. In the complex the bare K^+ is stripped of its water of solvation and is completely enclosed in the interior of the antibiotic. The valinomycin-K^+ complex presents only the hydrophobic parts of its structure to the exterior, and, in this protected basketlike form, the potassium ion may be easily extracted from water into organic solvents, or it may penetrate into lipid bilayers. In this way K^+ can be carried across cell or mitochondrial membranes which are otherwise relatively impermeable to the ion. Carrier

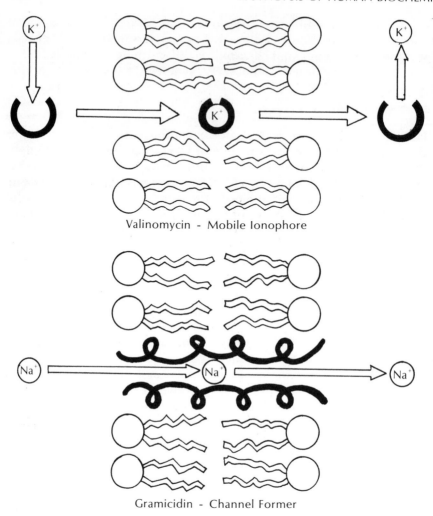

FIG. 4–5. Ionophore actions in membranes—the mobile ionophore (valinomycin) transfers potassium ions in a hydrophobic cage which migrates through the bilayer; the channel-former (gramicidin) produces a transient hydrophilic pore which allows the passage of sodium ions through the bilayer.

ionophores like valinomycin, or those which are specific for other cations such as Ca^{2+}, have been of great value in studying the mechanisms and energetics of ion transport, and in investigating the effects of alterations in ion concentrations on biologic phenomena. The second category of ionophore is the channel-former typified by gramicidin. This antibiotic selectively promotes the movement of sodium ions through membranes, although its action is not as completely specific for Na^+ as valinomycin is for K^+. In this case there is no 1:1 lipid-soluble complex formed with the ion, nor does the antibiotic act to allow the extraction of Na^+ into fat solvents. Instead, gramicidin, when added to lipid bilayers, promotes the conductance of sharp pulses of sodium ions across the membranes. Since the occurrence of these bursts of ion movement is not influenced by lowering the temperature below the point where alkyl chains of the bilayer are fluid (a procedure which drastically reduces movement of valinomycin-K^+ complexes through the membrane), it is inferred that gramicidin forms a pore or a channel through the lipid bilayer. Such a fixed channel structure would not be affected by "freezing" the bilayer, as would the mobile carrier-ionophore. Models constructed from two molecules of gramicidin with hydrophobic residues projected outward would produce a hydrophillic channel of the appropriate internal dimensions to allow passage of a sodium ion, and would ensure that the channel would be the correct overall length to span a lipid bilayer. The occurrence of bursts of Na^+ conductance is consistent with the spontaneous associations– dissociations of gramicidin dimers to alternately open and close sodium channels across the bilayer.

Transport Proteins

Simple diffusion accounts for the movement of only a small number of molecules (*i.e.*, water, urea) through biologic membranes. For most ions and low-molecular-weight metabolites there exists a series of proteins with the ability to assist or facilitate the transmembrane passage. In the situation where substances move downhill in the energetic sense, that is from regions of high to lower concentrations, the process is termed *facilitated diffusion*. When energy input is required to move the substances against the chemical potential gradient from low to higher concentrations the process is termed *active transport*. Properties of transport proteins have to accommodate three general phenomena of membrane translocation: first, they must contain specific binding sites for the transported substance on either, and sometimes both, sides of the membrane; second, they must provide either a mobile carrier or selective channel system across the bilayer; finally, in the case of active transport they must undergo some fundamental change in binding affinity, conformation, or orientation in the membrane when converted from nonenergized to energized states. At present the study of transport proteins and their mechanisms of action have not progressed much beyond the work done by Michaelis & Menten some 50 years ago concerning enzyme catalysts. Because of the important ramifications mechanisms of membrane transport hold for human physiology, metabolism, and biochemical genetics it is of paramount importance to press on both with basic research and clinical applications. As in so many other areas of science the spin-offs from one approach have and will continue to promote advances in the other.

The parallel between enzymology and the study of transport proteins is an apt one because transport proteins behave like enzymes in many ways, and the transport functions of membrane proteins may be associated with catalytic activities. The rate of simple diffusion across a membrane will show a direct, straight-line relation to the concentration of diffusing molecules. However, when a protein facilitates the rate of movement the process exhibits saturation kinetics. The relation is strictly analogous to that of the Michaelis–Menten curve for an enzyme-catalysed reaction, with a limiting Vmax value corresponding to saturation of the carrier mechanism, and a series of Km values for closely similar molecules correlating with their respective affinities for binding sites on the transport protein. Other structural analogues of transported substances may also complex to the binding sites, or interfere with the carrier mechanism, and thus provide antagonists of transport systems analogous to competitive and noncompetitive inhibitors in enzymology.

One of the most important and most fully studied examples of facilitated diffusion systems is the transport of glucose. This type of system accounts for the entry of glucose from the blood into skeletal muscle and adipose tissue; both the Km and Vmax values are increased several-fold by insulin in these tissues. A comparable, but insulin-insensitive system, operates for transport of glucose into erythrocytes from blood plasma. The carrier is quite specific for certain hexose and pentose sugars, which compete with one another when present in mixtures relative to their Km values. Some sugars, such as fructose, have Kms so much above normal physiological concentrations that they are not effectively transported by the glucose carriers, while some sugar derivatives or analogues may block the normal carrier function and act as inhibitors. In this case, as for most passive, facilitated processes, the transport of glucose inward is driven by the metabolic consumption of the sugar in glycolysis or the hexosemonophosphate pathway.

A different driving force is required for tissues, such as kidney or intestinal mucosa, that must function to transport glucose against the concentration gradient. In such cell types energy-dependent, active transport is required to move the sugar molecules to regions of higher concentration, and thus may be distinguished from passive, facilitated processes by sensitivity to respiratory inhibitors (such as cyanide) or uncouplers of oxidative phosphorylation (such as dinitrophenol). Despite this crucial difference the active transport of glucose bears several similarities to facilitated diffusion in the erythrocyte. The uptake of sugars by intestinal epithelium also exhibits competition among molecules depending on their Kms, and their selective uptake rates determined by effectiveness of interaction with the carrier system. Saturation of the mediated transport mechanism is particularly striking in the renal tubules. Glucose in the glomerular filtrate is efficiently reabsorbed by the tubular cells unless the normal Vmax capacity is exceeded (as in diabetes mellitus when blood sugar concentrations exceed the saturation level), or the Vmax is decreased below normal blood glucose levels (as in kidney damage, or as in the presence of poisons of the glucose carrier such as phlorhizin). In both instances glucose will escape the normal reabsorption process and appear in the urine (glucosuria).

Although mechanisms of carrier action, and particularly the bioenergetics of active transport, are imperfectly understood in several instances the transport of sugars, amino acids, and ions (*i.e.*, potassium) appear to be associated intimately with movements of sodium ions. The energy-linked translocation of Na^+ by the so-called sodium pumping system has been frequently implicated as a driving force for the permeation of such molecules. Accordingly a great deal of interest is being directed towards active Na^+ transport and its role in general transport processes, and, in particular, its role in maintaining ion gradients of excitable tissues. Again much of this work, for reasons of convenience, has centered on energy-dependent Na^+ and K^+ movements in erythrocytes, and the associated hydrolysis of ATP, the so-called Na^+/K^+ ATPase activity.

It was noted earlier that active transport processes in intact cells are characteristically sensitive to inhibitors of respiratory energy production. However, this is not universally true; in the erythrocyte, for example, there is no mitochondrial oxidative metabolism, and the maintenance of ion gradients depends on glycolysis. The ultimate source of energy in both aerobic and anaerobic cells is ATP. This can be shown conclusively with the valuable model system provided by the erythrocyte "ghost." When placed in very dilute solutions erythrocytes swell osmotically, eventually discharging their contents of hemoglobin and glycolytic enzymes when the membrane ruptures. The pale ghost membranes may be washed free of intracellular contents, and when placed into solutions of higher ionic strength will reseal and contract to the original cell shape. It was observed that such resealed ghosts would carry out normal active transport of Na^+ and K^+ against concentration gradients only if ATP had been placed in the resealing medium so that it was trapped on the inside. Coincident with the movements of ions, it was readily shown by measurement of inorganic phosphate release that the ATP was hydrolysed to ADP. Because of its dependence on sodium and potassium ions this catalytic action of the transport system is referred to as Na^+/K^+-ATPase, and is present in the plasma membrane of all cells that carry out active transport of these ions. A similar Ca^{2+}-activated ATPase is present in the membranes of sarcoplasmic reticulum, the compartment for sequestration of calcium ions in muscle tissue.

Generally cells maintain a high internal potassium ion concentration, a necessity for efficient operation of glycolysis, protein synthesis, and several other enzymatic processes, although extracellular fluids contain little K^+. Conversely the external Na^+ levels are high, and the cell expends a considerable amount of its energy reserves in pumping out the sodium ions that leak into the cell to maintain low internal Na^+. The ATPase protein, which is responsible for these ion movements, has been studied from the standpoint of its asymmetry in the membrane using erythrocyte "ghosts," and it has been isolated by detergent extraction of the membrane "battery" of the electric eel. These investigations have revealed that the system consists of a complex of molecular weight about 250,000, comprising a glycoprotein subunit, presumably facing outward, of 50,000 molecular weight, and two catalytic subunits that span the membrane, each of 100,000 molecular weight. In order to react with ATP the catalytic subunit must bind Na^+ on the inner facing surface (Fig. 4– 6a); in the process the protein is phosphorylated and ADP is released (Fig. 4– 6b). The phosphorylated protein will release the phosphate group as inorganic P only when it binds K^+ on the external surface (Fig. 4– 6c); K^+ binding is competitively antagonized by the cardiotonic steroid, ouabain, when the latter interacts at the outer side of the membrane.

The mechanism to account for these phenomena is still speculative. Presumably the energized state of the protein involves a series of conformational changes such that Na^+ is preferentially released from the external surface. The movement of K^+ inward seems to be a consequence of the outward Na^+ translocation, and is probably in competition with active uptake of sugars and amino acids; in many cases active transport of the latter substances is also inhibited by ouabain. A variety of models have been constructed that imply the movement of ion-binding sites, or the rotation of the ATPase protein through the membrane in the manner of a revolving door. The present evidence for the strictly maintained asymmetry of the protein shown by ouabain binding or by ferritin-conjugated antibodies, as well as the general energy considerations arguing against flip-flop of proteins through lipid bilayers, seems to make such rotating models implausible. In the absence of contradictory evidence it may be more realistic to visualize a transmembrane ion channel system, similar to that noted earlier for gramicidin-induced Na^+ movement, as an interim model. The Na^+/K^+-ATPase, in addition to containing asymmetric binding sites for the two ions at either end of the hydrophilic channel, should undergo changes in conformation to "open" or "closed" positions during the energy cycle. Future experiments are needed to establish the nature of the channels for movement of ionic species through the membrane.

Anesthetic Actions

In the action of opium narcotics previously described, the primary event is selective combination of the drug with a receptor at the surface of the nerve membrane resulting in blockage of nerve impulses in pain centers of the nervous system. Such drugs were historically known to exert analgesic effects. Early surgeons relied on mixtures of opiates and alcohol (laudanum) to provide partial relief from the tortures of amputations and other major operations, which had to be conducted at breakneck speed to minimize the trauma and suffering to the conscious patient. The advances of modern surgery and dentistry could never have developed without the

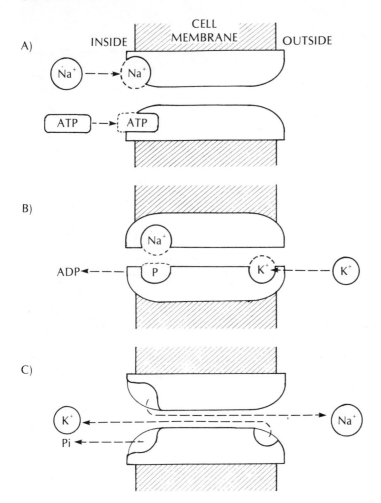

FIG. 4–6. Model of a transport protein, the Na^+-, K^+-ATPase system for movement of cations across membranes by energy-requiring, active transport: (A) ATP and Na^+ bind to sites on the internal surface of the protein; (B) cleavage of ATP energizes the transport protein which may bind K^+ on the external surface; (C) energy is released as the phosphoryl-protein is hydrolysed with movement of the two cations across the membrane in opposite directions.

discovery of the much less selective membrane-active drugs that produce local or general anesthesia.

While only slight modifications of the chemical structure of morphine (e.g., from the l- to the d-stereoisomer) may eliminate its effectiveness, the anesthetics include such a wide range of compounds that specific interac-

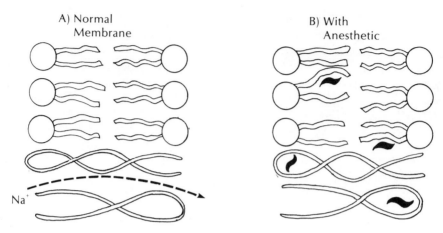

FIG. 4–7. Anaesthetic effects on membranes—the anaesthetic molecules (shown in black) interact with hydrophobic regions of lipids and proteins in membrane bilayers to expand their volumes and prevent the transmembrane flux of sodium ions.

tions with membrane components are unlikely. In fact the only common factor of chemical properties among general anesthetics (ether, chloroform, halothane) seems to be the degree of lipid solubility of the substances. The ability to induce anesthesia correlates well with the ability of agents to enter into the hydrophobic core of the nerve-cell membrane. For a divergent series of anesthetic agents it is known that the effective concentration leads to occupation of about 0.3% of the volume of the membrane, despite wide variations in shape and size of the agents.

Clues to the mode of anesthetic action have arisen from studies of the effects on membrane properties. Many of these effects may be demonstrated in non-neural membranes (e.g., erythrocytes), and even in artificial membranes (e.g., liposomes). The most fundamental effect is a disordering of the lipid bilayer; apparently, the insertion of low amounts of anesthetic into the hydrophobic core increases the fluidity of the alkyl chains. Associated with this fluidization of the bilayer is a dramatic expansion of the membrane volume, not by 0.3% as expected for the occupying molecules, but by 2 to 3%. This tenfold enhancement of membrane expansion is not observed with liposomes composed solely of lipids, so the crucial interaction must involve conformational changes of membrane proteins within their hydrophobic regions (Fig. 4–7). The anaesthetic effect on nerve membranes is interpreted in the following manner: expansion within the membrane is produced by this relatively nonspecific interaction of lipids and proteins with the anesthetic agent; the perturbation of those proteins associated with acetylcholine receptors, which are responsible for sodium ion passage, causes blockage of the ion channels; the prevention of ion movement across the nerve membrane thereby interferes with propagation of excitatory impulses and produces loss of conscious sensation.

The fact that membrane expansion is a key factor in anesthesia is illustrated by the effects of elevated pressure. The anesthesia produced by chloroform or ether on aquatic animals (*e.g.*, newts) may be completely reversed if atmospheric pressure in the tank is greatly increased (80–90 atmospheres). The reversal of anesthesia may be related to compression of the nerve membrane back to its normal state of order. Parallel studies with liposomes show that the same increase in pressure reverses the expansion of the membrane caused by the anesthetic and reduces fluidity of the lipid bilayer to that seen in the absence of the drug.

FUNCTIONAL MEMBRANES—ELECTRON TRANSPORT SYSTEMS

Associations of enzymes with membranes confer a number of unique organizing features upon cellular metabolism. Firstly, catalysis by membrane-bound enzymes may take place in the extracellular space, in the cytoplasm, or in the internal compartments of organelles depending on the orientation of the enzyme's active center in relation to membrane asymmetry. Secondly, *vectorial* metabolism, the property of cellular processes that establishes definite directionality with respect to the two sides of a membrane and thus concentration gradients or charge separations, may provide the driving force for active transport or the generation of electrical potentials. Finally, the membrane may provide a matrix for the physical association of enzymes in close proximity to one another in multienzyme complexes. These distinguishing features are exemplified by the vital oxidative systems in our cells which involve the transport of electrons by the cytochrome components of mitochondrial and endoplasmic reticulum membranes. This type of solid-state catalysis possesses obvious advantages of efficiency, particularly for sequential reactions.

Recent studies of mitochondrial oxidations have revealed a distinctive arrangement of electron carriers which is of fundamental significance for the associated energy transfers of oxidative phosphorylation. The cytochromes and other components of the electron transport system are mainly bound as integral proteins of the inner membranes and their invaginations, the cristae, of the mitochondria (Fig. 4–8). An exception is cytochrome c, a relatively small hemoprotein (molecular weight ~ 12,000), which is a peripheral protein. Cytochrome c is readily extracted from the mitochondrial inner membrane by salt solutions; this does not occur if the membrane is turned inside out. Other studies with specific antibodies to cytochrome c confirm that the protein is only accessible from the outward-facing surface (*i.e.*, the cytoplasmic side) of the inner membrane. Other cytochromes, such as cytochrome oxidase (molecular weight ~ 300,000), are intimately bound to hydrophobic regions of the inner mem-

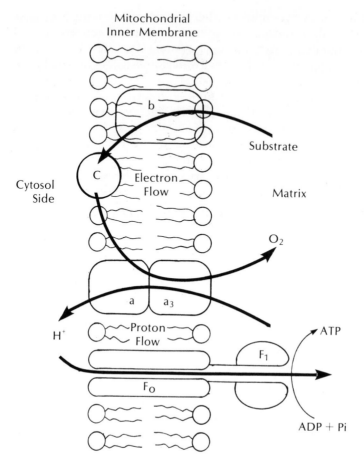

FIG. 4–8. Mitochondrial inner membrane and oxidative phosphorylation. Electron flow through the cytochrome chain $(b \rightarrow c \rightarrow a \rightarrow a_3)$ is vectoral and coupled to the outward movement of protons. Inward movement of protons back through the F_0 channel and the F_1-ATP synthetase drives the production of ATP.

brane and may be extracted only by the use of strong membrane-disruptive agents or detergents. The proteins themselves are nonfunctional as electron carriers unless they are allowed to reassociate with lipids similar to the phospholipids of the original membrane. Studies with antibodies and nonpermeating chemical labels have proven that the cytochrome oxidase protein is accessible to reagents on either the surface facing outward (*i.e.*, cytosol side), or the surface facing inward (*i.e.*, matrix side). Examination of the isolated oxidase protein has shown it to consist of several subunits, two of which contain a heme grouping designated a and

a_3. Reconstruction of multimeric protein as it appears in the inner membrane indicates that the heme a-bearing portion, which reacts with cytochrome c, is accessible from the outer surface, while the heme a_3, which reacts with oxygen, projects into the inner matrix space. Presumably the other protein subunits, which are very hydrophobic, span the interior of the bilayer and maintain the interaction between the electron carriers, $a-a_3$, of the oxidase. The transmembranous flow of electrons from reduced cytochrome c to O_2 is accompanied by consumption of hydrogen ions on the inner surface of the membrane. In this way a proton gradient, more acidic on the outer surface, is established across the mitochondrial inner membrane coincidental with the movement of electrons in respiration. This is thought to be the principal driving force for the synthesis of ATP at the third coupling site of the respiratory chain. Analogous transmembrane or vectorial electron flow with concurrent proton movements out of the mitochondria may be detected at the other two coupling sites, but the exact topology of the electron carriers has not been as well established as it has been for cytochrome oxidase.

Although cytochrome c may be shown to form lipoprotein complexes with phospholipids it appears that linkages to the mitochondrial membrane for cytochrome c are mainly through polar regions. The attachments must be largely electrostatic since they are so readily dissociated by salt solutions. On the other hand the linkages that bind cytochrome oxidase are hydrophobic and involve the nonpolar fatty alkyl residues of the membrane bilayer. Moreover it has been observed that phospholipids that associate with cytochrome oxidase in the membrane fall into the category, noted earlier, of the annulus or boundary structure. The evidence from spin-labelled lipid probes shows that the phospholipid molecules that surround the cytochrome oxidase protein have much less mobility than those that are freely associated in the bilayer proper.

The electron transport systems of endoplasmic reticulum have nothing to do with cellular respiration or oxidative phosphorylation. The cytochromes in isolated microsomal membranes bear a resemblance to the b cytochromes of mitochondria, and are also capable of transferring electrons between reduced nicotinamide coenzymes (NADH or NADPH) and cytochrome c. However, in the case of the b cytochrome in endoplasmic reticulum (designated cytochrome b_5), the electron transfers are involved in the production of unsaturated fatty acids. Another cytochrome, termed P_{450} (for the maximum absorbance of its complex with carbon monoxide), functions to detoxify certain drugs by hydroxylation in endoplasmic reticulum; administration of these drugs, such as the barbiturates, causes an induced synthesis of the hydroxylation system and an increased proliferation of endoplasmic reticulum membranes.

The cytochrome b_5 protein has been studied in considerable detail. It is a typical integral protein buried in the membrane, and it is released intact (molecular weight $\sim$ 17,000) only by treatment with detergent solutions. However, a fully active, water-soluble portion bearing the heme group (molecular weight $\sim$ 12,000) may be clipped from the membrane by mild digestion with a proteolytic enzyme (trypsin). The remainder of the cytochrome b_5 molecule, which is not essential for its function, remains buried in the membrane and contains the C-terminal sequence of 50 amino acids which are predominantly hydrophobic in character. It is presumed that this hydrophobic peptide spans the membrane acting as the handle for attachment of the reactive N-terminal sequence which projects into the cytoplasm.

RECONSTITUTED SYSTEMS

It has been a traditional scientific research technique, exemplified by the triumphs of organic chemistry, to prove a structure of an unknown first by analysis of its parts, and then by synthesis of an identical structure from simple starting materials. Such experimentation has not been accomplished for living systems, with the exception of the reassembly of certain viruses from their components. Multienzyme systems have been reconstructed from enzymes and coenzymes, but most investigations of cellular structures have been predominantly analytical. As of late the approach to synthesis, or reconstitution, of membrane structure and function from component parts has been dramatically successful, mostly as a result of the natural tendency of the lipids to form bilayers.

The purified Na^+/K^+-ATPase protein has been successfully reincorporated into phospholipid liposomes to produce a reconstituted ion-pumping system. The orientation of the protein in the artificial membrane is opposite to that found for intact erythrocytes; thus Na^+ must be added to the outside surface of these vesicles, and K^+ must be added to the inside surface. When ATP is subsequently applied on the outer surface, Na^+ is pumped in and K^+ is pumped out of the liposomes, while ATP is hydrolysed to ADP and inorganic P. As might be expected from the inverted orientation, ouabain has no effect on these reactions unless it is trapped on the inside of the vesicles. Stoichiometry of the reconstituted transport system is the same as for natural cell membranes (Na^+ transport : K^+ transport : ATP hydrolysis = 3 : 2 : 2).

Great strides have been made in the reconstitution of the respiratory chain and the associated ATP synthesis of oxidative phosphorylation. The cytochrome oxidase protein isolated by detergent extraction from the mitochondrial inner membrane can be taken up into phospholipidvesicles. Under these circumstances the enzyme, which is catalytically inactive in the absence of lipids, is just as effective in transferring electrons from reduced cytochrome c to oxygen as it is in the intact mitochondrial inner membrane. However, the orientation of the oxidase protein in the liposomes is not selective, and the cytochrome c can interact on either side of the membrane. If the reduced cytochrome c is introduced only inside the sealed vesicles there is a vectorial movement of electrons from inner to outer surface of the membrane with an associated influx of hydrogen ions. This is consistent with the chemiosmotic hypothesis of proton movements during respiration, although the orientation is inside out in comparison with that for the intact mitochondrion. The reconstituted system also exhibits the phenomenon of respiratory control (*i.e.*, a low rate of electron flux is observed which is stimulated several-fold by the addition of an uncoupling agent, such as dinitrophenol). The uncoupling agent presumably acts as a protonophore to carry the hydrogen ions back across the liposome bilayer barrier as a lipid-soluble acid, and thus dissipates the back pressure on the oxidase system resulting from the buildup of protons inside the vesicles.

Similar reconstructions may be performed to insert other segments of the respiratory chain into artificial membranes, and the whole sequence of electron transfers from NADH to O_2 may be reconstituted from the appropriate flavoproteins, cytochromes, benzoquinones, and so forth. It is also possible to harness the proton gradients generated by the asymmetric electron flows in such systems as energy sources. For example, in the liposomes, previously described, bearing cytochrome oxidase, the mitochondrial ATPase protein designated F_1 may be added on to the outer surface. (Again this is the reverse orientation to that seen for the mitochondrial inner membranes where the F_1 protein may be shown to project inward into the matrix by electron microscopy or by studies with specific antibodies.) In this system the oxidation of trapped, reduced cytochrome c provides the proton gradient required to drive the ATPase in reverse, and hence allows the reconstitution of ATP synthesis associated with the third coupling site of oxidative phosphorylation.

CELL – CELL COMMUNICATION

The cell theory has dominated our concepts about living organisms for over a hundred years. It makes the assumption that individual cells are self-contained and lead independent lives. The membranes of cells serve to maintain the integrity and separateness of individual cells, and yet we are becoming increasingly aware of the role of membrane components in fostering the interdependence of cells. The ways in which cells may communicate with one another involve both direct physical junctions that are visible, and chemical interactions at a molecular level. In some cases there may be transfer of substances between cells through permeable channels linking their cytoplasmic spaces. From recent investigations of cellular aggregation phenomena it is apparent that carbohydrate residues of membrane glycoproteins and possibly glycolipids play the key role in determining the selectivity of such reactions at the cell surface. Since these recognition sites may prove crucial for the correct alignment and association of cells in tissue differentiation, for the determination of immunologic reactions of cells, and for the normal constraints on cell growth occasioned by contacts between cells, it is important to understand the chief biochemical features of the carbohydrate-containing components of membranes.

GLYCOLIPIDS AND GLYCOPROTEINS

The glycolipids found in animal tissues contain a variety of sugars complexed to the long-chain nitrogenous base, sphingosine. The nitrogen is invariably esterified with a fatty acid residue with the consequence that glycolipids bear a superficial similarity to phospholipids: the two long-chain alkyl substituents project into the hydrophobic interior of membrane bilayers; the polar sugar substituents are exposed to the aqueous environment on the exterior surfaces. Simple cerebrosides, which abound in the myelin sheaths of nerves, contain a single sugar—generally galactose, or glucose if the cerebrosides are located in certain other tissues. The sugar moiety is transferred to the sphingosine portion by its activated uridine coenzyme, UDP-galactose or UDP-glucose, in the presence of the appropriate transferase enzyme. More complicated glycolipids arise by the sequential transfers of sugar moieties from uridine nucleotide derivatives of glucose, galactose, and N-acetylgalactosamine to the sphingolipid backbone. Transfer of one or more molecules of the acidic sugar, sialic acid, by its activated cytidine nucleotide derivative leads to the production of the class of glycolipids known as gangliosides. As the name implies, gangliosides are also abundant in nervous tissue, but in the membranes of neurons rather than in myelin as is the case for the galactocerebrosides.

In addition to the simple sugars and their N-acetylamino derivatives found in glycolipids, the glycoproteins may contain mannose and fucose as well as sialic acids. Reactions of these carbohydrate moieties have been demonstrated by histochemical and enzymatic labelling procedures on the outer surfaces of plasma membranes of a number of different cell types. Specific proteins isolated from plant sources (lectins) bind selectively to certain of these sugar molecules, and are useful in localizing different glycoprotein types in cell membranes. These studies, and others involving treatment of cell surfaces with proteolytic enzymes which strip off most of the carbohydrate residues as small glycopeptides, have established that the protein-bound sugars extend in short oligosaccharide chains mainly on the outward facing N-terminal segment of membrane glycoproteins (Fig. 4– 9).

The sugar chains are anchored to the polypeptide chain by glycosidic links either to the amide-N of an asparagine molecule, or to the hydroxyl-O of a serine or threonine molecule. As for the glycolipids, activated sugars are added to the membrane glycoproteins as their UDP coenzymes by the appropriate transferase enzymes. These reactions occur predominantly in the membranes of the Golgi apparatus, and it is believed that Golgi mem-

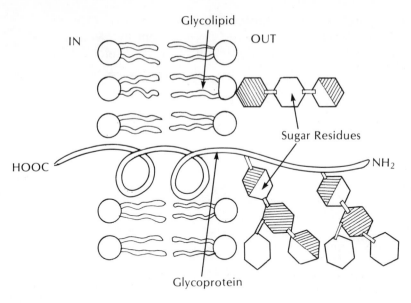

FIG. 4–9.   Membrane glycolipids and glycoproteins. The sugar residues
project from the outward facing surfaces of the membrane bilayer.

branes are transitional forms between intracellular and plasma membranes that complete the formation of characteristic glycoprotein structures on the outer cell surface. Generally, it is a N-acetylated amino sugar
that is added first in the anchor position; sequential addition from UDP
derivatives of mannose, galactose, and amino sugar residues occurs onto
the projecting hydroxyls of the anchor sugar by glycosidic bonds much in
the way that a glycogen chain is propagated. Branching within the
oligosaccharide chain may also occur. Finally, the chain, containing any-
where from three to about a dozen residues, is terminated; the short chains
are generally those attached to serine or threonine by O links, while the
more complex structures, often bearing fucose or sialic acid residues at
their outer ends, are commonly attached by the N-asparagine type of link.

IMMUNE RESPONSES

The immunologic reactions of cell surfaces (important in determining the
compatibility of cells for blood transfusions or for tissue transplantations)
have been defined by the observation of carbohydrate-containing antigens.
Model studies in which glycolipids or glycoproteins are incorporated into
liposomes have proven particularly useful for the controlled investigation
of antibody–antigen reactions in membranes. Just as in the case of the
interactions of lectins at plasma membrane surfaces, the observation of
competitive-binding interference by free sugars has furnished considerable

Erythrocyte Surface Sugars	Blood Group	Serum Antibodies

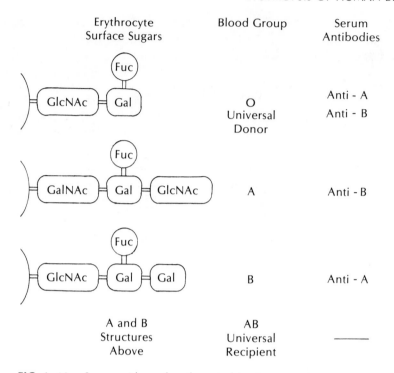

	O Universal Donor	Anti - A Anti - B
	A	Anti - B
	B	Anti - A
A and B Structures Above	AB Universal Recipient	———

FIG. 4–10.   Sugar residues of erythrocyte blood-group substances.
Fuc = fucose; Gal = galactose; GalNAc = N–acetyl-galactosamine;
GlcNAc = N-acetylglucosamine.

information about which of the carbohydrate moieties are specifically implicated in reactions of different membrane antigens. The definition of molecular components of antigenic groups on cell surfaces has been delineated most thoroughly for the blood group antigens on the erythrocyte plasma membrane.

A related series of glycolipids provides the rationale for the ABO group of antigenic characters in the membrane (Fig. 4–10). The A antigen has a short oligosaccharide sequence terminating in N-acetylgalactosamine, the sequence for cells of the B group terminates in galactose, while O group cells lack both of these terminal sugars and exhibits no antigenicity. The molecular genetics of these variations is explained on the basis of the production of specific sugar transferase enzymes as the gene products. The A individual exhibits a transferase that selectively uses UDP-acetylgalactosamine for the final sugar transfer on the membrane, while the transferase from those in the B group will only use UDP-galactose. The AB group possesses both transferases and thus both the A and B surface antigens, while the O group has neither enzyme nor either A or B determinants.

The use of liposomes containing varying antigenic determinants has proven useful for studies of the mechanisms involved in cytolysis by immunologic agents. Thus it is possible to prepare immunologically sensitive liposomes by incorporating erythrocyte membrane glycolipids into bilayers composed predominantly of phospholipids and sterols. The lysis of such sealed membrane-bounded structures may be measured by the release of water-soluble markers (*e.g.*, radioactively tagged glucose) trapped in the interior. On addition of the appropriate antiserum plus the nine serum protein components that are termed "complement" to the antigen-bearing liposomes, formation of immune complexes on the liposomal surface results in damage to the model membrane and consequent leakage of the trapped glucose (Chapter 1). Early reports suggested that the lesions consisted of fairly large holes (10 nm in diameter), but recent work suggests that smaller pores or channels (about 2 nm) may account for the cytolytic response. The model system is particularly useful for studies of both humoral and cell-mediated effects because a number of different lipophilic antigens of either natural or synthetic origin may be introduced, and because it is possible to study the effects of variations in membrane fluidity through changes in fatty acid or sterol composition in relation to the immune response of the membrane.

MEMBRANE JUNCTIONS

It is becoming apparent that there may be direct physical bridges between cells in addition to communication by transmembrane movements of signals and surface receptor responses. The most well defined intercellular connections are the desmosomes, areas of cell-to-cell adhesion with fibrous proteins extending across the space between two cell membranes. Although the cells adhere to one another their membranes are separated and their cytoplasmic spaces do not communicate. Another type of connection between cells is the so-called tight junction. In this structure each contiguous cell in a sheet is joined to its neighbors by a seam that seals the outer layers of the bilayers. Thus there is no extracellular space between the cells, and the sheet of cells is capable of forming a completely impermeable partition between two fluid compartments. (An example is the so-called blood–brain barrier in which the blood vessel cells of the endothelial lining, unlike those in other body locations, are linked by tight connections to form a continuous protective shield to prevent entry of toxic substances into the brain.) Still another type of intercellular connection is referred to as the gap junction, and by contrast with the other two connections previously mentioned it provides a direct linkage between the cytoplasm of two adjoining cells. The extracellular spacing is tighter than that provided by desmosome links yet looser than that provided by tight junction connections. However the most important feature is that the gap junctions form a series of channels connecting epithelial cells of the liver,

kidney, skin, and so forth. By observing the course of fluorescence in adjacent cells following the microinjection of fluorescent compounds of
varying dimensions it may be shown that small molecules (up to 1000
molecular weight) can pass through the channels without reaching the
extracellular space between cells.

CELL – CELL RECOGNITION

In associating with one another to form the tissues of the body, cells exhibit
the ability to distinguish one cell type from another and to adhere selectively to other cells based on surface membrane properties. In addition to
the integral components of the outer face of the bilayer, the cell surface
membrane also includes peripheral glycoproteins, polysaccharides, and
extracellular ions such as Ca^{2+}, which are responsible for communicating
with neighboring cells.

Contact Inhibition

A valuable technologic advance for the study of cellular interactions is the
recent development of procedures for the growth of various cell types in
artificial media. These cultures of animal cells develop and divide when
supplemented with appropriate nutrients, but generally growth continues
only until a limiting concentration of cells is reached. For example, such
cells may multiply and spread out on the surface of the culture vessel until
they form a continuous layer of cells that abut one another on all sides.
Cessation of growth may coincide with the cells touching one another, and
for this reason has been termed contact inhibition, or density-dependent
inhibition of growth. The mechanism of this effect is unknown but undoubtedly involves a concerted action of surface components of the plasma
membrane, which act both as exterior receptors for adjacent cells and as
transmitters of signals to the interior to shut down cell division. The parallels with the receptor-transmitter functions of the adenyl cyclase system
are obvious. Moreover, establishment of contact between cells is associated with an increase in concentration of intracellular cyclic AMP at the
same time as DNA synthesis is arrested. Both qualitative and quantitative
differences between the glycoprotein or glycolipid components of the cell
surface have been detected in comparing the rapidly-growing cells with
those which are arrested by density-dependent inhibition.

Malignancy and Cell Transformation

The implications of intercellular reactions and properties of the cell membranes in relation to cancer has recently received much attention. Tumor
cells are characterized by run-away growth and division, and by devia-
tions of invasive tendencies and other social reactions with their neighbors. Some of these aberrations are apparent from *in vitro* studies of cancer
cells; an experimental tool which has been of great value in such work is

the use of oncogenic viruses that are capable of promoting a transformation from normal to malignant cells in cultures. It has been established, based on the *in vitro* experiments, that in most cases the transformation process is characterized by interference with the contact and adhesion phenomena observed in normal cells. Thus the transformed cells lose the property of contact inhibition and instead continue to grow and produce multiple layers of cells. Many types of cancer cells exist, and it is difficult to make generalizations concerning the membrane alterations that occur in these malignant cells. In many cases the recognition and adhesive properties among cancerous cells are defective, and the intercellular communications are drastically reduced. Although such changes in surface properties are undoubtedly secondary manifestations of the malignant transformation, their study is of importance for the understanding of how membrane-mediated signals may govern both the growth rates and the invasive tendency of tumor cells.

CELL FUSION—CELL HYBRIDS

The union of cells to form syncitia with several nuclei contained in a single large cytoplasmic space (polykaryocyte) is a rare event under physiological conditions. A typical example is the differentiating muscle cells (or myoblasts), which fuse together to form the multinucleated ribbon structures (or myotubes) of mature muscle. In the test tube it is also possible to trigger the joining of cytoplasmic spaces between cells by application of agents that cause their membranes to meld. When the two cells so united are from different species the resulting partnership is termed a hybrid cell, and it will display some genetic characteristics of each of the species of origin.

The phenomenon of cell fusion is not completely understood (Fig. 4–11). Cells growing in culture can be induced to fuse with the adsorption of certain virus particles, but the same viruses that are inactivated by ultraviolet light can also promote cell fusion, indicating that infection and the replication of virus is not necessarily related to the fusion stimulus. Moreover, a number of chemical agents, including highly surface-active cytolytic lipids, such as lysolecithin, are capable of inducing fusion of cells when applied just below the cytotoxic level. The principal event in the fusion stimulus appears to be a localized disturbance of the normal arrangement of the plasma membrane bilayers of apposed cell surfaces (Fig. 4–11b). This disturbance is accompanied by extrusion of cytoplasmic enzymes, including lysosomal hydrolases, to attack surrounding glycoproteins of the cell membranes, and by modification of cytoplasmic microtubules on the interior, so that the two plasma membranes may coalesce in the region of the displacement to form a single cell (Fig. 4–11c).

Use of the fusion process with mixed cultures of cells from various species has led to the production of stable genetic hybrids that can survive

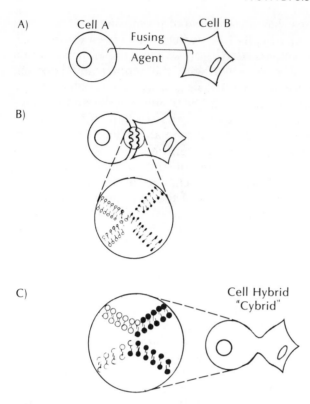

A) Cell A   Cell B
              Fusing
              Agent

B)

C)   Cell Hybrid
                               "Cybrid"

FIG. 4–11.   Fusion of cell membranes to produce cell
hybrids (cybrids).

and maintain the chromosomal characteristics from each of the two parental species in varying proportions. Interspecies hybrids of human cells and other animal cells ranging from mice to mosquitoes have been produced in this way. Such monstrous matings may conjure images of biology perverted to demonology. In reality these chimeric forms are not quite demons, but are more than just curiosities. Surface antigens and other species-specific markers, such as characteristic cytoplasmic proteins, can show the degree of admixture of genetic determinants in the compound nuclei of the new hybrid cells. Using selective staining procedures to identify the species of origin of the surviving chromosomes, one can then determine which chromosome, or even which part of a chromosome, carries the gene for a particular protein found in the hybrid. In this way it has become possible to provide highly detailed genetic maps of the 46 human chromosomes. The hybridization technology promises to open many new venues for studies of cell surface antigens, virus infections, and malignant transformations of cells.

INTRACELLULAR COMPARTMENTS

A number of analogies have been used to characterize the cell and its functions. The medieval vision of a tiny man, or homunculus, inhabiting the fertilized egg has given way to our modern views of DNA as the pluripotent passenger and director of development. Yet the metaphor of cell as organism is not without value when we consider that individual cells must encompass all those activities of the whole body (*i.e.*, reproduction, respiration, movement, digestion, excretion, etc.) in a microcosm; and as the human organism has differentiated organ systems for these specialized processes so the individual cells of the body have differentiated organelles.

In several instances the organelles exist as well-defined forms (*i.e.*, nuclei and mitochondria). As for the discrete organ systems of the body (*i.e.*, brain or liver), these are the easiest to isolate and study in their natural functional states, and their internal spaces are also quite readily distinguished from the rest of the cytoplasm. In other cases organelle systems may be more diffuse, and like certain vascular tissues of the body (*i.e.*, the blood vessels and lymphatics) may have pleimorphic and interconnecting structures that render them more difficult to separate as intact forms making the topologic considerations of the spaces they occupy much more complex. (Examples of such organelles are the tubules and cisternal spaces of the endoplasmic reticulum and Golgi networks.) Finally, there is a cellular analogue of the body's connective tissue, sometimes loosely termed the cytoskeleton. Rigid microtubular elements maintain compartmental shapes, and contractile microfilaments play a role similar to that of human musculature in evoking changes in compartmental shapes within the cells.

ENDOCYTOSIS AND EXOCYTOSIS

In a preceding section we dealt with transport mechanisms that allow small molecules to cross cell membranes. Bulk transfer of substances either into or out of cells may also take place. Here our metaphor of the cell as homunculus breaks down, for there are no pre-formed orifices for the cell's digestive apparatus. Instead, large particles or bulk quantities of liquids to be taken up by cells must be engulfed by the plasma membrane by the mechanisms of *phagocytosis* or *pinocytosis;* together these uptake events are termed *endocytosis*. The significance of these processes varies for different cell types; it is of particular importance for the neutralization of bacteria and other foreign bodies by certain phagocytic cells of the reticuloendothelial system. The first step involves adsorption of substances on the external surface. There follows an energy-requiring process whereby the plasma membrane invaginates to surround the engulfed substances and to internalize them within the cytoplasm as membrane-coated vesicles termed *endosomes*. Further steps in the digestion of such substances are detailed in the later section on lysosomal functions. Similarly, the ejection of macromolecules, or waste products, which are too large to penetrate the cell's membrane require similar, but reversed processing, termed *exocytosis*. Such events are of greatest significance for secretory cells of the exocrine or endocrine glands, or neural cells involved in chemical transmission. The steps in packaging substances for bulk extrusion by membrane-encapsulated secretory vesicles are dealt with in the following section on functions of the Golgi apparatus.

Both endocytosis and exocytosis entail extensive turnover and exchange of intracellular membrane components. When extracellular particles become encapsulated portions of the plasma membrane are depleted, and in order to maintain the cell's normal surface area new membrane obviously must be synthesized. Formation and assembly of plasma membrane or the membranes surrounding secretory vesicles undoubtedly contribute significantly to the energy requirements for these processes. The newly formed components do not lead to net accretion of membranes, but rather to a cycle of renewal and breakdown fueled by the cell's energy sources, and readily demonstrated by the incorporation of appropriate radioactive membrane precursors. For example, all nucleated cells in the body show continual uptake of ^{14}C-labelled glycerol or ^{32}P-phosphate into their membrane phospholipids even in the absence of growth, indicating that a dynamic equilibrium or steady state of synthesis and degradation of membranes must occur at all times. Stimuli that evoke increases in secretory phenomena accelerate the labelling of certain membrane components above the resting levels, and similar increases of phospholipid turnover are seen in phagocytosis. As noted in the accompanying sections, current concepts of membrane replacement and renewal are based on a view of continual ebb and flow of the phospholipids, glycolipids, and proteins among endoplasmic reticulum, Golgi apparatus, plasma membranes, and lysosomes of the cell. In its lifetime, as in ours, the cell as homunculus will pass through many stages from birth through differentiation to senescence, and each of its membrane components may be recycled many times within that lifetime.

GOLGI AND SECRETORY PHENOMENA

In many cells there is a complicated arrangement of smooth membranes that generally enclose materials destined for secretion from the cell. This system of cisternae and vesicles is termed the Golgi complex or apparatus (named after its discoverer). Stereographic studies by serial sectioning or thick sectioning have revealed that the internal canals of the Golgi are interconnected and probably link with internal channels of rough and smooth endoplasmic reticulum membranes. The Golgi membranes are most extensively developed in actively secreting cells of exocrine or endocrine glands, and in neurosecretory cells making them of prime importance in the secretory phenomenon. In the acinar cells of the pancreas, for example, the Golgi apparatus may be distinguished as a highly oriented stack of membranous disks lying between the endoplasmic reticulum at the base of the cell and the vesicles which store secretory products towards the apical membrane bordering the acinar space. The lumens of the endoplasmic reticulum, Golgi, and secretory vesicles provide a continuum that is topologically exterior to the cytoplasm, and thus protects the cell interior from potentially hazardous secretions.

A type of time-lapse photography, pulse-chase labelling, and autoradiography, may be used to show the sequence of events in secretion. Briefly, the experimental procedure involves the incubation of small pieces of glandular tissue in a solution containing a ^{3}H-labelled amino acid (*e.g.*, leucine), which is quickly incorporated into newly synthesized secretory proteins. After a few minutes the tissue is removed to a second solution containing unlabelled amino acids, which are subsequently incorporated into the proteins too. In this way the radioactive protein, produced by the short pulse of the first incubation, is chased through the cell by the subsequent slug of unlabelled protein formed in the second incubation. Pieces of tissue may then be removed at various times during the chase incubation, and sections are laid down on photographic emulsion. After developing the film, exposed areas from the adjacent radioactive tissue may be lined up with their structural counterparts as seen in the electron microscope, hence a self- or autoradiograph of labelled regions in the cell.

This technique used with pieces of pancreas has demonstrated a progression over time of the labelled secretory proteins. At the shortest intervals the label is confined to rough endoplasmic reticulum at the basal ends of the cells where polypeptides are presumably being synthesized on the membrane-bound ribosomes (Fig. 4–12). In the next minutes the label is seen migrating through channels of smooth-membrane endoplasmic reticulum into the cisternae of the Golgi system. Within one hour the label is transferred to the zymogen granules, and subsequently it is detected in the acinar spaces leading to the pancreatic ducts during secretion. Each part of the transfer process is energy-dependent, and addition of respiratory inhibitors (*e.g.*, antimycin) during the chase will cause a back-up of the label at that point. Continuing protein synthesis seems to be unnecessary for the transfer process since addition of inhibitors of translation (*e.g.*, cycloheximide) affect only the pulse labelling but not the events occurring during the chase.

Support for such a model of secretion has been obtained in studies of a number of exocrine and endocrine glands, as well as liver and neural cells. Events at the level of the ribosome have been clarified somewhat. Only those proteins destined for use within a cell are synthesized on free ribosomes, while secretory, or "export" proteins, are formed by the ribosomes that are complexed with endoplasmic reticulum membranes as detailed in

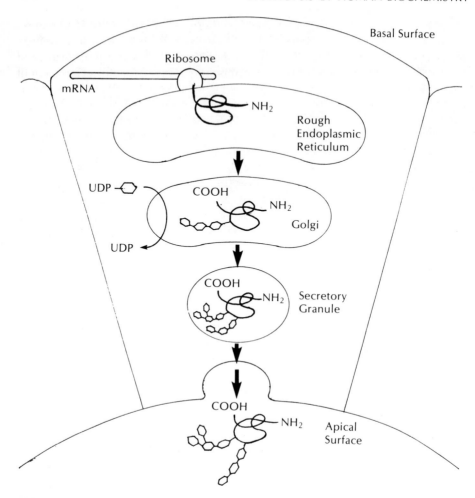

FIG. 4–12. Processing of secretory proteins. The initiation of protein synthesis occurs on ribosomes bound to rough endoplasmic reticulum with the polypeptide conveyed to the lumen. Sugar molecules are added in the Golgi apparatus, and membrane-bound proteins are transformed to secretory granules that fuse with the cell membrane and extrude the secretory protein.

Chapter 5. An interesting feature of this process is the demonstration that the first 30 to 40 amino acids of newly synthesized secretory proteins are initially buried within the membrane. It is hypothesized that this N-terminal tail of the growing polypeptide directs the ribosome to an attachment site on the membrane, and on entering the hydrophobic core it forms a tunnel through the lipid bilayer. As synthesis proceeds the protein is elongated to extend through the bilayer into the lumen of the endoplasmic reticulum where some sugar residues may be attached (in the case of glycoproteins). However, most of the sugars are added after the protein dissociates from the endoplasmic reticulum and passes into the cisternae of the Golgi complex (Fig. 4–12). It is possible to separate membranes of the Golgi type from those of the endoplasmic reticulum by fractionation of carefully disrupted cells using differential centrifuging techniques. Biochemical studies show distinct differences in chemical composition and enzyme activities of the two membrane types. Most of the glycosyl transferase enzymes, for example, are recovered in the isolated membranes from the Golgi complex. This is consistent with the results of pulse-chase studies in which ^{3}H-labelled sugars (*e.g.*, galactose) are used instead of amino acids. Initially the label is mainly associated with glycoproteins of Golgi membranes. With time the labelled sugar is seen to progress through secretory proteins and membrane proteins of the secretory granules. Later the label is detected outside the cell, either as secretory glycoproteins, or as integral components on the outside surface of the plasma membrane.

Several problems are unresolved concerning the final stage of secretion, which is the exocytosis by fusion of secretory granules with the plasma membrane. For instance, both the trigger and the specificity of the fusion process are not fully understood. In epithelial cells, such as those of the pancreas, secretory products are always discharged directionally from the apical surface. There must be a mechanism whereby secretory granules recognize the appropriate membrane surface in order to direct the correct orientation for secretion. A possible involvement of the microtubule system in migration of the secretion products through the cell is suggested by the observation that colchicine, an agent that disrupts microtubule structures, inhibits secretion by a number of cell types. Possibly the microtubule system forms a framework within the cell to interconnect secretory vesicles and guide their movement to the apical plasma membrane.

The Golgi apparatus has also been implicated in production of secretory mucopolysaccharides and in the addition of sulfate groups in mucus-secreting cells. In liver and intestinal cells the passage of the VLDL and chylomicron lipoproteins out of the cells can also be traced through the intermediary way-station of the Golgi tubules. In the lactating mammary gland there is marked hypertrophy of the Golgi membranes, which play a role in the secretion of milk proteins (although in this instance the lipid moieties are extruded by a separate mechanism involving direct budding of fat droplets surrounded by a coating of plasma membrane). Thus the Golgi system seems to play quite a major role in packaging secretory products, although there may be alternative processes in certain cases. During the transformation of the contents passing through the Golgi lumen, it should also be noted that the Golgi membranes themselves are undergoing metamorphosis toward a composition intermediate between that of endoplasmic reticulum (where membrane phospholipids and proteins chiefly originate) and the plasma membrane (where the coating from secretory vesicles must terminate). Such considerations suggest that the Golgi membrane is a transitional state in the maturation of cell membranes. In light of its abundant glycosyl transferase content, it seems likely that events in the Golgi system may play a key role in the differentiation of integral components of plasma membranes, in the development of the characteristic external glycoproteins, and possibly in the development of glycolipid determinants of the cell surface.

LYSOSOMAL FUNCTIONS AND DISEASES

Lysosomes, as their name implies, are the repositories of lytic enzymes in cells. Fortunately for us such powerful digestive catalysts are encased in a protective membrane, much in the same way that the intestinal casing provides a lumen for the digestion of foodstuff that is topologically on the outside of the body. Under certain toxic or degenerative influences the lysosomal membranes may leak their enzymes into the cell, or the membrane may engulf cytoplasmic components; in these instances cell contents can be digested by *autolytic* or *autophagic* processes involving the lysosomes. Following the ingestion of extracellular materials, the lysosomes may also combine with the foreign substances and conduct their breakdown by *heterophagic* processes.

In general, the lysosomes contain a fairly wide range of hydrolytic enzymes: acid phosphatase, acid ribonuclease, and acid deoxyribonuclease, which cleave phosphodiester links of the nucleic acids; a number of enzymes capable of hydrolysis of glycoside links; lipases and phospholipases, which hydrolyse ester links; proteases and peptidases, which hydrolyse peptide links. Many of these lytic enzymes show maximal activity under acidic conditions below the normal intracellular pH value. It is probable, moreover, that there are wide differences in the complement of enzymes in the lysosome populations of cells depending on tissue of origin, and the physiological or developmental state of the cells.

As is the case for the digestive enzymes destined for secretion from cells, it now appears certain that the synthesis of the lysosomal enzymes occurs on membrane-bound ribosomes of the endoplasmic reticulum, and that the newly synthesized enzymes in an inactive or precursor state traverse the route through the complex network of the Golgi complex. The membrane-enclosed body formed in this way is termed the *primary lysosome* (Fig. 4– 13). By a process of membrane fusion it is able to meld with various types of endosomes that arise from phagocytic or pinocytic processes described earlier. The organelle that results from the fusion is termed a *secondary lysosome*, and it is only at this stage that the hydrolytic enzymes become activated to conduct their digestive functions. In this way the macromolecules taken up by endocytosis are mixed with and subject to the action of a wide variety of hydrolases. The small, water-soluble reaction products (amino acids, sugars, nucleic acid bases, etc.) may then dif-

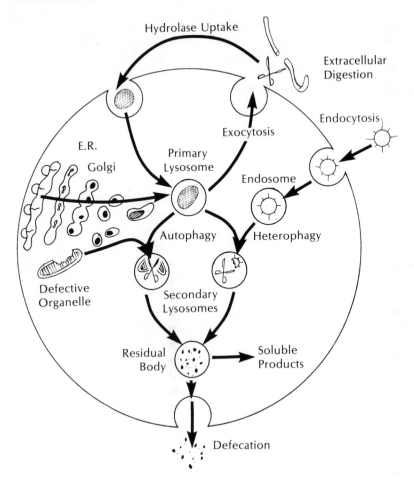

FIG. 4–13. Primary and secondary lysosomes. Primary lysosomes may fuse with intracellular particles (autophagy) or those brought in from outside (heterophagy), with digestion occurring in the resultant secondary lysosomes.

fuse into the cytoplasm for metabolic use. Any indigestible material will remain encapsulated within the membrane, the resultant vesicle being termed a *residual body*. Analogous with exocytotic secretion, undigested contents of such vesicles may be eliminated from the cell following fusion of the residual body with the plasma membrane.

Several involutional or resorptive changes in tissues have been attributed to the actions of lysosomal digestion. For example, the osteoclasts, which are involved in remodelling bone by resorption of the calcified matrix components, are especially rich in the hydrolytic enzymes that are associated with lysosomes. In this instance, the hydrolases are extruded into the extracellular spaces from primary lysosomes where some breakdown of matrix occurs; the material may undergo further attack following uptake into the secondary lysosomes. This two-stage process involving exocytosis of lysosomal hydrolases for extracellular digestion and phagocytosis of partly degraded matrix for intracellular digestion is probably of general significance for bone and cartilage resorption. In the uterus lysosomal hydrolases have also been implicated in endometrial breakdown during menstruation, and during the period immediately following birth when postpartum involution of the organ occurs. A significant feature of the uterine lysosomal system is its sensitivity to steroid hormones. The synthesis of lysosomal hydrolases is dependent on maintenance of estrogenic hormones. Moreover, studies of the release of lytic enzymes from the isolated lysosomes have shown a greater susceptibility to autolytic stimuli in animals treated with estradiol. Similar effects were observed with testosterone in the accessory sex organs of male animals. In addition to the sex hormones, vitamin A and vitamin D have a disruptive effect on the lysosomal membranes of a number of tissues, while the anti-inflammatory steroid hormone, cortisone, has a counteracting stabilizing influence.

The lysosomes of endocrine tissues may play a part in modulating release of hormones from the glands. A direct role is seen, for example, in thyroid tissue where the active hormones, thyroxine and triiodothyronine, are released from the polypeptide storage form, thyroglobulin, only after the latter has been endocytosed by follicular epithelial cells and hydrolysed by the proteolytic enzymes in secondary lysosomes. In the pituitary it appears that lysosomal hydrolases can suppress secretion of some polypeptide hormones by digesting the contents of excess storage vesicles. Such interactions between lysosomes and hormones may prove to be of general importance in the modulation of endocrine effects on the body.

There are several categories of diseases in which the lysosomes are key participants—the most prevalent and important category to clinical medicine being infections caused by microbial agents. The mobile phagocytic white cells of blood, as well as the fixed macrophages of the tissues, form a primary line of defense against such parasites by endocytosing the foreign particles to form phagosomes. There is an adaptive increase of the lysosomal digestive enzymes accompanying the uptake of such foreign organisms into the phagocytic cells. Fusion of the phagosomes with primary lysosomes (producing secondary or phagolysosomes) exposes the infectious bacteria or viruses to the digestive actions of the acid hydrolases. The ensuing hydrolysis of microbial macromolecules generally contributes to the inactivation and eventual death of the invading microbes. In some instances phagocytes are less successful in carrying out digestion of the invaders: certain bacteria (e.g., the tuberculosis bacilli) may possess outer protective coatings that render them resistant to lysosomal hydrolases; some viruses may also escape attack, as is the case with the reovirus whose double-stranded RNA is not susceptible to ribonuclease. Such organisms may replicate within the lysosomes, eventually destroying the cells they infect. In other cases pathogenic microbes may release toxins that damage the lysosomal membrane: some bacteria release streptolysins, which are believed to lead to release of the lysosomal hydrolases, causing destruction of mitochondria and other host cell structures; certain viruses (*e.g.*, picornia) generate production of cytotoxins in the host that cause rupture of the lysosomes and rapid autolysis by similar means.

The autolytic response is not confined to microbial infections. Subjecting cells to sublethal doses of ionizing radiation or cytotoxic chemicals (*e.g.*, chlorinated hydrocarbons) can also result in damage to the lysosomes with leakage of hydrolytic enzymes into the cytoplasm. In such situations the digestive process serves to confine the toxic actions to a focal area of injury, and to clear the site of damaged cells preparatory to regeneration of healthy tissue. There is also much suggestive evidence that enhanced lysosomal activity may be an accompaniment, if not causative agent, in the inflammatory processes of rheumatoid arthritis. Increased numbers, as well as enhanced lability of lysosomes, may be observed in rheumatoid synovial tissues; this suggests that autophagic digestion contributes to the degradation of normal matrix components in the joint (*e.g.*, mucopolysaccharides and associated proteins). Some of the observed ameliorative effects of cortisone in joint disorders may be related to its noted ability to reduce lysosome fragility.

Another major group of pathological conditions involving the lysosomes are the hereditary storage diseases. These are sometimes referred to as the inherited lysosomal diseases, and in all cases where the etiology has been established, they can be attributed to the deficiency or abnormal processing of a particular hydrolytic enzyme that is normally localized in lysosomes. As a consequence, the substrate for the missing hydrolase accumulates within the lysosomes of affected organs. The membrane-bound, undigested material (generally a sphingolipid or a mucopolysaccharide) forms a characteristic residual body. With time the cytoplasm becomes progressively congested with the abnormal storage granules until cellular function is impaired. If the defect lies within cells of the nervous system, as is frequently the case, the patient exhibits severe progressive neurologic impairment that is irreversible and often fatal in the early years of childhood.

The sphingolipidoses are a group of rare familial diseases, each disorder being characterized by deposition of particular sphingosine-containing lipids in certain tissues of the body. Not only is there considerable variation with respect to the organs affected, but there is also a good deal of variation among the diseases concerning the pathological features, severity of their effects, and age of onset. It is now clear that the sphingolipids, which are being continually synthesized in cells, undergo turnover by a complex pathway of catabolic reactions each catalysed by a specific lysosomal hydrolase. In Tay–Sachs disease, for example, absence of the hydrolase that splits a hexosamine residue from glycolipids in the nerve cells leads to accumulation of the ganglioside substrate, GM_2, in the lysosomes. The absence of the hexosaminidase enzyme, accumulation of GM_2, and presence of typical onionlike residual bodies in affected neurons are characteristic of the syndrome. Clinical signs of the disease appear in infancy and include progressive mental impairment, paralysis, blindness, and the presence of a readily recognized, cherry-red spot in the retina accompanying degeneration of the macula. The degeneration of nerve cell bodies and secondary demyelination generally lead to early death. Although several other sphingolipidoses (Krabbe's disease, Sandhoff's disease, and metachromatic leucodystrophy) have primary effects on the nervous system, there are others, such as Neimann–Pick disease, which exhibit major involvement of the visceral organs as well. The adult or chronic form of Gaucher's disease, as the name implies, has a late and slowly progressive accumulation of glucocerebroside in the spleen, bone marrow, and other reticuloendothelial cells of the body; only the infantile or acute type of the syndrome exhibits a primary involvement of the neurons and severe neurologic impairment. Patients suffering from Fabry's disease accumulate excessive quantities of ceramide trihexoside in the intestine, lymph, and kidney resulting in the development of uremia.

Deficiencies of specific lysosomal hydrolases that normally contribute to catabolic turnover of acid mucopolysaccharides have been demonstrated in tissues from patients with the familial syndromes, which were once known collectively as gargoylism in reference to the frequent grotesque distortion of the victims' facial features. Distinctions based on genetics, biochemistry, and clinical features have revealed the presence of a half-dozen different syndromes of mucopolysaccharidosis in what had once been thought of as a single disease. Afflicted individuals, who may show mild to severe skeletal deformities, mental retardation, cardiac disease, or clouding of the cornea, also show variability in the nature and extent of accumulation of sulfated mucopolysaccharides in the tissues and in the urine. Definition of the biochemical lesions in the mucopolysaccharidoses has followed a novel path that provides some tantalizing clues about normal features of lysosomal enzymes. This approach exploited the growth of fibroblast cells from patients with the diseases, and the ability of substances added to the culture medium to correct the abnormal chemical pathology. Cells from the skin of afflicted patients, when cultured in medium containing ^{35}S-labelled sulfate, accumulated excessive amount of radioactive mucopolysaccharides in comparison with normal cells. This accumulation could be prevented if the cells were grown in the presence of extracts from normal tissues, which apparently contributed "corrective factors" to the culture medium. These factors were subsequently identified as enzymes capable of hydrolysing a particular linkage in the mucopolysaccharide chain, and correction of the defect was thus effected by replacement of the enzyme which was missing from the deficient cells, presumably by endocytosis into the lysosomes.

Two interesting features of the mucopolysaccharidoses were clarified by such studies. First, the two categories of gargoylism which were originally separated based on genetic and clinical determinations could be shown to compliment one another in the cell culture assay. Hurler's syndrome, which is inherited as an autosomal recessive trait and shows early evidence of mental retardation and opacities in the cornea of the eye, was distinguished from Hunter's syndrome, which is inherited as an X-linked recessive trait and shows later, less severe evidence of mental retardation and no corneal clouding. A mixture of fibroblasts taken from patients with the typical symptoms of Hurler's and Hunter's syndromes were able to interact in culture to effect a mutual correction of the excess accumulation of mucopolysaccharides. Thus different factors were defective in the two diseases, and cells from one genotype could secrete enough of the missing enzyme into the culture medium to correct the deficiency in the other cell type. The factor missing from the Hurler patients was isolated from normal tissues and identified as a glycosidase that cleaves terminal iduronic acid residues from sulfated mucopolysaccharides, while the deficient enzyme in Hunter patients was characterized as the hydrolase (sulfatase) which normally splits off the sulfate group. In a similar fashion a group of mucopolysaccharidoses were differentiated, and the enzyme defect defined for each syndrome. The severity of clinical manifestations ranges widely, from the pronounced skeletal abnormalities but absence of mental impairment in Morquio's syndrome, to the pronounced mental retardation but minor skeletal changes in San Fillipo's syndrome, to the milder pathology of Scheie's syndrome which exhibits marked clouding of the cornea but few abnormalities of either the skeleton or nervous system.

The second aspect of the study was the discovery of a further class of lysosomal diseases in which there is an anomaly in the processing of the enzymes required for mucopolysaccharide breakdown. This class is exemplified by I-cell disease, a syndrome with severe skeletal deformities and mental defect reminiscent of Hurler's syndrome. Fibroblasts from the patients lack a number of lysosomal hydrolases involved in both mucopolysaccharide and sphingolipid catabolism, although the missing enzymes may be detected in the culture medium. Since the defect may be corrected by addition of enzymes from normal individuals to the cells, it appears that in the I-cell syndrome the enzymes are synthesized normally but lack a recognition marker which is necessary for the normal uptake of the hydrolases into the lysosome. In the normal course of events, digestive enzymes destined to be taken up by the lysosomes of cells are produced in the endoplasmic reticulum in a precursor transport form that bears the characteristic marker, a phosphorylated sugar residue. The latter, mannose-6-phosphate, is missing from precursors of lysosomal enzymes in the I-cell syndrome, leading to defects of several catabolic functions in these patients. From a clinical standpoint, the importance lies in the further elaboration of the recognition markers, with the prospect of modifying missing lysosomal enzymes to produce high uptake forms that may be applied to correcting the abnormal storage diseases.

SUGGESTED READINGS

Atkinson PH, Hakami J: Alterations in glycoproteins of the cell surface. In Lennarz WJ (ed): The Biochemistry of Glycoproteins and Proteoglycans, pp 191– 239. New York, Plenum, 1980

Barski G, Belehradek J Jr: Cytoplasmic membranes in somatic cell interaction and hybridization. In Jamieson GA, Robinson DM (eds): Mammalian Cell Membranes, Vol 5, Responses of Plasma Membranes, pp 284– 306. London, Butterworths, 1977

Capaldi RA: The structure of mitochondrial membranes. In Jamieson GA, Robinson DM (eds): Mammalian Cell Membranes, Vol 2, The Diversity of Membranes, pp 141– 164. London, Buttersworth, 1977

DeMello WC: Intercellular Communication. Plenum, New York, 1977

Emmelot P: The organization of the plasma membrane of mammalian cells: Structure in relation to function. In Jamieson GA, Robinson DM (eds): Mammalian Cell Membranes, Vol 2, The Diversity of Membranes, pp 1– 54. London, Butterworths, 1977

Favard P: Membranes of the Golgi apparatus. In Jamieson GA, Robinson DM (eds): Mammalian Cell Membranes, Vol 2, The Diversity of Membranes, pp 108– 140. London, Butterworths, 1977

Finean JB, Coleman R, Michell RH: Membranes and Their Cellular Functions, 2nd ed, New York, John Wiley & Sons, 1978

Frazier W, Glaser L: Surface components and cell recognition. Annu Rev Biochem 48:491– 524, 1979

Gregoriadis G: The liposome drug-carrier concept: its development and future. In Gregoriadis G, Allison AC (eds): Liposomes in Biological Systems pp 25– 86. New York, John Wiley & Sons, 1980

Hakomori S – I: Glycosphingolipids in cellular interaction, differentiation, and oncogenesis. Annu Rev Biochem 50:733– 764, 1981

Heidmann T, Changeux J-P: Structural and functional properties of the acetylcholine receptor protein in its purified and membrane-bound states. Annu Rev Biochem 47:317– 358, 1978

Hughes RC: Membrane Glycoproteins—A Review of Structure and Function, London, Butterworths, 1976

Kelly RB, Deutsch JW, Carlson SS et al: Biochemistry of neurotransmitter release. Annu Rev Neurosci 2:399– 444, 1979

King AC, Cuatrecasas P: Peptide hormone-induced receptor mobility, aggregation, and internalization. N Engl J Med 305:77– 88, 1981

Kolodny EH: Lysosomal storage diseases. N Engl J Med 294:1217– 1220, 1976

Lester HA: The response to acetylcholine. Sci Am 236(2):106– 118, 1977

Nicolson GL, Poste G, Ji TH: The dynamics of cell membrane organization. In

Poste G, Nicolson GL (eds): Cell Surface Reviews, Vol 3, Dynamic Aspects of Cell Surface Organization, pp 1– 73. Amsterdam, North-Holland, 1977

Op den Kamp JAF: Lipid asymmetry in membranes. Annu Rev Biochem 48:47– 72, 1979

Pagano RE, Weinstein JN: Interactions of liposomes with mammalian cells. Annu Rev Biophys Bioeng 7:435– 468, 1978

Pearse BMF, Bretscher MS: Membrane recycling by coated vesicles. Annu Rev Biochem 50:85– 101, 1981

Pitt D: Lysosomes and Cell Function, New York, Longman, 1975

Quinn PJ: The Molecular Biology of Cell Membranes, London, MacMillan Press, 1976

Rubenstein E: Diseases caused by impaired communication among cells. Sci Am 242(3):102– 121, 1980

Sabatini DD, Kreibich G: Functional specialization of membrane-bound ribosomes in eukaryotic cells. In Martonosi A (ed): The Enzymes of Biological Membranes, Vol 2, Biosynthesis of Cell Components, pp 531– 579. New York, Plenum, 1976

Sharon N: Carbohydrates. Sci Am 243(5):90– 116, 1980

Silverstein SC, Steinman RM, Cohn ZA: Endocytosis. Annu Rev Biochem 46:669– 722, 1977

Snyder SH: Opiate receptors and internal opiates. Sci Am 236(3):44– 56, 1977

Snyder SH, Childers SR: Opiate receptors and opiate peptides. Annu Rev Neurosci 2:35– 64, 1979

Sweadner KJ, Goldin SM: Active transport of sodium and potassium ions: Mechanism, function, and regulation. N Engl J Med 302:777– 783, 1980

Tedeschi H: Mitochondria: Structure Biogenesis and Transducing Functions, New York, Springer-Verlag, 1976

Wattiaux R: Lysosomal membranes. In Jamieson GA, Robinson DM (eds): Mammalian Cell Membranes, Vol 2, The Diversity of Membranes, pp 165– 184. London, Butterworths, 1977

Weissman G, Claiborne R (eds): Cell Membranes—Biochemistry, Cell Biology and Pathology, New York, HP Publishing, 1975

Whaley WG: The Golgi Apparatus, New York, Springer-Verlag, 1975

Wikström M, Krab K, Saraste M: Proton-translocating cytochrome complexes. Annu Rev Biochem 50:623– 655, 1981

Wilson DB: Cellular transport mechanisms. Annu Rev Biochem 47:933– 966, 1978

Zwaal RFA: Some aspects of structure– function relationships in biological membranes. In Dumont J, Nunez J (eds): Hormones and Cell Regulation, Vol I, pp 1– 14. Amsterdam, North Holland, 1977

Chapter Five *Molecular Genetics*

Although many physiological processes have yielded to modern biochemical analysis, no single event has had such a profound impact on biology and medicine as the definition of mechanisms for the transmission and expression of inherited characteristics at a molecular level. The explosion of the science of molecular genetics has seemingly conferred almost godlike power on mankind, allowing him to understand and perhaps ultimately control the destiny of his own and other species. Beginning three decades ago with the description of the double helical structure of DNA, genetic research has demonstrated the processes involved in replication, transcription, and translation of the nucleic acid code words from the genotype into the polypeptide sequences of the phenotype. This research has culminated in the present day understanding of the actions of hormonal and nutritional inducers, environmental mutagens, and carcinogens that modify and alter functions of the genes. Terms such as "cloning," "recombinant DNA," "genetic engineering," have escaped from the laboratory to the sociopolitical arena much as "natural selection," "evolution of species," "descent of man," did in Victorian times. Molecular genetics and its implications have presented almost as much of a challenge to our collective wisdom as have the consequences arising from the development of atomic physics.

Part of the dilemma in representing the current state of this important field is the continual rapid progress being made in molecular genetics which renders the account almost obsolete as it is being published. The scientist or clinician who wishes to keep abreast of advances in biochemical genetics must follow the accounts of weekly science news magazines as assiduously as he would the sporting news. A second difficulty is the fact that most of the hard data is directly based on studies with simple prokaryotic organisms, namely bacteria. While we may presume many universal attributes of biologic systems (*i.e.*, the genetic code) it is becoming increasingly apparent that the complex eukaryotic organisms, such as ourselves, will exhibit fundamental differences in kind in the way that their genomes are structured, expressed, and regulated. In the following sections the genetic mechanisms of eukaryotic organisms are stressed, although in many respects these mechanisms are less clearly comprehended than those of the prokaryotes, which have been more thoroughly investigated.

CHROMOSOME STRUCTURE

In bacteria the genome is embodied in a single, bare molecule of nucleic acid. In cells of higher organisms, such as human cells, there are several levels of complexity in the organization of the genetic material. First of all the total length of DNA molecules is several orders of magnitude larger, and the nucleic acid polymers are associated with a number of different proteins that modify the structure and activity of the genome. In addition to these chemical differences the DNA of eukaryotes is compacted into discrete beadlike structures, known as nucleosomes, and the latter are further coiled and packaged into the dense sticklike bodies called chromosomes. The human germ cells of ovary or testis each contain 23 such bodies, 22 autosomal chromosomes and one sex chromosome. The sex chromosome in the prefertilized ovum is designated as the X chromosome and in the sperm cells it is designated as either an X chromosome or the smaller Y chromosome. The latter carries only information regarding the differentiation of accessory male sex organs to the offspring, but the X chromosome, in addition to carrying information regarding female sex characters, specifies other cellular functions. Consequently the fertilized egg, and all somatic cells of the resulting progeny, will contain 46 chromosomes—22 pairs of autosomal chromosomes, plus the 2 sex chromosomes, X + Y in the male or X + X in the female. In the female only one of the X chromosomes is active, the second X being randomly inactivated to produce the dense mass of chromatin termed the Barr body (so called after its discoverer). Females, therefore, are chromosomal mosaics; the cells are randomly developed from one or the other germ line as two sets of clones, one expressing the maternal X genome, the other the paternal X genome.

The chromosomes may be visualized as entities only during cell division: at mitosis in somatic cells when chromosomes are duplicated and 46 (the diploid number) are passed on to each daughter cell; or at the reduction division of meiosis in germ cells when only a single set (the haploid number) are transmitted to each developing egg or spermatozoon. Separation of chromosomes in late metaphase of division is effected by spindle fibers attached beside the region of union, or centromere, between pairs of chromosomes, at a structure known as the kinetochore. If the cells are dividing in culture medium, application of the crocus alkaloid, colchicine, will prevent the assembly of tubulin protein subunits from forming the microtubular elements of the spindle, and hence it will freeze the action at metaphase. In this way a number of human cells can be arrested during division in order to provide a complete picture of an individual's chromosome complement or karyotype. The characterization of the karyotype enables verification of the sex of an individual and detection of certain anomalies of transmission of the autosomes or sex chromosomes in syndromes such as Mongolism. Individual chromosomes are very distinctive in size, shape, position of the centromere, certain banding features, or presence of satellite bodies. Modern techniques have enabled the mapping of various chromosomes to pinpoint the localization of human genes specifying particular proteins or functions.

MORPHOLOGY AND COMPOSITION

While the chromosomes exhibit a rigid and reproducible shape during mitosis, during interphase the chromosomes are not individually distinguishable. The DNA-protein complex known as chromatin is visible under light or electron microscopy as a highly diffuse network of pale fibers termed *eu*chromatin. Euchromatin condenses and becomes dark-staining in a few regions associated with the nuclear membrane, and is termed *hetero*chromatin. Other visible areas of denser chromatin fibers are seen as a layer surrounding the subnuclear organelles, the nucleoli, which are attached to the peripheral heterochromatin by a stalklike structure. The major nucleic acid found in the core of the nucleolus is RNA, which is organized in granular or fibrillar form; it is now clear that this is the site of synthesis and assembly of RNA components of the cellular ribosomes.

The chemical composition of isolated chromosomal material shows some variations depending on methods of isolation, species, and stage of the cell cycle. Some preparations contain variable amounts of phospholipids, but these probably accompany the dense perinuclear chromatin fragments as contaminating fragments of nuclear membranes. The major components are DNA, RNA, and proteins of various types. The DNA has a constant contribution (C) on a per nucleus basis that depends solely on whether the cell has undergone meiosis (C = haploid amount), mitosis (2XC = diploid amount), or is approaching mitosis after the S phase of the cell cycle when DNA synthesis has reached completion to provide the doubling of chromosomes for two daughter cells (4XC = tetraploid amount). As a proportion of the chromosomal mass DNA is between 15 to 30% of the total. Most of the remainder is made up of proteins, with a variable but minor amount (generally less than 10% of total mass) as RNA.

The proteins have been traditionally classified into two categories: the histones, a group of low-molecular-weight, basic proteins (isoelectric points > 10); the nonhistone proteins, a rather heterogeneous group which provide the remainder of the chromatin protein. The histones are few in number (only 5–7 discrete bands are usually seen on electrophoresis), are very similar throughout eukaryotic species (plant histones show only minor differences in amino acid sequences from their animal counterparts), and are maintained as a fairly constant component of chromatin through the cell cycle relative to DNA (on a mass basis the histone/DNA ratio is close to 1.0 in interphase or metaphase chromosomes). The basic properties of histones result from the preponderance of the basic amino acids, arginine and lysine, in their polypeptide chains. In the neutral pH ranges within cells the nitrogen substituents on projecting side-chains of these residues will be protonated, and hence the histones, under physiological conditions, will bear an excess of these positively charged basic groups. The five common histones have been numbered: H1, the largest, which is very rich in lysine residues; H2a and H2b, which both contain slightly more lysine than arginine residues; H3 and H4, which both have somewhat more arginine than lysine residues.

HISTONE COMPLEXES WITH DNA

Histone proteins are capable of forming electrostatic linkages with negatively charged phosphate groups of DNA molecules because of the preponderance of their positively charged side-chains. In fact, calculations show that the number of basic groups of chromatin in the histones is equivalent to the number of acidic groups in the DNA, suggesting that each phosphate group could be neutralized by formation of such complexes. However, examination of the amino acid sequences shows that the basic groups of histones are distributed in an irregular fashion precluding total neutralization of DNA phosphates. Moreover, although titration studies and binding properties of DNA in chromatin to various dye molecules reveal the masking of the polynucleotide chains by histone complexes there are large stretches of DNA that are not covered by histones in this way.

Early studies suggested that some functions of the DNA (*e.g.*, transcription to produce messenger RNA) could be inhibited by electrostatic binding to histones in chromatin, and various schemes were proposed whereby release of histones (*e.g.*, by acetylation of their basic nitrogen groups) could unmask fully active DNA to explain the mechanisms of gene regulation. Current work, discussed in a later section, suggests that it may rather be some proteins among the more heterogeneous group of nonhistone proteins that act to control DNA functions. The histones play a major structural role in the folding, bending, and supercoiling of DNA molecules that is required to compact the several meters of extended double helices within the confines of the nucleus in a single human cell. The packages of DNA show quite distinct stoichiometries with respect to the numbers of the various histone molecules involved, and though much remains to be learned the outlines of a regular repeating structure are clearly discernible.

NUCLEOSOMES

This term has been coined to refer to the periodic histone-DNA aggregates visualized when chromatin is examined under the electron microscope and are seen to be spaced along chromatin fibers like beads on a string. The nucleosome units may be isolated following carefully controlled digestion of chromatin to rupture the linkers and release the individual beads of the chromatin necklace. Physical and chemical analysis of the repeating units demonstrates that each has a spherical diameter of approximately 100 nm, contains a stretch of DNA about 200 base-pairs in length, has one molecule of histone H1 associated with it, and has two molecules of each of the other histones. The histones, H3 and H4, can be isolated from chromatin in the form of the tetramer, $(H3)_2$ $(H4)_2$, and the further addition of equimolar amounts of H2A + H2B in the presence of DNA can reconstitute the dimensions and periodic properties of chromatin fibers. Current models, which attempt to reconstruct the repeating nucleosome structures picture the DNA helix as coiled around a globular aggregate of eight histone molecules (Fig. 5– 1); presumably the latter associate with the phosphate groups of the DNA through their NH_2 terminal segments, which are richest in basic amino acids, and associate with each other through their more hydrophobic middles and COOH terminals. The path of the DNA molecule is not known precisely but appears to follow one and three-quarter turns, or 150 base pairs, wound about the core particle of 100 nm diameter; the remainder of the DNA, about 50 base pairs, that runs between core particles is known as "linker" DNA. This arrangement contributes to the tight packaging of DNA in the chromosome.

Present studies exclude histone H1 from the globular core structure, and assign it a more exposed location on the exterior surface of the nucleosome (Fig. 5– 1). This is consistent with the selective removal of H1 from chromatin by mild salt extraction or brief trypsin digestion, its ready accessibility to specific anti-H1 antibodies, and the lack of requirement for H1 in the reconstitution experiments to produce the core particles of nucleosomes. The H1 component has been shown to be unique among the histones in other ways as well. The other histones that form the backbone of nucleosomes, in particular H3 and H4, are strongly conserved throughout evolution with respect to amino acid sequences (only 2 or 3 amino acid residues vary among widely different eukaryotic species). Histones of the H1 class are much more heterogeneous, and multiple forms are detected that differ qualitatively between species and quantitatively between tissues of the same species; there is no immunologic cross-reaction of the purified isohistones of one species with those isolated from the H1 class of

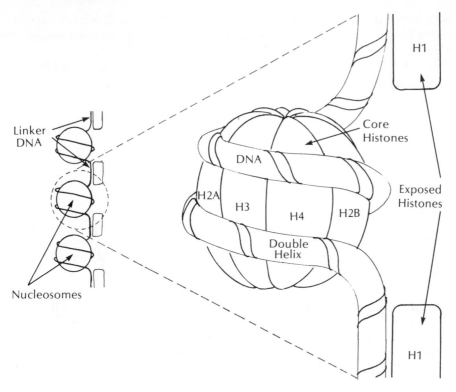

FIG. 5–1. Structure of nucleosomes. The core histones (H2A, H2B, H3 and H4) are enwrapped by the DNA strand; the exposed histone (HI) is bound to DNA strands linking the nucleosomes.

another species. The H1 component seems to associate with the short stretch (50 base-pairs) of DNA, which acts as the "linker" region between nucleosomes. Interest in this exposed and variable histone component has been stimulated by observations of losses of certain isohistones of the H1 class from tumors, and the demonstration of enhanced phosphorylation of H1 during peak periods of DNA synthesis and tumor growth, thus implying a key regulatory function in cell division for the H1 component.

FUNCTIONS OF THE GENE

The essential actions of genetic material in our cells may be viewed as two discrete processes: (1) transmission of a set of instructions governing the production, assembly, and operation of the chemical machinery of our own body cells, or of the germ cells, for succeeding generations of our descendants (this aspect is embodied in DNA replication, whereby exact copies of the genes are formed at mitotic or meiotic divisions of cells); (2) accomplishment of those instructions at the appropriate times in relation to cell differentiation and in response to physiological demands (this phase is exemplified by DNA transcription and translation, when the gene is expressed by the synthesis of complementary RNA molecules followed by the synthesis of the specific protein molecules required for morphology and activity of the body).

The replication of DNA within eukaryotic chromosomes follows the same rules of complementary base pairing, and uses analogous types of DNA polymerase enzymes as those used for bacteria and other simple organisms. However, presence of the histone and nonhistone proteins of chromatin may modify the reactivity of eukaryotic DNA as a template for polymerization of new DNA or RNA molecules. There is also a whole new set of controls imposed on gene expression in relation to triggers of cell division, and responses of multicellular organisms (*e.g.*, humans) to hormonal and other regulatory agents. In the following sections some of these special aspects of eukaryotic gene functions are considered, with special reference to the mechanisms of regulation of DNA actions.

DNA AND ITS REPLICATION

Three discrete DNA polymerase enzymes designated α, β, and γ have been characterized in animal tissues based on differences in molecular size, susceptibility to $-SH$ agents, and so forth. The α and β forms are both found in the nucleoplasm; only the α and γ forms increase notably during the stages of cell division when DNA synthesis is progressing indicating that these are the enzymes involved in replication processes required for duplication of genetic material. Some evidence suggests that the β form, which predominates in nondividing cells, is specifically involved with the repair of DNA molecules that have been damaged by exposure to ultraviolet irradiation or noxious chemical agents. Deficiencies of this DNA repair system can lead to skin lesions, cancer, or other pathological sequelae as detailed in a later section. The γ form is detected in the cytoplasm, and is specifically involved in the replication of the circular DNA molecules found inside mitochondria.

It may be calculated that operation of the α-DNA polymerase at its maximum rate would require several weeks to carry out complete duplication of human chromosomes if it started its action at a single site on each DNA molecule, as is the case for bacterial or virus replication. Instead, it is now clear that DNA reduplication in the chromosomes of eukaryotic cells proceeds simultaneously in both directions and from several different initiation points, or replication forks, along the length of DNA molecules. The process may be visualized by fixing the DNA duplex strands at the moment of replication and spreading the molecules on an electron microscope grid. The regions of duplication are indicated by the separation of parent duplex strands into eyelike bulges at multiple locations along the DNA molecules

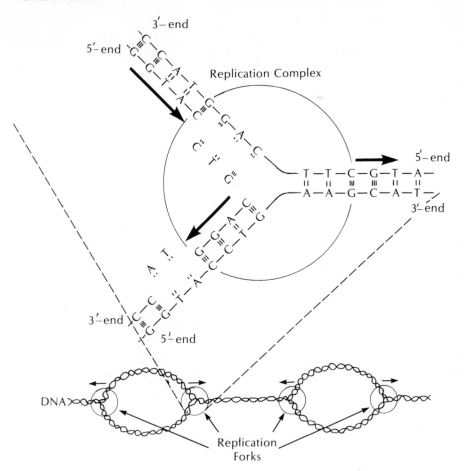

FIG. 5–2. Synthesis of DNA by action of DNA polymerase at a replication fork. The top parental strand is replicated with a continuous new strand, while the bottom strand is replicated as smaller fragments which are subsequently joined.

(Fig. 5–2). In this way several thousand DNA polymerase molecules, each forming a replication site and moving at only a few micrometers per minute, can account for the duplication of the several centimeters of total DNA in a human cell in a matter of minutes rather than weeks.

Another fascinating complication of the human genome is the presence of great stretches of DNA with no known functions in relation to coding for gene products. In contradiction to the bacterial or viral genomes where economy is the byword, there is a profligate dispensation of DNA in chromosomes of eukaryotic organisms. In simpler viruses, for example, the genes have been found to overlap, so that the tailend of a DNA sequence that specifies a certain protein A also acts to provide the message for the beginning or NH₂ terminal sequence of protein B. In sections of bacterial

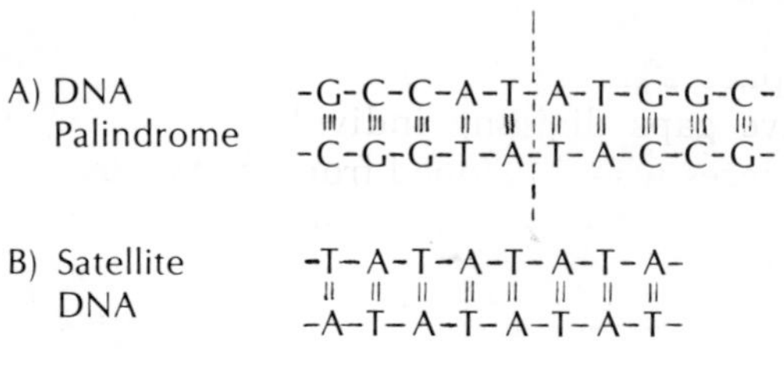

A) DNA   -G-C-C-A-T-A-T-G-G-C-
    Palindrome ‖ ‖ ‖ ‖ ‖ ‖ ‖ ‖ ‖ ‖
               -C-G-G-T-A-T-A-C-C-G-

B) Satellite   -T-A-T-A-T-A-T-A-
   DNA   ‖ ‖ ‖ ‖ ‖ ‖ ‖ ‖
               -A-T-A-T-A-T-A-T-

FIG. 5–3. DNA structures: (A) palindromic
sequences; (B) satellite DNA sequences.

chromosomes a series of proteins with related functions may be encoded
by DNA sequences that are contiguous, and that are controlled by the same
starting signal genes (described in the next section under the operator gene
concept). In chromosomes of the higher animals, however, it is apparent
that there is a gross excess of DNA relative to the amount that is required
to provide a single copy of each RNA molecule or polypeptide gene product
in the cell. In some cases this is owing to the redundancy of certain genes
that are present in multiple copies. But even allowing for these redundant
genes there is still a great excess of DNA. Portions of the nucleotide sequences that do not encode cellular RNA or polypeptide sequences may be
involved with regulatory mechanisms affecting gene action, the eukaryotic
analogues of the bacterial operator genes. In addition, there appear to be
very large segments of DNA with no likely role in the coding process *per se*.
Some of these gap or spacer segments have very short repeating sequences
that, because of the multiplicity of complementary interfaces between the
two DNA strands, may be readily recognized by their tendency to reanneal
together rapidly when the strands are artificially separated (Fig. 5– 3).
Other curious sections read the same backwards as forwards, analogous to
the palindrome attributed to Napoleon: "Able was I ere I saw Elba." The
significance of the short repeating segments (termed satellite DNA), or the
backward–forward sections (termed palindromic DNA) is still to be determined; possibly such features contribute to the formation of loops,
kinks, or cross-links of DNA strands that are essential in chromosomal
packing, or they may act as recognition signals for interactions with DNA
or RNA polymerases, steroid-binding proteins, nonhistone proteins, or
other modulatory factors of gene action.

As noted in the next section, the operon model of concerted gene action, which was proposed from studies of prokaryotes, may not readily
apply to regulation of eukaryotic chromosomes. The presence of very large
gaps between the unique coding segments of DNA for individual polypeptides makes it highly improbable that a single operator gene could regulate the multiple transcriptions through a single RNA polymerase initiation. Moreover, the mapping of human chromosomes has shown that
polypeptides of related functions (for example sequential enzymes of a
metabolic pathway) are often encoded on separate chromosomes.

An additional perplexity is the recent discovery that eukaryotic chromosomes not only possess spacers in between recognizable gene sequences, but that the latter have gaps dividing individual genes into pieces. This concept of genes in pieces was developed from studies on the synthesis of egg albumin in the oviduct of laying hens, a model system that has been immensely valuable for the study of mechanisms of endocrine regulation of gene action as described in an earlier section (Chap. 3). The protein end-product determined by the ovalbumin gene contains 386 amino acids requiring a DNA sequence of 1158 nucleotides. Isolation of the gene and determination of the actual sequence of bases in relation to the sequence of amino acids of the albumin led to the surprising discovery that it contained 7700 nucleotides, a length far in excess of that needed for coding, and that most of the excess nucleotides occurred in between the segments required to code for the polypeptide sequence. In short, the genetic code for the protein was interrupted by sizable stretches of apparently noninformational DNA. This has been confirmed for the human genes coding for hemoglobin and the immunoglobulins, which also have *inter*vening or *"in*tron" sections of DNA nucleotides inserted between the *ex*pressed coding or *"ex*on" sequences. To be efficacious in protein synthesis, either the exon portions must be selectively transcribed, or else the intron sections must be selectively eliminated after transcription; as noted in the following section the evidence suggests the latter mechanism. The significance of genes-in-pieces doubtless will be debated for some time; it is plausible, as postulated by some biologists, that the exon segments may represent domains of folded polypeptide sequences which, by appropriate selection pressures during evolution, might be assorted and assembled to give rise to unique protein functions during emergence of new species. From the standpoint of evolutionary changes, redundant genes and genes-in-pieces may have furnished flexibility for experiments by nature in the generation of increasingly complex eukaryotic organisms.

TRANSCRIPTION AND ITS REGULATION

There are four basic types of ribosenucleic acid that are produced from templates of DNA in the nucleus: ribosomal RNAs (rRNA) of which there are 4 classes based on size; transfer RNAs (tRNA) which are the smallest molecules of about 25,000 molecular weight; messenger RNAs (mRNA) of varying number and sizes; heterogeneous nuclear RNAs (hnRNA) of similar number but considerably larger size than mRNAs. These RNA species are synthesized by different RNA polymerases, which exhibit different sensitivities to inhibitors.

The ribosomal RNAs are produced from several different loci in the chromosomes, and it is clear that each cell has many copies of the DNA sequence required as template. In the nucleus these sections of the chromosomes may be seen to come together forming the specialized region known as the nucleolus where the rRNAs are synthesized. A particular

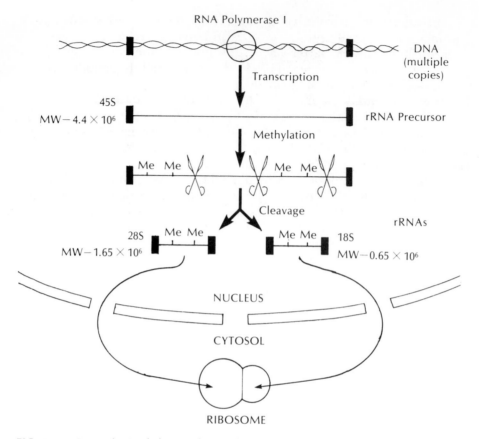

FIG. 5-4. Biosynthesis of ribosomal RNAs by the action of RNA polymerase I. The original RNA transcript is processed in the nucleus by methylation and cleavage to smaller fragments.

RNA polymerase, designated I is associated with nucleoli and transcribes from the DNA a single long stretch of linked ribosomal RNA; the latter is tailored into the individual rRNAs of the ribosomes by specific cleavage reactions with ribonucleases creating the desired lengths (Fig. 5-4). Both prior to and after the cleavage reactions a number of riboses and bases of the rRNA sequences are covalently modified in the nucleolus by the addition of methyl groups. The latter substituents undoubtedly contribute to stability and rigidity of the final rRNA products. Synthesis of rRNA by RNA polymerase I is characteristically sensitive to low concentrations of the antibiotic inhibitor, actinomycin D.

A somewhat similar process occurs for the transcription of tRNAs, which also have many reiterated copies of the specifying DNA sequences. A different enzyme, RNA polymerase III, reads off the templates for the various tRNA molecules as somewhat larger precursors, and appropriate nu-

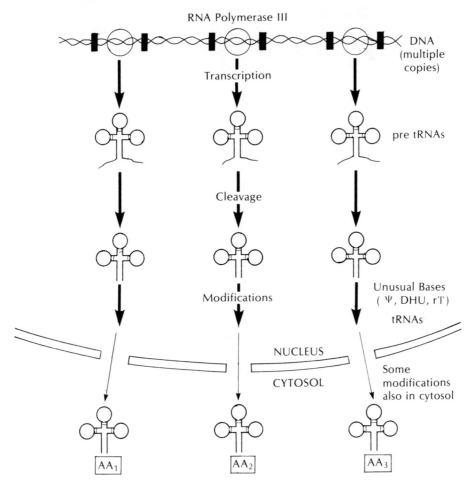

FIG. 5–5. Biosynthesis of transfer RNAs by the action of RNA polymerase III. The original transcripts are subsequently modified by cleavage and introduction of unusual bases.

clease action splits off the extra 20 to 30 nucleotide residues from the pre-tRNAs. The unusual nucleotide bases (methylated bases, pseudo-uridine) that are so characteristic of tRNAs are subsequently introduced by post-transcriptional modifications (Fig. 5– 5). Recent work suggests that these alterations may be of significance in control of cell differentiation since the tRNA methylation reactions are markedly activated by steroid hormones, and characteristic differences of methylated tRNAs have been detected in tumor tissues.

A separate mRNA molecule, complementary to each structural gene DNA sequence, is produced for each polypeptide that is synthesized by a cell. A number of these specific mRNA molecules have now been isolated in

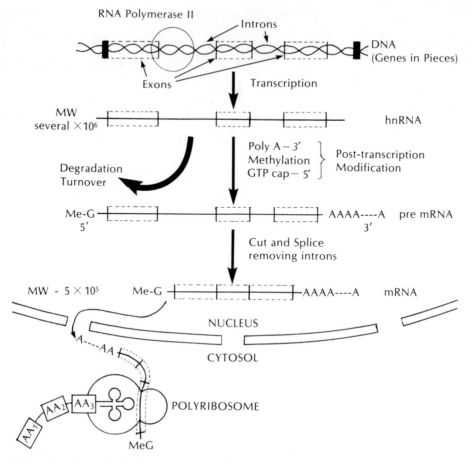

FIG. 5–6. Biosynthesis of messenger RNA by the action of RNA polymerase II. The primary transcript is modified by addition of the poly A tail, the cap structure, and methylation, then subjected to cutting and splicing to join the exon sequences.

a pure state, and several have been fully sequenced. Transcription of mRNA involves a third enzyme, RNA polymerase II; its function may be distinguished by its high sensitivity to the inhibitor, α-amanitin. RNA polymerase II synthesizes high molecular weight compounds termed hnRNA. These molecules seem to serve as precursors, albeit rather inefficient ones, for the mRNA molecules, which subsequently emerge from the nucleus into the cytoplasm (Fig. 5–6). The inefficiency of the conversion is demonstrated by the observation that most of the hnRNA molecules undergo rapid turnover, that is they are quickly broken down again in the nucleus, with only a few percent of the total amount synthesized escaping to play the role of genetic messengers in the cytoplasmic polysomes.

As is the case for rRNA and tRNA maturation, the mRNAs of eukaryotic organisms must undergo very extensive modification after they are released from the chromosomes. This is a far cry from the simpler prokaryotic systems where transcription and translation of genetic messages occur in one compartment of the cell, and the reactions of protein synthesis may be initiated at one end of a mRNA molecule while the other end is still attached to the DNA template. First of all, in eukaryotic cells the original transcript from the gene, the hnRNA, is up to fivefold longer than the mRNA, and thus there must be extensive tailoring by nuclease action to pare it down (Fig. 5–6). Secondly, the mRNA molecules that have been isolated generally have been found to have a long string of from 20 to 200 AMP molecules (poly A) attached on their 3'-OH ends. (Exceptions to this are found in histone mRNAs which lack the poly A tail). Addition of the extra AMP residues probably occurs shortly after transcription because poly A chains are attached to many giant hnRNA molecules in the nucleus. Two different functions have been attributed to the poly A segment: (1) it may be involved in aiding the movement of messengers out of the nucleus (addition of 3'-deoxyadenosine, or cordycepin as it is called, inhibits the production of the poly A tail and also blocks the transport of newly synthesized mRNAs to cytoplasmic ribosomes); (2) it may serve to protect eukaryotic mRNAs from premature degradation in the nucleoplasm and cytoplasm (in contrast with prokaryotic mRNA, which turns over very quickly, the eukaryotic messengers persist for hours or days and tend to break down only as they age and undergo concomitant shortening of their poly A segments). Other modifications of the mRNA molecules occur after shortening of the hnRNA precursors, and include the introduction of a few methyl groups, and a characteristic GMP residue which is added onto the 5' terminus of the messenger, and which also becomes methylated. This feature, which is unique to mRNA molecules, is referred to as the 5' cap, and plays a key role in the primary attachment of the messenger to the ribosome.

The presence of poly A sequences at the 3' end of hnRNA and mRNA molecules has facilitated their isolation, the former from nuclei, the latter from polysomes in the cytoplasm. The process used for isolation is the attachment of a complementary poly T or poly U segment to an inert support, which will thereby have a strong affinity for poly A-containing RNAs. Poly A-containing RNAs may thus be separated from other RNA species, and then eluted from the support by lowering the ionic strength. By such procedures a number of mRNA species have been obtained in homogeneous states. The yields are greatest from tissues that actively synthesize a particular protein, for example hen oviduct, which produces massive quantities of ovalbumin, or reticulocytes, which concentrate on accumulating hemoglobin. As noted earlier, isolated ovalbumin mRNA has been thoroughly characterized, and it has been possible to select the separate mRNAs coding for the α- and the β-globin polypeptide subunits of hemoglobin by the appropriate technical subterfuges. The mRNA for ovalbumin reading from its 5'-end contains: the methylated guanosine cap plus a 64-long noncoding region ending with the initiator codon of translation, AUG—the 1158-long coding region—a 637-long 3' non-coding region starting with the terminator codon, UAA—the variable poly A stretch at the 3'-OH end. Thus the molecule has considerable noninformational RNA. The intervening sequences (introns) totaling 5000 nucleotides which interrupted the coding region of the ovalbumin mRNA in its precursor form have obviously been eliminated. Recently other studies have shown that the 550 nucleotide intron, which interrupts the mouse β-globin gene in the coding sequence between amino acids 104 to 105, is transcribed in the nuclear precursor hnRNA, but it is not present in the finished mRNA sequence. Presumably a very exact cut and splice job is needed in hnRNA processing to remove these noncoding intronic sequences from the middle of the coding or exonic sequences of the messenger molecules.

Induction and Repression

Many of our current concepts about the control of transcription have been derived from or influenced by the operon model for the induction of enzyme synthesis in bacteria (Fig. 5–7). Although there is little evidence for the operation of an identical system in eukaryotic cells, the principles of this adaptive phenomenon in prokaryotes may be relevant to gene regulation in higher forms. In the bacterial system certain enzymes are responsive to environmental stimuli and are termed inducible. For example, presence of a substance in the growth medium may induce the synthesis of new enzyme to metabolize that substance. The gene for this inducible enzyme is normally present in an inhibited or repressed state that is incapable of producing the necessary mRNA. At least three regions of the bacterial chromosome are involved in the induction of enzyme synthesis: a promoter region that binds the RNA polymerase to the DNA of the gene (this binding may require cyclic AMP and an appropriate receptor protein, or some other activator); an operator region, between the promoter region and the structural gene for the enzyme, that controls the ability of RNA polymerase to read along the gene (the normal repressed state arises from the binding of a specific repressor protein to the operator thus competing with binding of the polymerase enzyme at an overlapping site); a regulatory gene that normally produces the mRNA to code for the repressor protein (the repressor has one binding site for a particular operator section on the genome, and an additional selective binding site for the inducer substance). When the inducer is added to the bacteria it is taken up and bound to the repressor protein. This combination causes a conformational change in the repressor molecule thus interfering with its binding to the operator. As a result the RNA polymerase may now read through the structural gene message for the enzyme, and the mRNA will be synthesized, at

A) Repressed

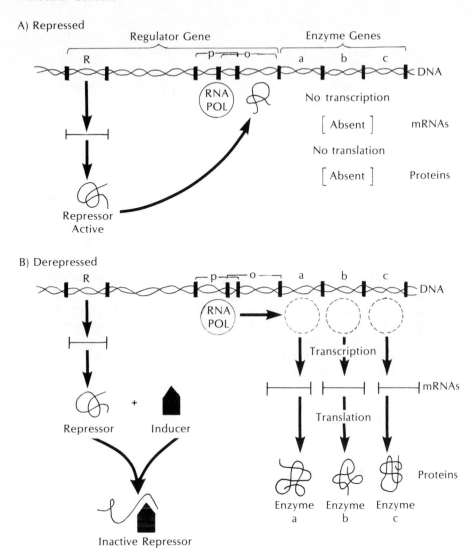

FIG. 5–7. Coordinate regulation of enzyme induction in procaryotic systems: (A) repression occurs through formation of a repressor protein which binds to DNA blocking transcription by RNA polymerase, (B) derepression occurs by an inducer molecule which binds to the repressor and thus inactivates it to allow transcription to proceed.

least so long as inducer molecules are present in high enough amounts to restrain the repressor from tying up the operator. This is why enzyme induction by such a system is termed derepression. The operon provides an economical autoregulatory mechanism. The increased enzyme resulting from induction metabolizes the inducer, and when the latter declines sufficiently synthesis of the enzyme, which is no longer required, will be shut down. The mechanism of derepression can simultaneously control a series of enzymes in a metabolic sequence in bacteria since the structural genes for other enzymes in the pathway are contiguous, and the action of a single repressor–inducer on the operator region preceding the gene sequence can regulate synthesis of all the enzymes.

Histone and Nonhistone Protein Effects

As noted earlier this type of coordinate repression of enzymes is quite unlikely in higher organisms, simply because of separation distances and nontranscribed gaps in between the genes. However, the search for eukaryotic analogues of bacterial repressors is a very active field of study. The most obvious contenders are the protein constituents of chromatin. It should be kept in mind that although the preceding account has stressed the organizational aspects of chromatin proteins in forming the chromosomes, particularly the histones, this structural role does not preclude a regulatory or modulatory function. In fact a good deal of evidence implicates the histone components as general repressors of gene action, while certain nonhistone proteins of the chromosomes may play the part of selective positive gene regulators. For example, it has been known that the diffuse euchromatin regions of nuclei are much more active in RNA synthesis than the condensed heterochromatin regions. The less dense euchromatin, which may be separated from heterochromatin by differential sedimentation, shows a higher content of nonhistone proteins and depletion of the histones. Observations on inactive genes in differentiated tissues suggest that the presence of repressors, possibly fairly nonspecific ones such as the histones, may block the initiation of transcription, while tissue-specific, nonhistone proteins may selectively recognize sequences of nucleotides of certain genes to unblock the synthesis of particular messengers.

Several types of chemical modification are known that may influence the way in which chromatin proteins interact with the DNA; these include acetylation or methylation of nitrogenous basic groups on the histones, or phosphorylation of serine hydroxyl groups in the histones and nonhistones. Addition of an acetyl group to the ϵ-NH_3^+ of a lysine residue will neutralize the charge to make the histones less basic, while the methylation of lysine or arginine nitrogens will increase the size of the side-groups without altering the overall charge. The formation of phosphate esters will render those regions of the proteins more acidic, and may induce more subtle, but no less important, changes of protein conformation. By altering the strength of the electrostatic or other interactions of such proteins with the DNA molecules it is not difficult to conceive how these chemical modifications could convert condensed inactive regions of heterochromatin into the more open configuration of euchromatin, which is accessible to serve as a template for transcription. Acetylation–deacetylation, methylation–

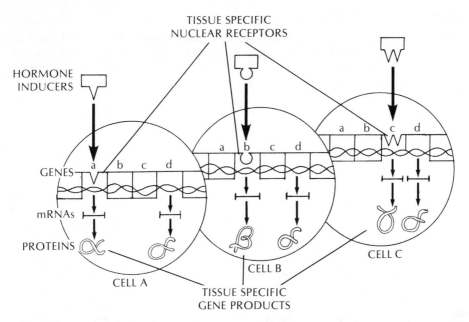

FIG. 5–8. Model for production of tissue-specific proteins by selective activation of nuclear receptors in eukaryotic systems.

demethylation, phosphorylation–dephosphorylation—these cycles of changes in nuclear proteins and the activities of the enzymes responsible for these changes are of continuing interest. Mention has been made of the correlation between histone phosphorylation and DNA replication. Accelerated phosphorylation of certain nonhistone proteins is associated with transcription in lymphocytes that are stimulated to grow. Within minutes of the stimulus the proteins become phosphorylated in the cytoplasm and are transported into the nucleus where their binding to DNA correlates well with the stimulation of RNA synthesis.

Nuclear Hormone Receptors

Another type of DNA-binding protein that has attracted much attention as a gene regulator is the nuclear hormone receptor (Fig. 5– 8). As noted in Chapter 3, a number of such receptors have been detected in target cells for estrogens, androgens, glucocorticoids, and thyroid hormones; the steroid receptor proteins have been most thoroughly studied. As for the case of the bacterial repressor, steroid receptors show two types of binding; one to the effector agent, which is the particular hormone, the second to a certain region of the genome. The detection of the receptors is readily followed by their very specific and high-affinity binding of radioactive tritium-labelled hormone. Radioautographic studies of cells treated with the labelled hormone show that these receptors are localized to the nucleus, and presumably are associated with certain regions of the genome which are associated with regulation of gene action. The second type of binding may be through a tissue-specific acceptor protein already complexed with the DNA, possibly one of the nonhistone chromatin proteins, since hormone-loaded receptors seem to associate selectively with chromatin isolated from cells of the target tissue. The alteration in chromatin structure that ensues is accompanied by a marked increase in the activity of RNA polymerase II, presumably reflecting activation by the hormone–protein receptor complex of appropriate DNA segments as templates for messenger synthesis. It is also of note that during this rapid response of increased RNA synthesis (as is observed, for example, in isolated uterine nuclei following the administration of estrogen) there is a parallel increase in acetylation of the histone H4 components of chromatin. Possibly the accelerated activity in transcription requires a disorganization of the normal compact nucleosome configuration in order to allow access of RNA polymerase II to the DNA.

There are now several instances in avian and mammalian species where the induction of synthesis of a particular protein by hormone or developmental stimulus is accompanied by a measurable increase of the corresponding mRNA in appropriate tissue cells, and by an increase of the incorporation of labelled precursors into the messenger. The results thus far suggest that increased protein synthesis in such situations is based on a primary response of increased RNA synthesis. However, one should be aware that eukaryotic mRNA species have a much longer existence than their evanescent bacterial counterparts, which normally disappear within minutes after they have been used for protein synthesis. In higher species the messengers may be very stable, surviving for many days on occasion, before they are broken down and recycled. Future aspects of gene control in eukaryotic cells may reveal whether factors regulating the rate of degradation of mRNA species may be important. Moreover, it may be of equal, or in some instances, greater importance to consider factors that influence the use of already existing messengers at the level of translational control of gene action.

PROTEIN SYNTHESIS

The ultimate expression of the genome lies in the structures and activities of cellular proteins. The business of interpreting the polynucleotide code of each gene in terms of the polypeptide sequence of each protein is the task of the ribosome and its associates. In the eukaryotic cell this process of translation of the genetic code is accomplished at a distance from the nuclear events of transcription and modification of messenger RNA. The ribosomes in the cytoplasm must perform the dual role of recognizing and then binding together at one assembly point—the mRNA conveying the instructions and the tRNA molecules bearing the energized amino acids. A third category of nucleic acids, the rRNA molecules, participates with ribosomal proteins in these recognition events. They work with soluble

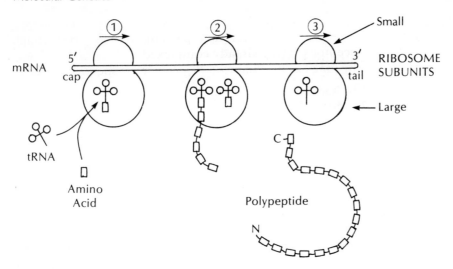

FIG. 5–9.   Three stages in the translation of messenger RNA: (1) initiation; (2) elongation; (3) termination.

protein factors (which promote initiation, elongation, or release of polypeptide chains) and GTP (guanosine triphosphates which facilitate formation and translocation of the assembly complexes) to form a remarkably efficient and precise translation machine.

A brief outline of polypeptide synthesis reveals just how complicated this bit of biologic machinery really is (Fig. 5–9):

1. The ribosome dissociates into a small (40S) plus a large (60S) subunit. The 40S subunit binds to the initiating tRNA molecule and the mRNA, then to the 60S subunit to produce the 80S initiation complex;
2. Sequential reactions of charged tRNAs at the A site of this complex bring appropriate amino acids in one by one to join to the elongating chain at the P site;
3. Introduction of a terminating anticodon for which there is no complementary tRNA molecule is followed by release of the finished polypeptide.

In the literature these three stages have become known as initiation, elongation, and release (or termination), and the appropriate cytoplasmic factors, in addition to a GTP energy source, are required at each stage. All of these processes depend on rigorously specific interactions of the components. We often stress the mRNA codon–tRNA anticodon interaction since it is the basis for the whole concept of colinearity of the genetic code with polypeptide sequence. Also of equal importance are the specific interactions between other tRNA regions with rRNAs and ribosomal proteins, and of mRNA regions beyond the coding sequence with ribosomal sites involved in initiation or termination processes. Of course the energy input from GTP hydrolysis, which aids in the binding of reactants and cofactors, and possibly in propelling the ribosome along the messenger in a unidirectional irreversible manner is also of prime importance. All of these events hinge on fidelity of recognition and a high degree of cooperation among proteins and RNA species to produce orderly conformational transitions of the tRNA and mRNA molecules. An understanding of these phenomena seems to be imminent as a consequence of the current rapid assimilation of knowledge about the intimate structural details of these components of protein biosynthesis.

Transfer RNA

The most thoroughly characterized reactants are the various tRNAs of which more than 100 different types from various species have now been sequenced. Purification of individual tRNAs has been facilitated by affinity binding through the anticodon segment to a specific immobilized complementary codon triplet, or by selective aminoacylation of the desired tRNA and chemical modification of the amino acid moiety to permit separation of the aminoacyl tRNA from all other uncharged tRNAs. A comparison of those sequences determined to date reveals common and variant features of the primary and secondary structures. The tRNA molecules are all relatively small (about 75 to 85 nucleotides), and show extensive double-stranded regions produced by complementary base-pairing. These double-stranded regions are interspersed with bulges of single-stranded sections where the molecule bends back on itself. This confers a highly convoluted shape, which in two dimensions is depicted as a clover-leaf, with the single-stranded loops forming the leaves and the stiff double helical stretches forming the interconnecting stems (Fig. 5–10). Some of the unusual tRNA bases are found in two of the loops—dihydrouridines (D), thymidine (T), pseudouracil (ψ)—and these initials are used to designate these loops. The anticodon is found in a separate loop; between the latter and the TψC loop is a variable loop from 4 to 21 nucleotides in length. On the 3′ end of the molecule, at the most remote position from the anticodon loop, is the sequence CCA; the terminal adenosine bears the activated aminoacyl group esterified to the 3′ or 2′-OH of the free ribose.

High resolution X-ray analysis of tRNA structure discloses many additional hydrogen-bonded structures in addition to the Watson–Crick base-pairing of the double helices. The three-dimensional picture that has emerged is one of an L-shaped molecule rather than the flat clover-leaf, since the D loop and TψC loop interact together by these tertiary links of H-bonds between the bases and riboses or phosphates of the backbones in adjacent chains (Fig. 5–10). The result is a molecule that is both compact and extended; compact so that the two tRNAs may be accomodated side-by-side on adjacent codons of the messenger, extended so that events involved with synthesis of the peptide bond may be carried on unimpeded at some distance from the organizing mRNA.

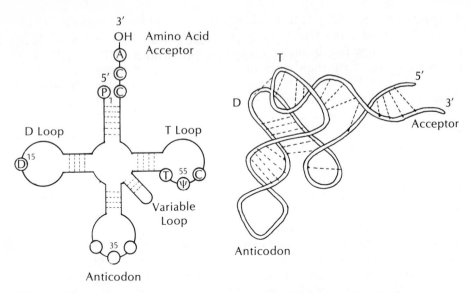

3'
OH   Amino Acid
         Acceptor

5'

P

D Loop   T Loop

D Loop   T Loop

Variable
Loop

35

Anticodon

Anticodon

Acceptor

FIG. 5–10. Structure of transfer RNA: At left a two-dimensional "cloverleaf" representation with amino acid acceptor "stem" and the "loops" designated D, T, variable and anticodon; at right, a three-dimensional representation with bonding between the D and T loops.

A consideration of the genetic code words corresponding to the 20 amino acids that are incorporated into proteins, making an allowance for the nonsense codons which specify termination, predicts that there may be as many as 61 different tRNA molecules in the cell. This would also imply that there are multiple tRNAs for a given amino acid (isoacceptor tRNAs). However, the true number is probably lower since the anticodon of a single tRNA may complex with several codons differing only in the last base position. For example tRNA^{Ala} with anticodon IAC recognizes the three alanine codons GUU, GUC, or GUA. This lack of specificity of interaction at the 5'-end of the anticodon is termed "wobble." Moreover, there may be more than one tRNA for a given codon. Note that tRNA$_i^{Met}$ has the same anticodon as tRNA$_m^{Met}$ but the former only recognizes the AUG codon at the beginning of the coding sequence of mRNA to initiate polypeptide synthesis, while the latter specifically acts to insert a methionine residue in the middle of a polypeptide chain. Thus the total number of different tRNA species seems at present to be indeterminate (probably in the range of 40 to 60 per cell).

Ribosome Structure

Continuing investigations are yielding fresh clues about the complex architecture of the eukaryotic ribosome and its interactions with tRNAs and mRNAs. Each of the subunits contains several dozen different proteins,

designated S1, S2, S3 . . . S31; L1, L2, L3 . . . L41 for components of the
small (40S) and large (60S) subunits respectively. Although RNA is an
equal contributor to ribosomal mass there are only 4 separate species of
rRNA, designated by their sizes in terms of sedimentation numbers 5S,
5.8S, 18S, and 28S; only the 18S species is found in the 40S subunit.
Sequences and possible secondary structures have been determined thus
far for only the two smaller components, both of which show evidence of
substantial double-stranded segments with hair-pin bends, and both of
which seem to have nearly identical nucleotide sequences in various animal species. The larger rRNAs with several thousands of nucleotides and
molecular weights approaching the millions show more interspecies variability of base composition. Most detailed structural work on ribosomes
has been done with the smaller less complex bodies isolated from bacterial
cells. In these prokaryotic ribosomes the use of chemical cross-linking,
sequencing techniques, and sophisticated electron microscopic visualization of components has afforded a detailed picture of the prokaryotic
rRNAs, proteins, and their interactions in reconstituting the active organelles. Work is continuing to provide the same reconstruction of the
eukaryotic ribosome. It seems likely that much of the additional complexity of eukaryotic ribosomes may be related to the presence of regulatory
components, which are probably more important for the control of the rate
and fidelity of protein synthesis in the animal cells at a translational level.

Initiation

The earliest recognition event in protein synthesis is the activation of the
amino acids. Each has a highly specific activating enzyme, termed
aminoacyl-tRNA synthetase, which is capable of transferring the amino
acid to any of its isoacceptor tRNAs. The process occurs in two stages, with
ATP providing the energy for the transfer: first, the acyl group of the amino
acid is attached through a highly reactive phosphate ester to the adenylic
acid portion of the ATP and is accompanied by release of pyrophosphate;
next, the aminoacyl adenylate, while still attached to the active center of
the synthetase enzyme, conveys the aminoacyl group to the 2' or 3'-OH of
free ribose on the adenosine at the 3' terminus of the appropriate tRNA.
Since the reaction is close to equilibrium and is freely reversible, the group
transfer potential of the resulting aminoacyl-tRNA is equivalent to that of
ATP. In order to provide the high degree of specificity for the amino acid
and for tRNA molecules it is apparent that the synthetase enzyme must be
a remarkable protein, with binding sites for aminoacyl group and the 3'
end of tRNA, and a recognition site for the anticodon region of tRNA some
80 Å distant. Alternatively, the synthetase may distinguish different tRNA
molecules based on subtle variations of three dimensional structure in the
tertiary configuration of the tRNAs. The synthetases of higher organisms
are not free in the cytosol, as in the bacterial cell. Instead they are found as
high molecular weight nondiffusible, complexes in the cytoplasm of

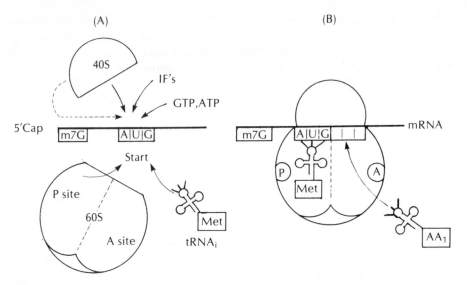

FIG. 5–11. Initiation of translation: (*A*) formation of initiation complex by combination of initiator tRNA and the two ribosome subunits at the "start" codon of mRNA; (*B*) entry of the first aminoacyl tRNA in the A site of the large ribosome subunit.

eukaryotes along with other protein factors. The sequestering of the synthetases in this manner may be of particular significance in the light of control processes in eukaryotic translation.

The second recognition event involves the particular aminoacyl tRNA species, which is responsible for initiation of the translation process. Because this tRNA becomes charged specifically with the amino acid, methionine, and because it can only react to incorporate the methionine at the N terminus or initiation site it is designated Met-tRNA$_i$. At least one soluble protein initiation factor (IF) and GTP are required for the first stage, which is binding of Met-tRNA$_i$ to the small 40S ribosome subunit (Fig. 5–11a). This stable initiation complex may form in the absence of mRNA, but the latter is required before the 60S subunit can combine with it to produce the functional 80S initiation complex (Fig. 5–11b). In the process, ATP plus several more IF's are involved, and the GTP is hydrolysed to GDP + P$_i$. As well as the necessity for recognition binding of AUG initiation codon at the beginning of the message to the anticodon of Met-tRNA$_i$, there is a requirement for the presence of a methylated G residue at the 5' cap of the messenger. Although the ribosomes are capable of accepting and translating messengers from other cell types, there is some evidence that selectivity exists between different mRNAs. Thus certain IFs in muscle cytoplasm can promote the differential use of messenger for myoglobin or for myosin synthesis. In reticulocytes the relative amounts of α-globin and β-globin chains of hemoglobin are maintained in a 1 : 1 ratio during translation despite the preponderance of mRNA for the former.

Elongation

Once the stable 80S complex of Met-tRNA$_i$ and mRNA with combined 40S + 60S subunits has been initiated, the aminoacylated tRNA with anticodon complementary to the codon adjacent to AUG may now react (Fig. 5–11b). This first stage of elongation, like the attachment of initiation tRNA, requires the presence of GTP and a soluble protein elongation factor (EF-1). It seems likely that the process whereby the initiator and other tRNAs are lined up in the appropriate niches of the ribosome needs a priming step—aminoacyl tRNA-protein factor (IF or EF) and GTP produce a ternary complex; reaction of this complex with the ribosome releases the protein factor, hydrolyses the GTP, and orients the tRNA in its ribosomal binding site.

Three features of the binding of tRNAs with the ribosome are noteworthy. First is the fact that the growing polypeptide chain remains attached to the previously bound tRNA on the peptidyl or P site of the ribosome, while new incoming aminoacyl tRNA binds at the adjacent aminoacyl or A site (Fig. 5–12a). Second is the observation that these bound tRNAs and the mRNA segment containing their codons plus several residues on either side are protected from nuclease action, implying that these reactants must be buried to a significant extent within the ribosome. Finally, a more extensive interaction of tRNAs with the ribosome must somehow reinforce or amplify the rather weak triplet base-pairing of codon with anticodon to account for the selectivity of translation, and the very low observed error frequencies that occur from mistakes in codon–anticodon recognition. Some evidence suggests, for example, that the TψC loop segment of tRNA binds to the 60S subunit by a complementary sequence of the 5SRNA component in the ribosome to confer added stability on the binding event at the A site.

The formation of the peptide linkage is catalysed by the ribosomal enzyme, peptide transferase, which is in the vicinity of the P site of the 60S subunit. In the process the $-C=O$ of the previously bound aminoacyl group links with the HN– of incoming aminoacyl tRNA, so that the chain grows, starting with methionine at the NH_2 terminus and moving toward the COOH terminus. When the polypeptide is about 20 residues in length the terminal methionine is generally lopped off by a selective peptidase (Fig. 5–12b). As is the case for tRNAs and mRNA bound to the ribosome the remainder of the growing polypeptide is immune to attack by proteases, and presumably is shielded within a crevice or tunnel in the ribosome.

One of the most difficult concepts to understand is the mechanism whereby the mRNA and the ribosome move relative to one another in the so-called translocation process of protein synthesis. This second stage of

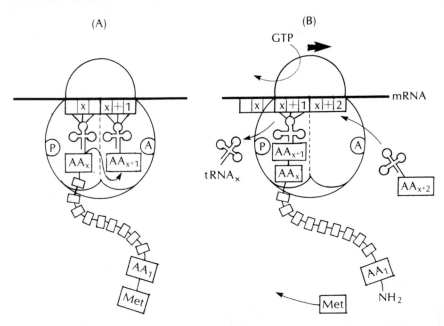

FIG. 5-12. Elongation of the polypeptide chain in translation: (A) movement of growing peptide to the A site coincident with linkage between amino acids on adjacent codons; (B) movement of the mRNA to convey the growing peptide to the P site and allow entry of the next aminoacyl tRNA in the A site.

the elongation phase results in movement of the polypeptidyl tRNA, still attached to its codon, into the P site, and exposure of the adjacent codon at the A site to allow entry of a new aminoacyl tRNA (Fig. 5- 12b). The soluble protein elongation factor (EF-2) is also necessary, and complexes with the ribosome in the presence of GTP. The GTP is hydrolysed, possibly to provide the energy requirement for movement, or for conformational changes of tRNAs and ribosomal proteins associated with the A → P shift and release of deacylated tRNA after formation of the peptide link. It is clear that synthesis of the polypeptide is driven by the high group transfer potential of the acyl ester in aminoacyl tRNA, while GTP hydrolysis is required chiefly to facilitate appropriate interactions of protein factors and RNAs and hence to promote the recognition events of protein synthesis at maximal rates consistent with fidelity of translation.

Termination

The mechanisms involved in the release of the polypeptide chain on reaching the end of the message are not clearly understood. The primary signal for termination events appears to be the presence of one of the codons UAA, UAG, or UGA in the A site (Fig. 5- 13a). Since there are no tRNA molecules with anticodons corresponding to these termination codons, elongation

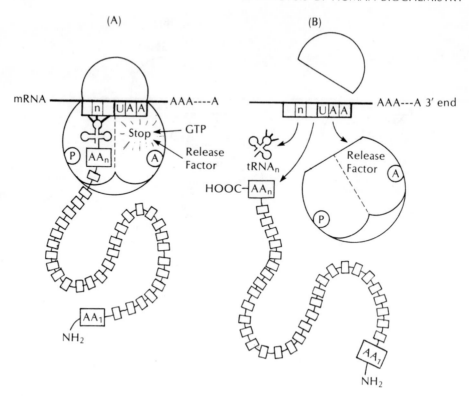

FIG. 5–13. Termination of translation: (A) further additions of amino acids prevented by arrival of the "stop" codon at the A site of the ribosome; (B) reaction of release factor and subsequent cleavage of finished polypeptide chain with dissociation of ribosome subunits from mRNA.

can obviously proceed no further. A soluble protein release factor (RF) plus GTP interact with the ribosome in the region of the A site. In the process the normal H_2N- acceptor for peptidyl transferase is somehow substituted by a H_2O molecule, so that the enzyme is converted to a hydrolase and aborts the synthetic sequence (Fig. 5–13b). The hydrolysis of GTP, which accompanies release of the completed polypeptide chain, seems to be involved with dissociation of the RF molecule, but many aspects of termination, such as the dissociation of deacylated tRNA and mRNA from the ribosome, remain to be clarified.

REGULATION OF GENE TRANSLATION

Reference was made previously to the ability of the ribosomal apparatus to discriminate among the many messenger RNA molecules that might be presented to it. This type of post-transcriptional control of protein synthesis is thought to be of particular significance for eukaryotic organisms, and there is a growing body of evidence concerning the types of signal molecules that may be involved in regulating the process of translation of the genetic message. In addition to the variations of the leader sequences on the 5' end of the mRNA, which may alter the rate at which these molecules are recognized to form initiation complexes, there are a wide variety of metabolites, hormonal agents, and other cytosol factors that have been implicated as modulators of the events occurring at the level of the ribosomes. In keeping with a basic premise for efficient control systems it seems likely that a major site of action for such regulators is at the stage of initiation of protein synthesis.

The simplest type of control system is the self-limiting aspect of protein synthesis with respect to supply of the amino acid precursors. Deprivation of any one of the essential amino acids prevents initiation of the translation process in mammalian cells in spite of normal production of mRNA and the presence of an adequate supply of ribosomal subunits. The defect induced by amino acid starvation appears to be associated with deficient charging of the tRNA molecules (the uncharged tRNAs possibly competing with the initiator tRNA for binding sites on the ribosome). This mechanism may apply to rapidly dividing cells, but in nondividing cells, such as the reticulocytes, protein synthesis is blocked by deprivation of amino acids at the polyribosome level during the elongation stage.

Hemin Control

The synthesis of proteins by isolated reticulocytes has provided a useful model for investigations of mechanism of translational control. The cells are highly specialized with respect to the major product of protein synthesis, hemoglobin, and the polypeptide component chains are thoroughly characterized. Reticulocyte extracts, when supplemented with hemin, will continue to synthesize proteins *in vitro* by reusing the endogenous globin mRNA molecules several hundred times. However, when deprived of hemin the synthesis of globin will proceed only for a few minutes before it stops abruptly. Cessation of globin synthesis is coincident with the formation of a protein inhibitor from its inactive precursor or proinhibitor. This 'hemin-controlled repressor' of globin synthesis appears to inhibit the binding of Met-tRNA$_i$ to the 40S ribosome, and to act catalytically since a single molecule of the inhibitor may inactivate up to a thousand ribosomes. The inactivation phenomenon requires the use of ATP or GTP, and one of the IF proteins becomes phosphorylated simultaneously. Thus the interpretation would be that hemin deprivation unmasks a normally latent protein kinase; the latter presumably acts as the repressor by specifically phosphorylating, and hence inactivating, one of the essential initiation factors that is associated with the ribosome.

The coordination of synthesis of proteins with that of their prosthetic groups, such as hemin, has obvious implications for cellular economy. Moreover, this type of translational repression is not limited to erythroid cells. Glucose starvation, absence of oxygen, and other conditions leading to energy deprivation also block protein synthesis in cells, and a number of different cell types have been found to produce associated translational inhibitors with similar properties to the hemin-controlled repressor of the reticulocyte. Thus this model of regulation of gene action at the level of translation may represent a general phenomenon in eukaryotes.

Viral Control

Another model of gene control with great significance for both medicine and cell biology is the perturbation of protein synthesis that results when cells become infected with viruses. Three associated phenomena are observed: first, virus infection often effectively shuts down the translation of messages produced by the host cell, while the viral mRNAs take over the ribosomes for synthesis of viral proteins; second, double-stranded RNA (dsRNA), which is found during replication of viral RNA genomes in infected cells, blocks the initiation of translation by a mechanism that is quite similar to the one for the hemin-controlled repressor system; finally, virus-infected cells produce glycoproteins termed interferons, which are able to induce an antiviral state in other cells by inhibiting the translation of viral mRNA.

Interference with host protein synthesis generally occurs early in the infective process and does not appear to require prior replication of the virus or synthesis of new viral proteins. This rules out the possibility that dsRNA itself is the effector molecule, and points rather to a pre-existing capsid protein of the virus as the inhibitory agent. The effect seems to be at the translation level, and may be associated with phosphorylation of ribosomal proteins. It is noteworthy that the transcription of host mRNA and its post-transcriptional modifications (adenylation, methylation, and capping) occur in a normal fashion during viral infection, and active host messengers can be obtained from virus-infected cells, so that there does not seem to be an increase in mRNA degradation. The intriguing aspect of this control process is its selectivity, since the viral messengers produced later may be translated by host-cell ribosomes that are ineffective in utilizing their own natural mRNAs. Part of the selective use of viral mRNAs may be attributed to their preponderant amounts during active transcription of the viral genome, and it has been observed in studies with isolated messengers that viral mRNAs are "stronger" and successfully compete with the host mRNAs for ribosomal initiation. The success of competition by more abundant, stronger messengers from the virus is accentuated when initiation factors are limited, and may be overcome by addition of excess amounts of initiation factors.

Double-stranded RNAs (dsRNAs), such as the replicative form of polio virus or synthetic poly I : C, appear to bring about an activation of protein kinases (as is the case for hemin deficiency in reticulocytes). Very small amounts of dsRNA are sufficient to block initiation, consistent with a catalytic mechanism, and coincidentally certain proteins of the ribosomes become phosphorylated as the translation mechanism shuts down. Although the mechanism of the dsRNA effect bears many similarities to that for the hemin-control repressor system there are important differences. In the first place the dsRNAs induce phosphorylation of several proteins. Secondly, dsRNA may also promote the formation of a small nucleotide inhibitor of translation, as well as the degradation of viral RNAs by activation of endonucleases. Finally, these effects of dsRNA are potentiated by, and inextricably entangled with, the effects of exposure of cells to interferon.

Interferon

Interferon, the antiviral glycoprotein, has been known to exist for some years but has not been well-characterized as yet; it is likely that it represents a family of related molecules with multiple sites of action. Induction of the synthesis and excretion of proteins capable of exerting antivirus protection on other cells has been observed in many cell types in response to virus infection, or to exposure to natural or synthetic dsRNAs. A major effect on cells treated with interferon is blockage of translation, particularly in this instance, the translation of viral mRNA. Interferon-treated cells exhibit induction of several protein kinase activities that are also dependent on the presence of dsRNA. The enzymes and their substrates are not fully characterized, but it appears that the interferon-induced translational control system involves, in part, a cascade of phosphorylation events culminating with the inactivation of one or more initiation factors. In addition a dsRNA-dependent reaction also gives rise to a small heat-stable inhibitor of translation in interferon-treated cells; this compound is an oligonucleotide containing three adenosine residues joined by 2', 5' phosphate ester linkages. Interferon also exerts effects in the absence of dsRNA; in this case the block in translation is observed to occur during elongation rather than at the initiation step. The elongation defect is manifested by blockage of ribosomes along the message and production of incomplete polypeptide chains; it is reversible by the provision of excess amounts of tRNA. The latter observation may provide a clue to the specificity of blockage in viral mRNA translation by interferon treatment of cells, particularly if the host message has a preponderance of codons corresponding to the more abundant tRNAs in its cytoplasm. Viral messages could be "stronger" in the formation of initiation complexes, but if their codons correspond to minor tRNAs they would lose in competition with host mRNAs at the elongation stages of translation in the presence of interferon.

Latent mRNA

Another type of post-transcriptional control of mRNA use has been attributed to the formation of complexes with specific proteins either in the nucleus or the cytoplasm. A number of these ribonucleoprotein particles (designated mRNPs) have been detected in various cell types. In immature reticulocytes, for example, the cytosol mRNP fraction contains mRNA as a silent, untranslated message which is converted to fully active globin mRNA on removal of the proteins. Thus the mRNP seems to serve as a method of storage of pre-formed messengers in a latent state. Small uridine-rich oligonucleotides, 20 to 30 residues long, have also been implicated as stabilizers of cellular mRNAs because they form double-stranded, untranslatable complexes with the messenger molecules along the 3'-poly A tails of mRNA molecules; these regulatory RNA molecules are sometimes referred to as translational control (tcRNAs). The role of both nucleoprotein and tcRNA complexes of mRNAs have been implicated in a variety of eukaryotic cells undergoing development. It has been known for several years that unfertilized ova carry pre-formed, dormant mRNA molecules. These maternal messengers are unmasked during early development following fertilization of the egg cell, and the synthesis of new histone proteins can be detected even when transcription is inhibited or when the nucleus of the ovum is removed by microsurgery. The histone mRNAs have been obtained from isolated mRNPs of unfertilized eggs, and can be shown to act as functional messengers for maternal histones when added to ribosomes. Fertilization may possibly provide the trigger to remove the masking agents of the latent mRNPs. Moreover, it is conceivable that a programmed schedule of binding and release of stored messages in response to appropriate hormones and other differentiation signals may play a key role in the timetable for the embryonic development of specific cell functions.

Polyamines

Another molecular species that is known to exert strong modulatory effects on protein synthesis in many organisms is the polyamine family of compounds, including spermine, spermidine, and putrescine (Fig. 5–14). The former are particularly abundant in male reproductive tissue as the names imply; putrescine, the decarboxylation product of the amino acid ornithine, is a precursor of the other two polyamines, and is generally found in smaller amounts in eukaryotic cells. Because of the protonation of the amino groups these compounds have a number of positive charges at intracellular pH values. Some of the effects of the polyamines are related to their ability as polycations to replace or augment the necessary functions of inorganic cations such as Mg^{2+} in the enzyme reactions of protein synthesis. In addition, the polyamines complex readily with nucleic acids through their phosphate groups, stabilizing both DNA and RNA structures against physical or enzymatic attack. The association is thought to be of particular significance for the structure of tRNAs. When isolated from a variety of cell types, tRNA molecules are found to have one or two molecules of a polyamine tightly complexed in their structures, and a number of physical measurements indicate that both spermidine and Mg^{2+} binding

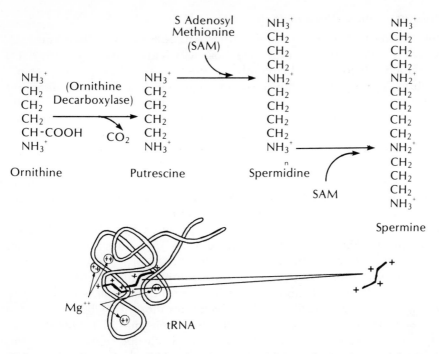

FIG. 5–14. Biosynthesis of the polyamines, spermidine, and spermine, and their complexes with transfer RNA.

contribute in a synergistic way toward the maintenance of the active tertiary configurations of the tRNAs (Fig. 5– 14). The polyamines have also been shown to influence many reactions of tRNA molecules, including the aminoacylation reaction and initiation and elongation processes on the ribosome. The stimulations by spermine or spermidine are most pronounced at low Mg^{2+} levels, and may therefore be most significant in cells when other cations become rate-limiting. Much interest is currently centered on the activities and regulatory properties of the ornithine decarboxylase and other related enzymes of polyamine biosynthesis in conditions of rapid cellular growth. In general those stimuli that induce the proliferation of cells, including tissue hypertrophy, embryonic or malignant growth, and administration of mitogenic agents or trophic hormones, are all associated with a several-fold increase in the activity of ornithine decarboxylase. The rise represents synthesis of new protein, which ceases when the tissue levels of polyamines increase, suggesting an autoregulatory negative feedback system with repression of synthesis of the decarboxylase enzyme by its end-products. It is apparent that the polyamines are ubiquitous and doubtless important modifiers of nucleic acid functions, mainly at the translational steps, but much remains to be clarified about their actions in living cells.

Membrane-bound Ribosomes

A noteworthy feature of protein synthesis in eukaryotic cells is the existence of two classes of ribosomes: (1) those which are free in the cytosol; (2) those which are bound to the membrane system of the rough endoplasmic reticulum. Much evidence has been obtained to indicate that the unbound ribosomes are involved in the synthesis of intracellular proteins, while bound ribosomes participate in the production of secretory proteins. This raises an additional complication in the regulation of eukaryotic translation—namely, how are the messages for the proteins destined for secretion from cells directed specifically to ribosomes that are attached to the endoplasmic reticulum? A second question related to this complication concerns the nature of ribosomal binding to the membrane, and the fundamental differences between ribosomes that produce free polyribosomes in the cytosol, and those that produce complexes with mRNA while associated with membranes. Answers to all of these questions are incomplete, but a general picture of the membrane-associated process is emerging.

The binding of ribosomes to the endoplasmic reticulum occurs through their large 60S subunits. Part of the binding is loose and readily reversed, and is associated with inactive ribosomes either before or after translation has taken place. The other portion of bound ribosomes is actively involved in the translation process, and may be released from the membrane only on interruption of protein synthesis by premature termination. By penetrating the lipid bilayer, the nascent polypeptide chain provides a link to the membrane (Fig. 5–15); thus a significant part of the growing protein is shielded within the membrane from proteolytic enzymes, and treatment of active rough endoplasmic reticulum with proteases leaves this protected nascent chain buried within the membrane. The further elongation of the polypeptide chain leads to extrusion on the side of the membrane opposite to the cytoplasm, namely into the lumen of the endoplasmic reticulum. Further processing of the projecting N terminus in the luminal space (Fig. 5–15) may involve removal of the hydrophobic amino acid residues of the "signal sequence," which favored penetration of the lipid bilayer, addition of hydrophilic groups (*i.e.*, sugars by enzymes of glycosylation), and tertiary folding of the chain, so as to anchor the active ribosomes on the membrane and ensure that the transmembranous movement of the secretory proteins is unidirectional.

Reasons for the selective use of certain messages by ribosomes while attached to membranes of the endoplasmic reticulum are still unclear. Part of the selectivity may be explained by the observation that certain mRNA molecules with long poly A tails may bind to the endoplasmic reticulum in the vicinity of their 3' ends. Thus the 5' leader sequence would be available for initiation of protein synthesis with small ribosomal subunits in the proximity of the membrane surface where large ribosomal subunits are already loosely bound. Alternatively, the initiation process

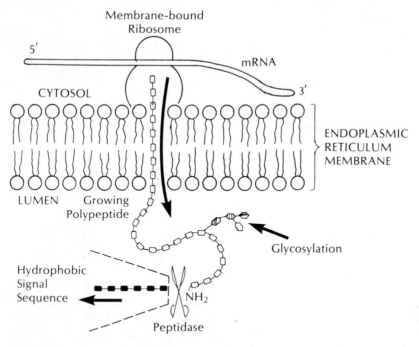

FIG. 5–15. Membrane-bound ribosomes and translation of a secretory protein across the lipid bilayer. Subsequent processing—addition of glycosidic residues, removal of transmembrane signal sequence—occur in the lumen of the endoplasmic reticulum.

and early stages of translation might occur with free 40S and 60S subunits in the cytosol until a certain length of the nascent polypeptide has formed. If the residues at the N terminus were sufficiently hydrophobic, the functioning ribosome-mRNA complex could then bind selectively with the endoplasmic reticulum, using both the hydrophobic interactions of the nascent polypeptide in the lipid bilayer and the electrostatic binding forces at the surface to cement the union of membrane with translation system. This concept, whereby a leading peptide sequence acts as the primary signal for binding to the endoplasmic reticulum, has been termed the "signal" hypothesis. It is supported by the observation that certain proteins which are produced for export from cells, such as the immunoglobulins, contain a preponderance of leucines and other hydrophobic amino acids near the N terminus. In other cases, of course, such signal sequences may be eliminated by proteolytic cleavage. In other cases it is suggested that hydrophobic residues may remain buried within the membrane. The generation of integral membrane proteins of endoplasmic reticulum and other cellular membranes may occur by such a process.

MUTAGENESIS

Defects or errors in operation of the genetic machinery may have drastic consequences either for the affected individuals or for their offspring and descendants. In the first instance alterations of the DNA of ordinary body cells result in the phenomenon of somatic mutation, a change in phenotype that is transmitted to daughter cells within the individual's life-time. Such a process may have beneficial attributes if harnessed, as has been proposed for the versatile response of immunoglobulin-secreting cells. However, deleterious mutations of somatic cells may also provide the trigger for transformation into malignant, cancerous growth, as noted further below. The second type of mutation involves the DNA of germ cells in the gonads. In this type of mutation the modification in genotype can be transmitted to the next generation and beyond in perpetuity. Although the potential also exists for improvement of the species, as is implied in the Darwinian view of evolution, helpful mutations are counterbalanced by the dangers of lethal or crippling defects in the gene products. (Some of these inherited lesions have already been discussed at length, as in the case of the aberrant hemoglobins [Chapter 1]). The inborn errors of metabolism, the inherited propensities to certain diseases (*i.e.*, diabetes or cancer), the genetically determined sensitivities to drugs, anaesthetics, or environmental chemicals, and other congenital anomalies associated with chromosomal breakage or dislocations illustrate the dangers that result from transmission of damaged DNA to succeeding generations. While some alterations could have beneficial effects, and while many minor changes in gene structure have neutral or trivial effects (*i.e.*, in noncritical spacer regions, reduplicated genes, or in areas of structural genes that are not crucial for the function of their RNA or polypeptide end-products), most gene mutation is viewed as undesirable and potentially perilous to ourselves and our descendants. The admonition about tinkering with other mechanical devices seems applicable for the human genetic machinery as well—"If it works don't fix it."

DNA ALTERATIONS

Mutagenic substances in the environment include a diverse group of chemicals that interact with the genome. There are two "general" alterations of DNA structure produced by mutagens: those in which a substitution of one base-pair by another is produced or, those in which a base-pair is either added to or subtracted from the chain (Fig. 5– 16). For example, a substitution would occur if one of the adenine residues were deaminated by a molecule of nitrous acid to form hypoxanthine. The latter would then pair with cytosine rather than thymine of the original complementary strand, and thus after two cycles of replication the A– T base-pair would be substituted by G–C. Cytosine may also be deaminated by nitrous acid to form

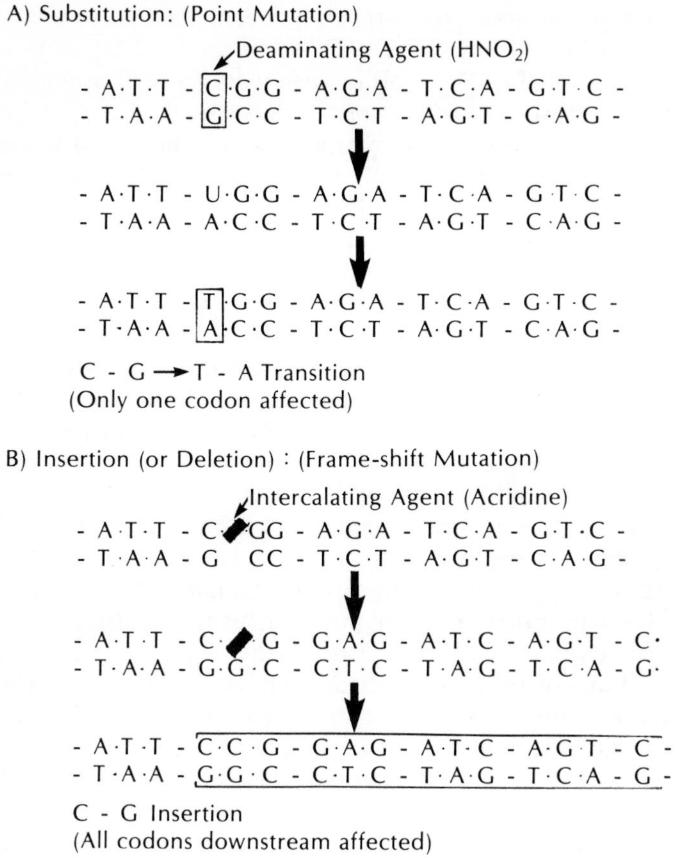

A) Substitution: (Point Mutation)

↙Deaminating Agent (HNO₂)

```
- A·T·T -[C]·G·G - A·G·A - T·C·A - G·T·C -
- T·A·A -[G]·C·C - T·C·T - A·G·T - C·A·G -
```

⬇

```
- A·T·T - U·G·G - A·G·A - T·C·A - G·T·C -
- T·A·A - A·C·C - T·C·T - A·G·T - C·A·G -
```

⬇

```
- A·T·T -[T]·G·G - A·G·A - T·C·A - G·T·C -
- T·A·A -[A]·C·C - T·C·T - A·G·T - C·A·G -
```

C - G ➙ T - A Transition
(Only one codon affected)

B) Insertion (or Deletion) : (Frame-shift Mutation)

↙Intercalating Agent (Acridine)

```
- A·T·T - C GG - A·G·A - T·C·A - G·T·C -
- T·A·A - G   CC - T·C·T - A·G·T - C·A·G -
```

⬇

```
- A·T·T - C   G - G·A·G - A·T·C - A·G·T - C·
- T·A·A - G·G·C - C·T·C - T·A·G - T·C·A - G·
```

⬇

```
- A·T·T - C·C·G - G·A·G - A·T·C - A·G·T - C -
- T·A·A - G·G·C - C·T·C - T·A·G - T·C·A - G -
```

C - G Insertion
(All codons downstream affected)

FIG. 5–16. Mutations of DNA: (*A*) by single base replacement with only one codon affected; (*B*) by insertion (or deletion) of a base to shift the reading frame of all codons downstream.

uracil, which would base-pair with adenine so that C–G would become T–A (Fig. 5–16a). Another instance of this type of substitution is the action of alkylating mutagens, such as ethyl methanesulfonate, which produce the O6-alkyl derivatives of guanine; the latter will mispair with thymine rather than cytosine during replication and hence the alteration from G–C of the original base-pair to A–T will occur. The analogue nucleoside 5-bromodeoxy uridine can masquerade for thymidine in replication, but once incorporated in DNA can induce guanine to mispair causing A–T to change ultimately to G–C in daughter cells. These changes of one purine–pyrimidine pair for another are termed transitions. Alternatively, the purine–pyrimidine pair can be substituted by a pyrimidine–purine pair (*e.g.*, A–T replaced by T–A), in which case it is termed a transversion. The mechanism of the latter substitution is not clear but such changes are commonly seen among spontaneous mutations. A mechanism to promote transversion could occur if a purine base were to pop out of alignment during replication, such that the DNA polymerase were to read a nearby pyrimidine in forming the base-pair. Mutations of the types G– C to C– G or T–A and A– T to T– A or C– G could arise by this process. The substitution types of mutation can result in production of mRNA molecules with a single codon altered, such that one amino acid in the translation product is substituted by another; the codon may be altered to a "nonsense" codon causing premature termination of the polypeptide chain; the alteration may occur in the "nonsense" codon, which normally signals termination, such that the mRNA is read beyond the – COOH end of the polypeptide. Examples of such mutations are cited among the variants of the human hemoglobins (Chap. 1). A second exposure of a mutated base-pair to the same type of substitution mutagen can regenerate or revert back to the original sequence (*e.g.,* A– T → G– C → A– T).

Rather than affecting a single codon, the addition or subtraction of a base-pair will affect every codon down-stream by shifting the frame of reference for reading the message during translation. For this reason such changes are referred to as frame-shift mutations. As noted below, such mutations may also back-mutate to normal, but with a very low frequency. The reversion may be enhanced by exposure of the mutant DNA to a frame-shift mutagen, but not to a substitution mutagen. Conversely, the substitution mutants may not be induced to revert by frame-shift mutagens. Insertion of a single base can occur if a spacer molecule of appropriate dimensions squeezes between two successive bases, or intercalates in the chain, so that the replication process will subsequently add an extra base in that position of the opposite strand. Flat heterocycles, such as the acridine ring, act as typical intercalating mutagens in this manner (Fig. 5– 16b). The intercalating agents may also cause a deletion mutation (*i.e.,* by a mechanism analogous to that proposed for transversion). In this case the mutagen molecules could prevent a base that is looped out during replication from looping back in by formation of flat stacks on either side of the extrahelical projection. Thus the replicated portion would be one-base short. Mutagens that cause a deletion may reverse effects from a previous insertion if they act in the same locus of the genome, presumably by bringing the codons back into register. Other types of mutagenesis not included in the above categories are produced by chemicals or physical factors such as heat, X-rays, and UV light that act to form interstrand cross-links between bases which will prevent normal DNA replication, or which will disrupt the strands, and possibly cause dislocation of large portions of the chromosomes.

Environmental Mutagens

Chemical mutagens present in the environment include aromatic hydrocarbons (benzpyrenes, benzanthracenes present in combustion emissions, cigarette smoke), halogenated hydrocarbons (vinyl chloride, chloromethyl ethers produced in polymer manufacture), nitrosamines (dimethylnitrosamine, nitrosopyrrolidine found in cured or smoked meats), and the mold metabolites known as mycotoxins (aflatoxin, ochratoxin generated by aspergillus fungi infecting improperly stored peanuts and grains), to name but a few of the synthetic and natural products to which humans may be exposed. In many cases such compounds are not directly mutagenic, that is they do not interact with DNA until they have been modified in some way by cellular metabolism. Moreover, there are instances, as noted earlier (Chap. 2), where the mutagen may be synthesized mainly within the body (*i.e.*, the nitrosamines which arise *de novo* within the stomach from dietary nitrite reacting with secondary amines). For these reasons attempts to identify the potential mutagens of human cells from among the myriad natural components in the foods we eat, or from among the pesticides, drugs, plastics, and by-products of industrial synthesis to which we may be subjected, present many difficulties. The ability of such agents to induce mutations in microorganisms with suitable mammalian activating systems (Ames test), or their ability to cause chromosome breakages or abnormal gene products in cells of higher animals have been indices employed to predict their potential damage to the human genome. There are, of course, uncertainties in making the extrapolation from such test systems to mutagenicity of somatic or germ cells in humans; uncertainties arise from species comparisons, relative toxic doses, variations of the intracellular conversions, and different rates of delivery to particular cells. Nonetheless, these laboratory approaches to identification of possible mutagenic chemicals are preferable to previous practices of releasing possibly toxic substances into the environment first and then conducting experiments with human populations by retrospective epidemiologic surveys.

Human Mutations

The incidence of spontaneous mutations in human populations has been estimated to be on the order of 5 to 10% based on extrapolations from other species and the occurrences of genetic disease. However, other data indicate that these calculations probably underestimate the aggregate of lesions that are difficult to detect or only mildly deleterious. In considering inherited defects of both dominant and recessive types, it is apparent that each individual undoubtedly carries several deleterious gene mutations without necessarily ever becoming aware of that fact. Of the recognizable syndromes with a familial basis over 1000 distinct genetic disorders have been identified in humans where the mode of inheritance is clearly known; there is also a like number where the mode of inheritance is less well defined. The bulk of these syndromes (*i.e.*, phenylketonuria, sickle cell anemia, galactosemia) are carried in the parents as recessive traits and are expressed in 25% of the offspring as a statistical average only in 1 per 10,000 or less in the population; the severe consequences and the potential for care or cure of these genetically inherited syndromes are such that these lesions have attracted much attention. Indeed, in the pediatric wards, and particularly in institutions for the retarded, 5% or more of the patients may be homozygotes suffering from recessive single gene defects. There is undoubtedly a high level of heterozygosity with respect to such cryptic recessive genes in human populations. In addition, it has been estimated that 1 in 200 live births may suffer from mutations of dominant genes producing various physical malfunctions or hereditary neoplasias; a similar number may suffer with inherited defects attributable to chromosomal mutations resulting in nondisjunction or translocations in autosomes and sex chromosomes (*i.e.*, Down's Syndrome, Turner's Syndrome, Klinefelter's Syndrome). The total disease burden from these two categories, therefore, involves 1% or more of the population.

DNA REPAIR

Just as the body has evolved processes for the healing of wounds and recuperation from other injuries, so too the genome has acquired mechanisms to neutralize the damaging actions of mutagens. These mechanisms are collectively known as DNA repair. They may involve a number of enzymes that are capable of recognizing and snipping out defective bases from the mutated gene and replacing them with a normal polynucleotide sequence. Simple strand breaks, such as those produced by ionizing radiations, may be rejoined directly by the enzyme DNA ligase. The abnormal bases (*i.e.*, uracil or hypoxanthine which are produced by deamination of cytosine or adenine respectively) and the alkylated bases would require the action of specific enzymes to excise the anomaly prior to reconstitution of the normal sequence; this process is referred to as excision repair, and it may be a very complex mechanism involving multiple enzyme systems. For example, some mutagens may produce simultaneous damage in both strands of the DNA molecule, as in the case of nitrogen mustards and other cross-linking agents, thereby blocking both replication and transcription. A collaborative action of several hydrolytic and synthetic enzymes will be required to correct such covalent alterations of the genome.

Much of our knowledge about the mechanisms and significance of excision-repair processes in human cells has been obtained by the study of restoration of DNA following damage resulting from exposure to ultraviolet rays (UV). By contrast with the simple ligation repairs that follow treatment with ionizing radiation by X-rays (repairs occur rapidly within 1–2 hours and involve only a few nucleotides or short patches), the repair of UV damage is complex, slower (1–2 days), and involves longer patches of 30 to 100 nucleotides. The nature of UV damage is formation of a cross-linkage within the same polynucleotide chain (intrastrand) rather than between two chains (interstrand). The linkages are produced between adjacent pyrimidine bases (*i.e.*, between two neighboring thymines to gener-

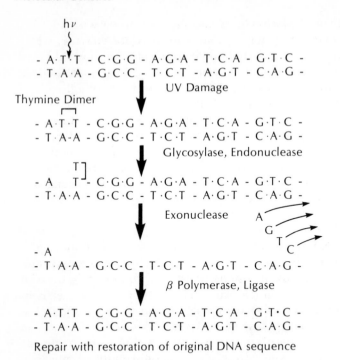

FIG. 5–17. Repair of damage to DNA by excision. The abnormal thymine–thymine dimer produced by ultraviolet irradiation is excised as well as a stretch of DNA below the damage, and the gap is subsequently patched.

ate thymine–thymine dimers) (Fig. 5–17). These lesions distort the DNA helix and block the replication process. If DNA polymerase is able to circumvent the dimer it will produce a gap, or a mismatch, opposite to the T–T region, and it will perpetuate an anomalous structure or a base substitution in daughter strands. Thus it is essential to the cell to remove the T–T segments prior to cell division in order to allow normal DNA replication to proceed, and to obviate the introduction of mutagenic errors in bypass replication. Although the alkylating agents and cross-linking mutagens that produce interstrand connections interact preferentially with chromatin in the linker regions between the nucleosomes, the intrastrand pyrimidine dimers are produced by UV throughout the cores of the nucleosomes as well. However, the core lesions in nucleosomes do seem to be repaired more slowly, suggesting that the organization and packaging of DNA with histone proteins in chromosomes may affect accessibility to the repair system. The excision process involves two snipping enzymes: an endonuclease cleaving the DNA on the 5′ end of the T–T dimer to produce a nick in the phosphodiester backbone; an exonuclease breaking the nicked strand downstream from the lesion to release the T–T dimer with adjacent nucleotides. Recent evidence suggests that a glycosylase enzyme, which may also be involved in elimination of alkylated or other abnormal bases, may first cleave the glycoside bond between the deoxyribose sugar and the 5′ pyrimidine of the dimer. The apyrimidinic site with the dangling T–T dimer would then provide the signal for attack by endonuclease. The deleted stretch is reconstituted by DNA polymerase using the intact complementary strand as template. The β-polymerase has been implicated in the repair process since it remains high in activity in differentiated cells, such as those of muscle or brain, which have retained repair synthesis capacity but have lost virtually all of the α-polymerase activity along with the ability for cell division and DNA replication.

Disorders of DNA Repair

The importance of DNA repair for normal maintenance and survival is illustrated by a number of hereditary defects in which serious diseases are associated with the biochemical lesions. First and foremost is the syndrome known as xeroderma pigmentosum. This rare, fatal disorder is inherited as an autosomal recessive trait with a frequency of about 1 in 200,000. The affected homozygotes show extreme skin sensitivity to sunlight with keratosis and hyperpigmentation progressing to multiple carcinomas and melanomas on exposed surfaces. The light-induced cancers of the skin develop in virtually every patient by age 20, and are the cause of early death. Cultured fibroblasts from afflicted individuals are unable to carry out the normal excision-repair process following exposure to UV. Genetic and biochemical studies have shown that there are several distinct forms of xeroderma, each with a different lesion of the repair mechanism. The sites of the presumed enzyme defects are not defined as yet but they seem to involve the early events of incision at the pyrimidine dimer, possibly at the level of the endonuclease or glycosylase, their activators, or associated factors which may facilitate access of repair enzymes to damaged DNA within the highly ordered structure of chromatin. Other genetic defects of DNA repair point up the diverse and discrete processes for correction of DNA lesions that are produced by different types of mutagens. Fanconi's anemia, a lethal bone marrow-deficiency syndrome, appears with the same frequency as xeroderma pigmentosum. While most homozygous individuals die at an early age from aplastic anemia, they also tend to develop leukemia and other forms of cancer; presumed heterozygotes for the condition also have a high incidence of neoplasia, and have been estimated to account for 5% of all deaths from acute leukemia. Cells from the homozygotes are capable of correcting UV damage but are unable to repair the lesions produced by interstrand cross-linking agents such as the nitrogen mustards. A third type of defective DNA repair is seen in ataxia telangiectasia, a recessive trait with an incidence of 1 in 40,000, that affects many systems including the brain (ataxia from cerebellar degeneration), skin (telangiectasia or permanent dilation of blood vessels), and immune system (sinopulmonary infection and lymphoreticular tumors as a common cause of death in early life). When affected individuals with neoplasms were subjected to radiotherapy they were found to be highly sensitive to ionizing radiation, such that many patients died from the treatment. Fibroblasts derived from patients with ataxia telangiectasia can rejoin single-strand breaks produced in their DNA by X-rays, but not more complex types of radiation-induced damage. The lesion appears to be at the level of short patch repair of the damage inflicted by X-rays or γ-rays; the UV repair system functions normally in cells from these patients. The heterozygous state is associated with a greatly increased predisposition to cancer. Because of the high incidence of the gene (1% of the population will be heterozygotes), it is estimated that at least 5% of all persons who die of malignancy below age 45 carry the gene.

CARCINOGENESIS

The malignant transformation of normal differentiated cells into undisciplined, invasive, rapidly dividing cancerous cells is both a perplexing challenge to medical scientists and a tragic circumstance to its many victims. Agents that cause this transformation are known as carcinogens, and they seem to be ubiquitous in the environment. The real tragedy of carcinogenesis is that it is largely self-inflicted, either from the unconscious exposure to toxic chemicals in air, water, or food, or often from the deliberate intake of harmful substances such as tobacco smoke. At the same time it is encouraging to note that many such major causes of cancer are potentially avoidable. In the long-run, the elimination of environmental carcinogenic agents will undoubtedly have a much greater impact on human survival by preventing malignancy than the discovery of therapeutic means to cure cancer following the appearance of a malignancy. The mechanisms whereby carcinogens act on cells, and the modification of the actions of carcinogens by other factors and by metabolic reactions are so complex that predictions about the effects of cancer-inducing substances are very difficult to make even on a statistical basis, and virtually impossible to make on an individual basis. Nonetheless, as the recent findings on the inherited lesions of DNA repair have demonstrated, it may soon be possible to identify those genetic factors that interact with environmental agents and accelerate the carcinogenic process. From such studies it should be possible not only to identify the most dangerous cancer-initiating and -promoting factors, but also to predict which individuals will be at highest risk and should accordingly take the greatest care to minimize their exposure to potential carcinogens.

Several attributes of normal cells are altered irreversibly as they undergo transformation into malignant cells, and many of these neoplastic alterations may be detected in animal cells growing in tissue culture.

Some of the parameters that are associated with the transformation of cells *in vitro* include anomalies of morphology, chromosomal karyotypes, and surface properties. Transformed cells become insensitive to restraints on cell growth and are able to multiply in media that would not support normal cells; for example, typical features of the transformation to malignancy include the ability of the cells to grow at high density, to grow in contact with other cells, and to grow in semisolid media that lack surfaces for anchoring the cells. The transformed cells can undergo an indefinite number of divisions; that is to say that they have acquired potential immortality under culture conditions by comparison with the finite life expectancy of normal somatic cells, and they attain the capacity to grow as autonomous tumors in susceptible host animals. Oncogenic tumor viruses, irradiation, and chemical carcinogens can effect the transformation process in parallel with their ability to produce cancer *in vivo*.

Development of a tumor is a protracted phenomenon that may require several months or years and may involve the interactions of diverse external agents with both genetic and epigenetic systems in the body. This process appears to be a multistage transition of a normal cell into a colony of cancerous cells which form the tumor. Experimental studies in animals and epidemiologic investigations in humans support the view that cancer arises from exposure to two classes of agents, initiators and promoters, acting in the appropriate sequence and over a significant portion of the individual's life span. The initiation event involves a single exposure to a carcinogenic substance, such as urethane (ethyl carbamate). As a result of exposure to the initiator there is an irreversible commitment of a cell to become transformed into a neoplastic cell. Promotion of the growth of transformed cells into a malignant tumor requires prolonged, multiple exposures to a different type of cocarcinogenic agent, exemplified by various esters of the macrocyclic diterpene, phorbol, which is derived from plants of the Croton family. Such promoter agents have no carcinogenic effect on their own, but they enhance and accelerate the development of tumors only after the initiation event has occurred. In practice it may be difficult to separate initiating and promoting activities. For example, polycyclic hydrocarbons (*i.e.*, dimethylbenzanthracene) are complete carcinogens, and act both as initiators and promoters of tumor formation. Cigarette smoke and other products of organic combustion may exert both promoting and initiating effects as evidenced by the return-to-normal incidence of lung cancer in smokers who have quit the habit for five years or longer.

Initiating actions of carcinogens necessarily involve an associated modification of the genetic material since this preparatory stage of carcinogenesis is irreversible and the neoplastic transformation is transmitted to daughter cells. Alterations to the genome by mutagenic agents or by incorporation of oncogenic viral genomic material, and interference with DNA repair processes are implicated in the etiology of various types of cancer. The correlation between the mutagenic and carcinogenic potency of various agents is very high. Some mutagens may fail to deliver their effects on tissues if they are extremely reactive or unstable in biologic fluids, or if they are not transported effectively within the organism to target cells. Alternatively some carcinogens are known to require tissue-specific conversion to active metabolites before they exert their effects. When the problems of targeting and activation are taken into account, the correlation between mutagenicity and carcinogenicity of initiating factors is greatly enhanced. The amino-azobenzene dyes, and certain other aromatic amines, act as specific liver carcinogens because they are converted *in situ* by hepatic microsomal enzymes to hydroxylated and also sulfated derivatives that are strongly mutagenic. Similarly it has been found that the polycyclic hydrocarbons, nitrosamines, and aflatoxins are inactive *per se* (precarcinogens), and are converted into highly active metabolites (ultimate carcinogens) within particular target cells. In general the ultimate carcinogens are strongly electrophilic derivatives of the parent compounds (diol-epoxides of the aromatic hydrocarbons, for example), which can form covalent adducts with DNA bases, and hence act as direct mutagens. Relative activities of the microsomal monoxygenase and other precarcinogen-activating enzymes will thus determine which tissue types may be affected selectively by various chemical agents, and also may be crucial in establishing which individuals in a population are most susceptible to the development of tumors.

Promoting actions of cocarcinogenic substances are dependent on the prior exposure of cells to initiators, and are reversible, nonmutagenic effects which accentuate the transformation and proliferation of precancerous cells. In contrast with the initiator molecule, which can produce its effect by a single hit, the promoter must be present continuously during the transition from a normal to a cancerous state. Many diverse epigenetic events have been observed to be affected by tumor-promoters. A number of membrane parameters, including glycoprotein components, turnover of phospholipids, and transport phenomena are altered by treatment of normal cells with phorbol esters. The promoting agents appear to induce aspects of the transformed phenotype in normal cells, such as the ability to grow at high cell densities and the induction of synthesis of enzymes (ornithine decarboxylase, plasminogen-activator) that are associated with expression of tumors. In addition to enhancement of the growth of cells in semisolid medium after the exposure to chemical, viral, or physical initiators of the neoplastic transformation, the promoters also block the terminal differentiation of cells thus destabilizing the normal limitations on tissue growth and allowing proliferation of transformed stem cells into a clone that eventually becomes the tumor.

Much effort has been devoted toward identification of environmental carcinogens with a view to restriction of their use and dissemination. Most concern has been directed to methods for detection of mutagenic or initiator types of agents, and such concern is warranted because of the single-hit nature of the lesions, along with the other deleterious effects of mutagens on human survival. The simple and inexpensive bacterial assays have obvious advantages for the screening of vast numbers of substances presently surrounding us; however, they may not entirely allow for the manifold metabolic manipulations that generate ultimate mutagens and carcinogens within mammalian cells. Tests on neoplastic transformability in whole animals and in cultures of human cells are required to supplement findings with the Ame's test and other microbial assays. Moreover, there is a growing awareness of the important role played by promoting agents, and procedures to assess the promoter or cocarcinogenic effects of agents that may have synergistic or potentiating effects on the classical initiators are being devised. Among the more significant and prevalent promoting agents are the dietary fats, which act to increase the incidence of human colon and breast cancer, possibly through the increased formation of intestinal bile acids or of fatty acid metabolites in tissues. On a more hopeful note it has been determined that many natural substances may exert anticarcinogenic effects either at the level of the mutagens (as noted in Chap. 2 for vitamin C and tocopherol inhibition of nitrosamine production), or at the level of the tumor-promoters (as noted for the vitamin A derivative, retinoic acid). Thus a combined awareness of our individual genetic risk factors, of external hazards, and of protective agents can provide the basis for private and public strategies to attack the cancer problem with the appropriate preventive measures.

SUGGESTED READINGS

Abelson J, RNA processing and the intervening sequence problem. Annu Rev Biochem 48:1035– 1069, 1979

Anderson WF, Diacumakos EG: Genetic engineering in mammalian cells. Sci Am 245(1):106– 121, 1981

Baxter JD, Recombinant DNA and medical progress. Hosp Pract 15(2):57– 67, 1980

Bostock CJ, Sumner AT, The Eukaryotic Chromosome, Amsterdam, North-Holland, 1978

Brandhorst BP: Heterogeneous nuclear RNA of animal cells and its relationship to messenger RNA. In McConkey EH (ed): Protein Synthesis, Vol II, pp 1– 67. New York, Marcel Dekker, 1976

Breathnach R, Chambon P: Organization and expression of eucaryotic split genes coding for proteins. Annu Rev Biochem 50:349– 383, 1981

Brimacombe R, Stoffler G, Wittmann HG: Ribosome structure. Annu Rev Biochem 47:217– 250, 1978

Burke DC: The status of interferon. Sci Am 236(4):42– 50, 1977

Cairns J: The cancer problem. Sci Am 233(5):64– 78, 1975

Chambon P: Split genes. Sci Am 244(5):60– 71, 1981

Chantrenne H: Regulation of protein synthesis at translation in eukaryotes. A brief review. In Dumont J, Nunez J (eds): Hormones and Cell Regulation, Vol 2, pp 1–13. Amsterdam, North Holland, 1977

Cohen SN: The manipulation of genes. Sci Am 233(1):24–33, 1975

DePamphilis ML, Wassarman PM: Replication of eukaryotic chromosomes: A close-up of the replication fork. Annu Rev Biochem 49:627–666, 1980

Devoret R: Bacterial tests for potential carcinogens. Sci Am 241(2):40–49, 1979

DiPaolo JA, Casto BC: Chemical carcinogenesis. In Gallo RC (ed): Recent Advances in Cancer Research: Cell Biology, Molecular Biology, and Tumor Virology, Vol I, pp 17–47. Cleveland, CRC Press, 1977

Drake JW, Baltz RH: The biochemistry of mutagenesis. Annu Rev Biochem 45:11–38, 1976

Fishbein L: Environmental sources of chemical mutagens. In Flamm WG, Mehlmann MA (eds): Advances in Modern Toxicology, Vol 5, Mutagenesis, pp 175–348. Washington, Hemisphere Publishing, 1978

Friedberg EC, Ehmann, UK, Williams JI: Human diseases associated with defective DNA repair. Adv Radiat Biol 8:85–174, 1979

Fuchs F: Genetic amniocentesis. Sci Am 242(6):47–53, 1980

Gefter ML: DNA replication. Annu Rev Biochem 44:45–78, 1975

Gelehrter TD: Enzyme induction. N Engl J Med 294:522–526, 589–595, 646–651, 1976

Hanawalt PC, Cooper PK, Ganesan AK et al: DNA repair in bacteria and mammalian cells. Annu Rev Biochem 48:783–836, 1979

Harris H: The Principles of Human Biochemical Genetics. 2nd Ed., Amsterdam, North Holland, 1975

Higginson J: Perspectives and future developments in research on environmental carcinogenesis. In Griffin AC, Shaw CR (eds): Carcinogens: Identification and Mechanisms of Action, pp 187–208. New York, Raven Press, 1979

Howard–Flanders P: Inducible repair of DNA. Sci Am 245(5):72–80, 1981

Isenberg, I: Histones. Annu Rev Biochem 48:159–192, 1979

Kornberg RD: Structure of chromatin. Annu Rev Biochem 46:931–954, 1977

Kornberg RD, Klug A: The nucleosome. Sci Am 244(2):52–64, 1981

Kurland CG: Aspects of ribosome structure and function. In Weissbach H, Pestka S (eds): Molecular Mechanisms of Protein Biosynthesis, pp 81–116. New York, Academic Press, 1977

Lake JA: The ribosome. Sci Am 245(2):84–97, 1981

Lane C: Rabbit hemoglobin from frog eggs. Sci Am 235(2):60–71, 1976

Lilley DMJ, Pardon JF: The structure and function of chromatin. Annu Rev Genet 13:197–233, 1979

Lodish, HF: Translational control of protein synthesis. Annu Rev Biochem 45:39–72, 1976

Maniatis T, Ptashne M: A DNA operator–repressor system. Sci Am 234(1):64–76, 1976

McCusick VA: The anatomy of the human genome. Hosp Pract 16(4):82–100, 1981

McGhee JD, Felsenfeld G: Nucleosome structure. Annu Rev Biochem 49:1115–1156, 1980

Miller JA, Miller EC: Ultimate chemical carcinogens as reactive mutagenic electrophiles. In Hiatt HH, Watson JD, Winsten JA (eds): Origins of Human Cancer, Book B, Mechanisms of Carcinogenesis, pp 605–627. Cold Spring Harbor Laboratory, 1977

Miller JH, Reznikoff WS (eds): The Operon, Cold Spring Harbor Laboratory, 1978

Ochoa S, deHaro C: Regulation of protein synthesis in eukaryotes. Annu Rev Biochem 48:549– 580, 1979

Ogawa T, Okazaki T: Discontinuous DNA replication. Annu Rev Biochem 49:421– 458, 1980

O'Malley BW, Towle HC, Schwartz RJ: Regulation of gene expression in eucaryotes. Annu Rev Genet 11:239– 275, 1977

Paterson MC, Smith PJ: Ataxia telangiectasia: An inherited human disorder involving hypersensitivity to ionizing radiation and related DNA-damaging chemicals. Annu Rev Genet 13:291– 318, 1979

Perry RP: Processing of RNA. Annu Rev Biochem 45:605– 630, 1976

Peto R: Epidemiology, multistage models, and short-term mutagenicity tests. In Hiatt HH, Watson JD, Winsten JA (eds): Origins of Human Cancer, Book C, Human Risk Assessment, pp 1403– 1428. Cold Spring Harbor Laboratory, 1977

Pitot HC: Biological and enzymatic events in chemical carcinogenesis. Annu Rev Med 30:25– 39, 1979

Revel M: Initiation of messenger RNA translation into protein and some aspects of its regulation. In Weissbach H, Pestka S (eds): Molecular Mechanisms of Protein Biosynthesis, pp 246– 323. New York, Academic Press, 1977

Revel M, Groner Y: Post-transcriptional and translational controls of gene expression in eukaryotes. Annu Rev Biochem 47:1079– 1126, 1978

Rich A, RajBhandary UL: Transfer RNA: Molecular structure, sequence, and properties. Annu Rev Biochem 45:805– 860, 1976

Sabatini DD, Kreibich G: Functional specialization of membrane-bound ribosomes in eukaryotic cells. In Martonosi A (ed): The Enzymes of Biological Membranes, pp 531– 556. New York, Plenum Press, 1976

Schimmel PR, Söll D: Aminoacyl-tRNA-synthetases: General features and recognition of transfer RNAs. Annu Rev Biochem 48:601– 648, 1979

Sheinin R, Humbert J, Pearlman RE: Some aspects of eukaryotic DNA replication. Annu Rev Biochem 47:227– 316, 1978

Smith-Kearny PF: Genetic Structure and Function, New York, John Wiley & Sons, 1975

Waring MJ: DNA modification and cancer. Annu Rev Biochem 50:159– 192, 1981

Weinstein IB, Lee L-S, Fisher PB et al: Cellular and biochemical events associated with the action of tumor promoters. In Miller EC, Miller JA, Hirono I *et al:* (eds): Naturally Occurring Carcinogens—Mutagens and Modulators of Carcinogenesis, pp 301– 313. Baltimore, University Park Press, 1979

Weissbach A: Eukaryotic DNA polymerases. Annu Rev Biochem 46:25– 48, 1977

Weissbach H, Ochoa S: Soluble factors required for eukaryotic protein synthesis. Annu Rev Biochem 45:191– 216, 1976

Wool IG: The structure and function of eukaryotic ribosomes. Annu Rev Biochem 48:719– 754, 1979

Wynder EL, Hoffmann D, McCoy GD et al: Tumor promotion and cocarcinogenesis as related to man and his environment. In Slaga TJ, Sirak A, Boutwell RK (eds): Carcinogenesis, Vol 2, pp 59– 77. Mechanisms of Tumor Promotion and Cocarcinogenesis. New York, Raven Press, 1978

Abbreviations and Acronyms

A adenine
ABP androgen-binding protein
ACAT acyl CoA-cholesterol acyl transferase
ACTH adrenocorticotrophic hormone
ADP adenosine diphosphate
Ala alanine
AMP adenosine monophosphate
apoA,B, etc apolipoproteinsA,B, etc
Arg arginine
Asn asparagine
Asp aspartic acid
ATP adenosine triphosphate
ATPase adenosine triphosphatase

C cytosine
C$_{1,2}$, etc factors of the complement system
CDP cytidine diphosphate
CMP cytidine monophosphate
CoA coenzyme A
CoASH coenzyme A, sulfhydryl form
COMT catechol-O-methyl transferase
CoQ coenzyme Q
CoQH$_2$ coenzyme Q, reduced form
CRF corticotrophin releasing factor
CTP cytidine triphosphate

Cys   cysteine
CySH   cysteine, sulfhydryl form
CyS-SCy   cystine, disulfide form

D   dihydrouridine
DHT   dihydrotestosterone
DIC   disseminated intravascular coagulation
DIT   diiodotyrosine
DNA   deoxyribonucleic acid
Dopa   dihydroxyphenylalanine
Dopamine   dihydroxyphenylethylamine
DPG   diphosphoglyceric acid
dsRNA   double-stranded RNA

EF   elongating factor of translation

FAD   flavin adenine dinucleotide
FADH$_2$ flavin adenine dinucleotide, reduced form
FDP   fibrin degradation product
FIGLU   formiminoglutamic acid
FMN   flavin mononucleotide
FMNH$_2$ flavin mononucleotide, reduced form
FSH   follicle stimulating hormone

G   guanine
GDP   guanosine diphosphate
GH   growth hormone
GHRF   growth hormone releasing factor
Gln   glutamine
Glu   glutamic acid
Gly   glycine
GMP   guanosine monophosphate
GTP   guanosine triphosphate

Hb   hemoglobin
HbA   adult hemoglobin
HbF   fetal hemoglobin
HbO$_2$, Hb(O$_2$)$_2$, etc   oxygenated forms of hemoglobin

HbS   sickle-cell anemia hemoglobin
HCG   human chorionic gonadotrophin
HDL   high-density lipoprotein
His   histidine
HMGCoA β-hydroxy-β-methylglutaryl CoA
hnRNA   heterogeneous nuclear RNA
HVA   homovanillic acid

IgA, IgD, etc   immunoglobulins A, D, etc
IF   initiating factor of translation
Ile   isoleucine

LATS   long-acting thyroid stimulator
LCAT   lecithin-cholesterol acyl transferase
LDL   low-density lipoprotein
Leu   leucine
LH   luteinizing hormone
LHRH   LH releasing hormone
LPL   lipoprotein lipase
Lys   lysine

MAO   monoamine oxidase
Met   methionine
MIT   monoiodotyrosine
mRNA   messenger RNA
mRNP   mRNA-protein complex particle
MSH   melanocyte stimulating hormone

NAD   nicotinamide adenine dinucleotide
NADH   nicotinamide adenine dinucleotide, reduced form
NADP   nicotinamide adenine dinucleotide phosphate
NADPH   nicotinamide adenine dinucleotide phosphate, re-
duced form

pCO$_2$ partial pressure of carbon dioxide
PGE, PGF, etc   prostaglandins E, F, etc
Phe   phenylalanine
Pi   inorganic phosphate

PNMT   phenylethanolamine-N-methyl transferase
Pol   polymerase
pO$_2$ partial pressure of oxygen
Pro   proline
PTU   propylthiouracil
p50   pO$_2$ corresponding to 50% saturation of hemoglobin

RF   releasing factor of translation
RNA   ribonucleic acid
rRNA   ribosomal RNA
rT$_3$ 3,3',5'-triiodothyronine (reverse T$_3$)

SAM   S-adenosylmethionine
Ser   serine
SHBG   steroid hormone binding globulin

T   thymine
TBG   thyronine binding globulin
TBPA   thyronine binding prealbumin
tcRNA   translational control RNA
TDP   thymidine diphosphate
Thr   threonine
TMP   thymidine monophosphate
TPP   thiamine pyrophosphate
TRH   thyrotrophin releasing hormone
tRNA   transfer RNA
Trp   tryptophan
TSH   thyroid stimulating hormone
TTP   thymidine triphosphate
TxA, TxB   thromboxanes A and B
Tyr   tyrosine
T$_3$ 3,5,3'-triiodothyronine
T$_4$ 3,5,3',5'-tetraiodothyronine (thyroxine)

U   uracil
UDP   uridine diphosphate
UMP   uridine monophosphate
UTP   uridine triphosphate

Val   valine
VLDL   very low-density lipoprotein
VMA   vanillylmandelic acid

ψ pseudouridine

Index

abetalipoproteinemia, 28
acetylation, of histones, 266–267
acetylcholine, membrane receptors for, 214–215
actinomycin D, inhibition of rRNA synthesis by, 260
autosomal chromosomes, 251
ACAT (acylCoA cholesterol acyl transferase), 27
acromegaly, hypersecretion of growth hormone in, 120–121
ACTH (adrenocorticotrophic hormone), 138–141
active transport
    ATP requirement for, 221
    of glucose, 220
    of Na$^+$, 220–223
acylCoA cholesterol acyl transferase (ACAT), 27
Addison's disease, deficiency of adrenal cortical steroids in, 149
adipose tissue
    effects of glucocorticoids on, 145
    effects of insulin on, 130
adrenal cortex, effects of ACTH on, 140–141
adrenal cortical hormones, 137–149
adrenal medulla, synthesis of epinephrine by, 188
adrenergic agents, mechanisms of action, 192
adrenocorticotrophic hormone (ACTH), 138–141
adrenogenital syndrome, 21-hydroxylase deficiency in, 142
albumin
    functions of, 18–20
    structure of, 20–22
aldosterone
    biosynthesis in adrenals, 142
    mechanism of action, 147–148
    in sodium reabsorption, 102
alkylating agents, as mutagens of DNA, 285–286

N-allyl morphine, antagonism of opiate binding by, 215
α-amanitin, inhibition of mRNA synthesis by, 262
amino acids, effects on glucagon and insulin secretion, 133
aminoacyl tRNA, synthesis of, 272–273
amylases, in carbohydrate digestion, 52
androgen-binding protein, in testicular functions, 151
androgens
    biosynthesis in adrenals, 142
    biosynthesis in the testis, 152–153
    mechanisms of action, 155–158
anemia
    in folic acid deficiency, 77
    in iron deficiency, 108
    in pyridoxine deficiency, 70
    in vitamin B$_{12}$ deficiency, 74–75
    in vitamin C deficiency, 83
anesthetics, effects on membrane permeability, 222–225
angiotensin
    in aldosterone production, 147
    in sodium reabsorption, 102
anterior pituitary
    polypeptide hormones of, 115–117
    secretion of ACTH by, 138–140
    secretion of gonadotrophins by, 152, 161
    secretion of growth hormone by, 117–119
    secretion of prolactin by, 121
    secretion of TSH by, 173–174
antibodies (*See* immunoglobulins)
anticoagulants, in treatment of thrombosis, 47–49
anticodons, binding to codons, 270–271
antidiuretic hormone (*See* vasopressin)
antigens, of cell membranes, 231
antihemophilic globulin (AHG), 39
antithyroid agents, 170, 182
Anturane (*See* sulfinpyrazone)